# DEVELOPMENTAL APPROACHES TO THE SELF

## PATH IN PSYCHOLOGY
Published in Cooperation with Publications for the
Advancement of Theory and History in Psychology (PATH)

*Series Editors:*
**David Bakan,** *York University*
**John Broughton,** *Teachers College, Columbia University*
**Miriam Lewin,** *Manhattanville College*
**Robert Rieber,** *John Jay College, CUNY, and Columbia University*
**Howard Gruber,** *Rutgers University*

---

**WILHELM WUNDT AND THE MAKING OF A SCIENTIFIC
PSYCHOLOGY**
Edited by R. W. Rieber

**HUMANISTIC PSYCHOLOGY: Concepts and Criticisms**
Edited by Joseph R. Royce and Leendert P. Mos

**PSYCHOSOCIAL THEORIES OF THE SELF**
Edited by Benjamin Lee

**DEVELOPMENTAL APPROACHES TO THE SELF**
Edited by Benjamin Lee and Gil G. Noam

# DEVELOPMENTAL APPROACHES TO THE SELF

Edited by
## Benjamin Lee
*Center for Psychosocial Studies*
*Chicago, Illinois*

and
## Gil G. Noam
*McLean Hospital and Harvard Medical School*
*The Clinical–Developmental Institute*
*Belmont, Massachusetts*

With the collaboration of
Kathleen Smith

PLENUM PRESS • NEW YORK AND LONDON

Library of Congress Cataloging in Publication Data

Main entry under title.

Developmental approaches to the self.

(PATH in psychology)
Includes bibliographical references and index.
1. Self—Congresses. 2. Developmental psychology—Congresses. I. Lee, Benjamin.
II. Noam, Gil G.      .III. Center for Psychosocial Studies. IV. Title. V. Series.
BF697.C576 1979a                           155.2'5                           82-22582

ISBN-13: 978-1-4613-3616-7      e-ISBN-13: 978-1-4613-3614-3
DOI: 10.1007/978-1-4613-3614-3

© 1983 Plenum Press, New York
Softcover reprint of the hardcover 1st edition 1983
A Division of Plenum Publishing Corporation
233 Spring Street, New York, N.Y. 10013

## ACKNOWLEDGMENTS

We would like to thank the following people for making this volume possible: Kathleen Smith, Lynn Clark, and Janine Poronsky for their efforts at coordination between Chicago and Cambridge and the demands of the two editors; Alexandra Hewer, Laura Rogers, Robert Kegan, Ann Fleck Henderson, Sharon Parks, Anne Colby, Steven Rolfe, and Paul Goldmuntz at the Clinical Developmental Institute in Belmont, MA. for their intellectual support; the colleagues at the Adolescent and Family Development Project of Harvard Medical School, Stuart Hauser, Allen Jacobson, and Sally Powers for their continued research input to our attempts to empirically investigate some of the concepts presented here; Maryanne Wolf, Larry Kohlberg, and Barney Weissbourd for being patient and critical listeners to the many versions of these papers.

# CONTENTS

# INTRODUCTION

Each of the three great schools of developmental psychology represented in this volume--psychoanalytic, cognitive-developmental, and Vygotskian--diverges in important ways. But more recent changes in each discipline have led to new possibilities for theoretical integrations. Each orientation has begun to focus upon the problem of "meaning construction", that is, how a person's subjectivity and consciousness is created through his interaction with significant others. Each discipline also discovered that as it switched to meaning and interpretation as the foci of their work, they had to reformulate and, in some cases, reject positions taken by their founding figures. The papers in this volume attempt to describe the newest developments in each of these fields and to foster a theoretical dialogue around the concept of the self. The papers in this book emerged out of discussions at a Conference on the Self, sponsored by the Center for Psychosocial Studies in Chicago.

For the psychoanalytic and cognitive-developmental approaches, we can observe a transition from what we call the biologism of both traditional Freudian and Piagetian memtapsychologies to a more "communicative-interactionist" point of view. Psychoanalysts have focused on the subjective experience of their patients as constituting a reality in its own right, and therefore have always focused upon problems of communication and interpretation. But Freud's emphasis on bio-sexual development led him to create a <u>metapsychology</u> in which the basic organizing principle is that of drive reduction. He viewed meaning as largely a means to further biologically based psychic needs and conflicts, and this is reflected in his treatment of language. The major function of language is to represent reality (as opposed to communication), and it is distortions in this function that forces one to look for unconscious motivations. If more complicated forms of thought arise through the structure of communication, it is only a short step to posit that more complicated affective and motivational processes are created and struc-

1

tured by communicative principles, rather than seeing meaning as only a transparent means for the expression of
affect.

This strong emphasis on bio-sexual development led to a
metapsychology which underestimated the importance of the
structure of meaning and communication in the development of
an experience-organizing ego and self. The ego, confronted
with powerful libidinal impulses, was attributed the place
of the "weak rider of the strong horse". American ego psychologists tried to overcome these theoretical shortcomings
by introducing the concept of an autonomous and conflict-
free sphere of ego functioning. They—and that is also true
for Erikson—still remained wedded to a tripartite model of
the mind in which the ego develops out of an initial ego-id
matrix and is confronted by superego demands and id forces.
However, Erikson and the proponents of British object relations psychology shifted the emphasis to psychosocial and
interpersonal aspects of development. Heinz Kohut's work
adds to these attempts at reformulation by introducing the
concepts of self and selfobjects as the constructors of an
overarching unity of subjective experience. He insisted
that empathic introspection and interpretation in the
clinical ("transferential") situation is the data generating
base of all of psychoanalysis, and is inherently communicative. Although still very controversial, Kohut's theory
links a developmental theory of social interaction with a
hermeneutic approach to subjectivity, relegating Freud's
bio-sexual drive theory to a secondary position. Michael
Basch's paper in this volume discusses basic themes in Heinz
Kohut's theory of the self. It integrates psychoanalytic
approaches with recent research in both the Vygotskian and
Piagetian paradigms. Starting with infancy, he traces how
affect and cognition are mediated through the development of
more complicated systems of communication into which the
child is socialized and analyzes how the therapist makes use
of these communication patterns in the process of psychotherapy itself. Basch concludes by arguing that the integration of psychoanalytic, Piagetian, and Vygotskian approaches provides the foundation for an operational definition of the self.

The cognitive developmental papers share with the psychoanalytic ones a turning away from the referential-nomin-
alism of their founding figure, Jean Piaget, to a more communicative interactionist approach. Most of Piaget's work

was concerned with the child's cognitive development in his interactions with physical objects, "mute" things. It became almost immediately apparent to later researchers that the cognitive structures developed in such interactions would not be adequate to deal with the child's communicative interactions with other agents. Kohlberg's theory of moral development advanced our understanding of social development and gave impetus to the field of social cognition. Theoretical formulations and empirical investigations have proven very successful and inspired a new generation of psychological research with a focus on the self, relationships between cognition and affect, and clinical orientations. Consequently, there has been a revival of interest in the theories of Mead and Baldwin who both maintain that self and other emerge only through social interactions that are essentially communicative.

The paper written by Noam, Kohlberg, and Snarey analyzes functional and structural theories of the self and proposes steps toward an integrated model of the self. These principles build on Eriksonian and social-cognitive theories. Embedded in this approach are new ways of studying the the relationship between cognition and affect, identity and biography, and new concepts of clinical integration centered around a life-span theory of the development of the self.

Werner van der Voort's paper demonstrates that there is a hidden social dimension to even the sensorimotor period of the child's development of object concepts. Through a fine-grained reanalysis of Piaget's own data, he shows that the first "Piagetian permanent object" is the hands of others, thus rooting even object development to social interaction.

Blasi argues for the importance of the self in creating any coherent epistemology. The critical inight is that any such epistemology must include a subject who makes judgments and differentially directs his attention to the world. The self becomes a process, not merely a self-regulating system of cognitions.

John Broughton presents a social cognitive perspective on the development of the self and identity during adolescence. He reviews the relevant work of Piaget, Kohlberg, and Erikson, and then points out their theoretical and empirical inadequacies in dealing with identity formation. His conclusion is that an adequate psychology of development must, ironically, go beyond psychology and view the self as trying to forge an identity in the crucible of the history of itself and the society in which it is embedded.

Robert Kegan, in "A Neo-Piagetian Approach to Object-Relations", builds on Piaget and Kohlberg and presents a developmental theory of self and object relations across the life cycle. Each stage of self and other is a new form of "meaning-making", a new construction of the subject's relations to significant others and to the self. His chapter is a statement of a social cognitive approach to the self as following an underlying developmental logic of self-other differentiations.

The last two papers are departures from both psychoanalytic and Piagetian concepts. Vygotsky clearly begins with a social systemic account of the origin of the "higher mental functions". In both the Kohutian and cognitive-developmental approaches, the structuring principles of psychological development lie in dyadic interactions involving two or more agents. This switch is a clear advance from the biological principles of Freudian drive reduction and the Piagetian analogies organic assimilation and accomodation and evolutionism. Vygotsky, however, insists that the higher mental functions arise through the child's differentiating different levels of a social institution, namely language, and that the principles which govern the structure and development of language (or for that matter, any social institution) are never reducible to either biological principles or principles of dyadic interactions. Vygotsky believed that ultimately psychological development followed dialectical principles which were derived neither from biology or social psychology, but from the interactions of social institutions over history. Language, being a social institution, is also subject to these laws, and, as the child uses language, he begins to transform his psychological processes along dialectical lines. The last two papers in this volume, introduce these Vygotskian themes and suggest ways in which they intersect with psychoanalytic and cognitive-developmental psychology.

In very important ways, a dialogue between these theory traditions will help advance a general developmental theory of consciousness. Vygotsky never developed a theory of the self or psychic conflict, while psychoanalytic theory is lacking a theory of adult development and has never articulated theories of social institutions. Cognitive-developmental approaches have tended to avoid the analysis of subjective phenomena, individual differences, and emotional development, and have only advanced a notion of the social up to the level of the dyad and group. Although this volume does not present an integrating theory which solves the shortcoming of each approach while maintaining their strengths, it does show that the most promising area of convergence is through the study of the developing social self as an analysis of meaning and communication.

Benjamin Lee and Gil G. Noam

# THE CONCEPT OF "SELF": AN OPERATIONAL DEFINITION

Michael Franz Basch

Chicago Institute for Psychoanalysis

Cultism is the curse of the thinking classes! There has been no shortage of studies of the self, of individual psychology, of human nature—call it what you will. Yet instead of progressing toward the common goal of an understanding of man as man, the proliferation of investigators has given rise to narrower and narrower viewpoints, each seeking to explain all of behavior through its particular expertise. As a result, the numerous schools of psychology form a veritable tower of Babel. Paradoxically, in psychology and its allied fields the quintessence of being human, the ability to communicate experience symbolically has been lost—and with it what Korzybski (1933) has termed "time binding", the ability to use and profit from the work of previous generations. Significant discoveries by one group, in one or another area of character formation, are unknown, misunderstood, or disregarded by others. Unnecessary duplication of effort is common, and the need to fit findings on the Procrustean bed of a favored theoretical construct undermines otherwise useful experimental efforts.

Seemingly there has been a surfeit of theorizing, and it sounds as if those who advocate a moratorium on speculation and renewed emphasis on uncontaminated observation and empiricism are correct. But it is not the tool, "theory", that is responsible for the chaos in psychology; rather, its improper use must be held accountable. Not less but more effort should be expended on examining not only extant theories of personality but their place in science at large.

## PSYCHOLOGY AND SCIENCE

The very existence of a relationship between theoretical and applied psychology and the other sciences is often questioned. It is often argued that the private nature of psychological experience precludes gathering scientific data. This implies that only what is public may be considered

7

scientific, and what is private must be considered to be unscientific. However, this is an error, based on a misconcep-
tion of the nature of reality, which must be corrected.  The
objective nature of physical reality has  long  been  challenged  by  philosophy, and Einstein's discoveries have convinced even the physical scientists of  the  relativity  of
their observations. It is now commonly accepted in the physical sciences that although the nature of the observation of
the  physical  world is very much dependent on the observer,
that fact does not make the operations any less  scientific;
this  tolerance,  however, has not been extended to the psychological sciences.  This may be because  even  though  the
physical  scientists can no longer claim certainty for their
observations, they are concerned with the material  (usually
tangible)  universe  while  the  psychologists are concerned
with the symbolic universe.  Both physical and psychological
scientists must learn that "reality" is independent  of  the
origin of the stimulus or its nature, and that the "real" is
that  about which valid predictions and/or post-dictions may
be made.  Validation of  predictions  and  post-dictions  in
science  is  through empirical experiment, but this does not
mean that the subject matter under consideration need be  of
the world of the senses.  There is a world of relations (for
example, the world of mathematics) which has a reality which
is  no  less  scientific  than  that of the world of matter.
Indeed the material world is a subsystem in the universe  of
relations.[1]

    The  practical  success of physical science has blinded
its practitioners to the fact that empirical experimentation
is a method of investigation within philosophy, just  as  is
speculation.   Scientists, for some time, seem to have operated on the assumption that science ranked  with  philosophy
and  had  replaced it--compounding that pretension by errone-
ously equating philosophy with the  speculative  method  and
that,  in turn, with religious mysticism.  The physical scientists, by narrowing their field of  investigation  to  the
material  aspect of the universe and by artificially and improperly reducing all phenomena to be studied to closed sys-
stems, postponed the day of reckoning  till  Heisenberg  and
Einstein came upon the scene. Indeed the physical scientists
captured  the  vocabulary  and equated their narrow concerns
with "Science" itself.  For them, only those activities that
could be reduced to material phenomena were worthy of  being
called  scientific;  they  conveniently  overlooked the fact
that by these criteria one could have a scientific  arithme-

tic but no mathematics, since only the truths of the former can be concretely demonstrated, the latter's being expressed in pure relationships.

As long as this unscientific (i.e., unimaginative) view of science dominates, a scientific psychology in which a comprehensive view of "self" is possible is a contradiction in terms. What is now required is (1) a stop to the attempt to have psychology ape the model of physical science, in the foredoomed hope of winning approval for its results by those criteria, and (2) a reexamination and redefinition of what is to be meant by "science" and "Science". A comprehensive epistemology, or philosophy of science, should not contradict what physical science has truly proven; it should encompass it and then go on to include and explain the vicissitudes of open systems—i.e., those in which information and its meaning determine the behavior of the system.

This necessitates a reexamination of the question of what is to be meant by the term "reality". As was mentioned, the physical scientists oversimplified the issue and equated the material, concrete universe—available directly or indirectly to the senses—with the "real". Thus, seeing an oasis is real; its hallucination is "unreal". The problems that such a viewpoint has created for perceptual psychology are only now beginning to be capable of resolution. It is suggested that these fundamental concerns must be dealt with before a psychological investigation can be begun. If this aspect is neglected, this study will flounder as has every other attempt at a unified theory of psychology.

The meaningless dichotomy of "objective" and "subjective" is based on the misguided notion of reality just mentioned. The question of the scientific nature of psychology has revolved around the issue of whether or not an introspective, and therefore supposedly "subjective", method could meet the standards of "Science".

Academic psychology generally answered this question in the negative, and sought to develop methods that emulated the criteria of observable replicability set by physics, chemistry, etc. The results obtained have been valuable in advancing our understanding of cognition, and to some extent in furthering our knowledge of perception and of the learning process. However, in spite of these valuable contributions, this approach failed--usually by default--to make headway in the areas of affect and human motivation.

From its inception, psychoanalytic psychology focused on the meaning of thought--i.e, its dispositional effect on future action. It has always maintained that introspection (the observation of what is conscious, and the predictive and post-dictive inferences therefrom) was as scientifically valid as observation based on sensory signals from the environment. Psychoanalysis has remained an investigative method and has never become a full-fledged psychological system because its explanatory theory is based on the same nineteenth-century model of perception and reality that hampers the rest of psychology--and all of science. Though psychoanalysis, unlike academic psychology, has never neglected affect and motivation in studying the meaning of behavior and has made considerable progress in understanding these factors, the lack of an acceptable scientific framework for all psychologies has prevented the mutually corrective influence that the various branches of the field could have exercised.

## PSYCHOANALYSIS: HISTORICAL CONSIDERATIONS

Psychoanalysis has been considered a method of treatment, a psychology, a research method, and even a Weltanschauung or philosophy of life. To a great extent, sometimes to the distress of many non-analysts, psychoanalysis is still intertwined intimately with the name and work of one man: Sigmund Freud. Freud's working life was a long one and his productivity was such that even today the depth and breadth of his ideas have not been fully assimilated. Indeed, until recently and especially in the English-speaking world, Freud and his work were not very accurately known. The German editions of Freud's work were many, and often confusing and contradictory in essential parts. Foreign translations were even more problematic. Freud tended to give rights to translation to those he felt were loyal scientific friends; and since these were not necessarily talented linguists, often the sense that Freud was trying to convey about his subject was lost. Similarly with the biographies of Freud: Jones' work, monumental though it is, gives us a picture of Freud as Jones saw him or would have had him be. In many respects, the Jones biographies (1953-57) missed the essence of the man; furthermore, the view that Jones has of Freud's work is very much colored by his (Jones's) rigidity and authoritarian proclivities. Other students looking at the same corpus of scientific writing can, and have by now, placed a very different emphasis on various aspects of that work. It was not until the '50s and early '60s that, volume by volume, a reasonably adequate and comprehensive translation of Freud's work became available in the Strachey Translation, the Standard Edition as it is called. So, in a sense, it is our generation that is privileged to have the first reasonably complete look at Freud's scientific offering and, paradoxically, only we who come after Freud departed life can develop a comprehensive understanding of his quest.

Prior to Freud's work on the hypnotic investigation and treatment of the psychoneuroses at the clinic of Jean-Martin Charcot, neurotic symptoms were not looked upon as psychological manifestations. Though Charcot and others had recognized that hysterical symptoms--no matter how bizarre--could be understood as reenacting, in some way, earlier traumatic events, they believed that such symptoms represented physiological aberrations of an inherently weak brain. Charcot accepted the position held by neurologists since the time of Descartes: that the activities of the cerebrum were synonymous with consciousness, and anything that was not conscious was not psychological. Since the origin of his symptoms were not part of the patient's consciousness and could be traced only through hypnosis, they were obviously not cerebral and, by the same token, not psychological in nature.

Freud, in a little-known essay (1888) and without any fanfare, disputed his teacher's dictum. No, he said, consciousness and the activity of the cerebral hemispheres could not be equated, and there was no reason for denying the appellation "psychological" to neurotic symptoms and their origin. Having said that, he had (without realizing it at the time, of course) founded what was to become the field of dynamic psychiatry and psychotherapy. As long as the origin of neuroses was to be found in physiological defects, interesting as the study of neurotic phenomenology might be, their cure had to be looked for in the detection and amelioration of brain pathology. Once considered as psychological these difficulties demanded a different explanation, one which Freud was not long in advancing; that neuroses were the outcome of a conflict between incompatible ideas--more accurately, a conflict between strong emotions aroused by some event and the adult attitudes or moral principles of the individual involved. Emotions were for Freud, as for Charcot, similar to electric charges that needed to be discharged by attaching themselves to ideas, thoughts, or concepts that would in turn lead to tension-relieving behavior. Although Charcot had focused mainly on traumatic neuroses and had succeeded in having patients recall under hypnosis what the event was that had upset them, Freud found (much to his surprise) that his patients always dated their difficulties to some sexual assault or experience in their early childhoood. Here he made his second major discovery (the first had been that neurotic symptoms were psychological events), when he realized that neurotic symptoms had a purpose. Unable to totally eliminate the gratification of the premature sexual stimulation or the wish for its repetition, while equally unable to acknowledge this need which contradicted moral principles and invited social sanction, the patient first distorted the need so he would not have to become conscious of it, and then expressed it in that distorted form. It then followed that the symptom could be eliminated if it were no longer necessary as a disguise for the wishes behind it. So, if the patient were helped to confront the origin of his symptom and faced its implications consciously, then it could be expected to disappear. This is more or less what happened in many cases.

It was in this way that what was to become the psychoanalytic method began in the practice of hypnosis. Eventually Freud found hypnosis to be tedious, often difficult to induce, and not as effective in many cases as he had hoped for originally. He began to have his patients, without their first being hypnotized, trace their symptoms by simply having them concentrate and speak freely of whatever occurred to them. This was not only effective but led to patients talking of matters not directly related to their symptoms, including the recounting of dreams. Freud soon found that dreams could be understood in the same manner as symptoms. Dreams expressed a conflict between unacceptable desires and the individual's self-concept which did not allow such sexual and aggressive notions to become conscious--i.e., to acknowledge them as his own. This led to the third and fourth major innovations that Freud brought to psychology. Although neurotic symptoms were limited to a small number of people, dreams were ubiquitous and universal. The recognition that dreams expressed conflicts similar to those of neurotic patients meant that all individuals were subject to uncon-

scious or non-conscious conflicts that needed to be worked
out and expressed in some fashion. Furthermore, it was possible to understand these unconsciously motivated expressions if, just as he was now used to doing with neurotic
symptoms, one understood dreams as statements made through
metaphor and analogy and did not take their seemingly confused form at face value.

Freud also learned that in a number of cases the supposed sexual assault suffered by neurotic patients in childhood was a figment of the imagination. At first chagrined
by having been "taken in" this way, he soon realized that
this was actually a significant finding. What was ubiquitous in neurotics was not the fact of premature sexual activity but the early interest in and excitement around sexuality. Nor was this limited to future neurotic patients; the
evidence of dreams showed that every child was in some way
affected similarly, had to prevent such thoughts from
becoming conscious, and as a consequence had to deal with
some degree of conflict in his later life around this issue.

It was now very clear that the equation between consciousness and mental life was a mistaken one. Everyone led
a psychological life on at least two levels: there was the
conscious one in which the origins and goals of behavior
were understandable, and at the same time there were memories and thoughts that were not only not in consciousness
but unavailable to it. Where the conscious and unconscious
life came into conflict, there was no doubt about which was
the stronger; the unconscious motive controlled behavior,
though it usually permitted itself to be rationalized by
consciousness that hid its true nature and purpose in some
way. Exceptions to this were the so-called errors of everyday life which clearly revealed their underlying motive to
everyone but the perpetrator, who couldn't understand why
everyone was either laughing at him or was angry with him.

Further advances were made with the recognition that
unconscious motives did not limit themselves in their
attempts at expression and fulfillment to neurotic symptoms,
dreams, or errors of everyday life. Any complex bit of behavior, when subjected to analysis, showed itself to be determined by many motives—some of which, often the essential
ones, were totally unknown to the subject. Again it was the
treatment situation that most dramatically brought out this
phenomenon and its implications. Freud noted that his pa-

tients seemed to lose sight of the reality of their relationship with him and at times dealt with him inappropri-
ately--reliving in their attitudes toward him fears, hopes,
and wishes that belonged to other times in their life and
had their origin in relationships to other people (most
often the patients' parents). They had transferred these
attitudes to him in the hopes of reliving the past so that
they might either repeat former gratifications, undo previous frustrations, or lay to rest old anxieties and fears.
Psychoanalysis proper began with the discovery of this
transference and the recognition that it was the analysis of
this phenomenon and its interpretation to the patient that
held the promised cure. As you can see, all the things that
Freud discovered are related to the meaning of human behavior, and it is in the investigation of goal-directed behavior and human motivation that psychoanalysis can make its
contribution to psychology and to science generally.

The impact of Freud's discoveries on Western civilization have been revolutionary. That much of our behavior is
shaped by the influence of the past on the present, and that
we are by no means in absolute control of how that past influences the present, is by now taken for granted. That
children are human beings, and that infants are persons in
the psychological sense who will be influenced by the transactions they have with their parents is also a commonplace;
and we tend to forget that it was Freud's work that made
this concept possible. There is no area in the arts, in politics, or in industry that has not been directly influenced
and shaped by Freud's monumental discovery of a human dimension that had not been dealt with before in systematic fashion. I would venture to say that if you listened to the ra-
dio and television broadcasts on any given day and checked
the newspapers, books, and magazines published on the same
day, Freud's name would come up thousands of times in one or
another context. Why then do we not find psychoanalysis
appropriately related to psychology generally, with productive interchanges taking place between it and the various
areas of that field?

FREUD'S SEARCH FOR A GENERAL PSYCHOLOGY

It is very difficult to picture oneself in Freud's

time, and imagine what it was like to live in a world not
yet changed by his insights. For many years, few people
paid attention to his ideas, and those who did were often
quick to ridicule them. Decades passed before a core of
interested students began to gather around him and form the
beginning of what was to become the psychoanalytic movement.
During its formative years, psychoanalysis was the work of
one man—an enforced isolation that affected its development
significantly. In some ways this was an advantage, since it
permitted Freud to proceed at his own pace, changing his
concepts and methods when necessary without coming to premature closure. However, it also meant that there was no real
opportunity for critical interchange with like-minded colleagues who might have corrected some of his fundamental
misconceptions—misconceptions which are still with us and
detract from the value and usefulness of Freudian psychology.

As I mentioned, Freud uncovered an aspect of mental
life that had not been scientifically examined before him.
The accuracy of his clinical observations is legendary; they
have been corroborated repeatedly by others using the psychoanalytic method and, as I will show later, have received
indirect confirmation from research in other fields. The
difficulties came in the area of formulating causal explanation for his findings.

Freud believed his insight into the meaning of behavior
obligated him to provide an explanation for the thought processes responsible for that behavior. He was not content to
demonstrate the existence of a subterranean form of thought,
the rules it followed, and its effect on behavior; he also
wanted to establish a general psychology based on his discoveries, a task for which he did not have the background.
He was by training a neurologist with extensive experience
in neurophysiological and neuroanatomical research. He
shared the Helmholzian viewpoint that all aspects of life
could ultimately be reduced to, and explained in terms of,
physics and chemistry. This meant that for psychology to
become a true science, mental life would have to be expressed as a brain physiology which ultimately would be
transformed into the basic physical sciences. Freud was not
the only neurologist challenged by the task of legitimizing

psychology by making it "scientific". Others, including his teachers Exner and Meynert, had proposed brain models to account for memory, thought, affect, and perceptions; and Freud adopted features from these earlier works to construct his own neurological psychology. His attempt in this direction was a manuscript, unpublished in his lifetime and found only a few years before his death, now called "The Project for a Scientific Psychology" (1895). It was written at a time when Freud had been working with neurotic patients for about ten years but had not yet achieved his understanding of dreams, thus before psychoanalysis had come into being. What is important about this is that the brain model (actually models) of the "Project", although it was to undergo a number of primarily terminologic modifications over the years, to this day serves as the basis for the explanatory theory advanced to account for the clinical findings of psychoanalysis.

The fundamental idea in the "Project"is that the process of thought is essentially a two-step operation. First, a stimulus in the form of an image is received and registered by the brain; this is perception. Then, if the percept is significant enough and there is no obstacle, this pictorial image becomes connected with verbal imagery, with language. Once a picture and the words describing it unite, the perception can be further manipulated logically by the brain --i.e., associated with other picture-word complexes in the interest of arriving at some goal. The hypothesis that the basic unit of thought is an image connected with a word-memory was first put forward in a neurologic work Freud published in 1891 entitled On Aphasia. Aphasics are individuals who have suffered vascular, mechanical or infectious brain damage which has left them with a partial failure of function. For example, such a patient may be able to identify by touch and correctly use objects placed in his hand, but will not be able to name them. Freud concluded that such pathologic phenomena were decompositions that reversed the process of normal development. That is, if perception and speech can be separated by disease, that indicates that originally these two were discrete functions synthesized during development. In other words, pathology reverses ontogeny. This assumption may explain why Freud never found it necessary to study the development of infants and children in any systematic way. He was always convinced that insight into normal development could be obtained by extrapolating from pathology.

Although Freud never mentioned it, the similarity between his formulation for the pathology of aphasia and that of the psychoneuroses must have been clear to him, reinforcing his convictions regarding the formation of the thought process. Like the aphasic, the neurotic seemed to be suffering from percepts that could not be expressed in words. Once the hypnotist enabled sensory memory to be joined with speech so that the patient could talk about the events precipitating his illness, the symptoms disappeared. Therefore, psychoneurosis must be a failure in normal thought-formation, due to the emotional stress of a traumatic experience. In any case, Freud always dealt with neurotic symptomatology as if it were a functional aphasia, and as if the analyst through his treatment restored the missing connections.

This concept of the thought process is fundamental in all of Freud's theory on the level of abstraction he called "metapsychology". He was to formulate a number of dualistic theories with which to account for the working of the brain or mind. As his clinical work progressed, he continuously found his explanations wanting and altered them in one way or another on an ad hoc basis. This gave rise to a very complex and, not surprisingly, unsatisfactory conglomeration of dualistic hypotheses and theories that continue to be the focus of much argument and many attempts at clarification. These dualisms--conscious and unconscious, primary and secondary process, unbound and neutralized energy, ego and id--can all be understood operationally as expressing in one way or another the postulated separation between sensory percepts and the words that describe them.

Something else to be learned from the "Project" is that, contrary to the commonly held belief in analytic circles, explanatory theory or metapsychology was not derived from psychoanalytic experience. Quite the contrary: Freud had worked out, before psychoanalysis was established, his theory of how the brain-mind worked; but he continued to try to use these explanations to encompass the clinical discoveries he was making. Freud was not trained in psychology. Had he been, he would have been familiar with developmental theories of perception, memory, and association that--inadequate as they were at the time--would nevertheless have given him a different perspective on his metatheory. He approached his findings with a physicalistic orientation, and accepted as a matter of course the necessity of explaining them in terms of brain mechanisms. Though he soon recognized the futility of his effort (given the state of knowledge about the brain in his day) and declared in The Interpreta-tion of Dreams (1900) that he would henceforth leave anatomy and physiology behind and utilize only psychologic explana-tions, all that happened—again contradicting the majority opinion in the field—was that the words changed. Instead of "brain", "mental apparatus" was substituted; but the new machine operated exactly like the old one in the "Project" and had the same shortcomings.

Though Freud said this "mental apparatus" was only a metaphor that he found useful in describing a process and was not to be confused with actual anatomy,it was clear that his work as a neurologist and his attempt to unite neurology and psychology in the "Project" of 1895 formed the founda-tion for these new theories. Indeed he himself, in discuss-ing the mental apparatus, often lapsed into neurological terminology; and he went so far as to say that eventually the actual organic substratum for these concepts would have to be found (see Basch 1975b, 1976a).

The operations of the mental apparatus were based on the following assumptions about general psychology: (1) The goal of the brain or mental apparatus is to avoid stimuli, or at least minimize them to the greatest extent compatible with the preservation of life. (2) Thought development is linear; and, though it is less complex, the infant's appre-ciation of self and the world is based on the same princi-ples of perception and cognition found in adulthood. (3) The intensity of thought, its affective component, is determined by the quantity of a postulated "psychic energy" which attaches itself to a wish and determines its influence on behavior. (4) Speech development precedes, and is essential for, thought. (5) True thought is equated with verbal logic and is made possible by a union of sensory, pictorial images with the more sophisticated and later-acquired word images. These premises, based on an adultomorphic view of infancy and childhood, are incorrect. In order to utilize psycho-analytic insight in formulating a psychology that will be scientifically valid as well as useful in practice, Freud's findings will have to be placed in the context of what we now know from developmental psychology, linguistics, and communication theory.

### THOUGHT AND WORDS

Things are seldom what they seem to be. Take the notion that we think with words: at first blush it seems reasonable enough. As I write this I am aware that the words form themselves first in consciousness and are examined, rejected, and revised there until finally approved and committed to paper. However, I have had the experience (not unique, I am sure) of lecturing and finding myself saying things that I didn't know I knew. How does this happen? In this instance the knowing seems to antedate the becoming conscious. Does that mean that inside my head there are, arranging themselves, words that I only later become aware of? What about the many complicated activities that we carry out every day--from tying our shoelaces to driving a car--that we have never put into words; are they the product of something other than thought? Infants don't speak; is their problem-solving activity the result of thinking or not? If not, what happens? If so, does that mean that there is wordless thought?

Man, anthropologists tell us, has possessed verbal language for only a few hundred thousand years at best; yet our race has been identifiable as human for millions of years. Were these non-verbal ancestors unthinking creatures; were their achievements not thoughtful ones? If words don't separate us from other clever problem-solving animals, what does? What makes human beings human? There must be something, for no other creature spends (or wastes) its time formulating (much less trying to answer) questions such as these. Anyway, do we psychotherapists need to have answers to such questions? Would it not be wiser to back off, to restrict ourselves to the area of practice, and to avoid issues that Freud himself could not deal with satisfactorily? For better or worse, that is not possible. We must continue to search for the solutions to such basic questions because our daily work as psychotherapists depends on how we answer them. Since psychotherapists work on the assumption that their patients' difficulties stem from some failure of or interference with their capacity to reason effectively, the development and nature of that function needs to be established. We do have the advantage today of advances in many areas of psychology and related fields that were not available to Freud; with these we can approach his clinical findings and explain them very differently. For example, how does it happen that the infantile, non-verbal years are the

most impressionable years?  Clinically it has been repeatedly demonstrated that the emotional climate and the events of
the  first  few months and years of life leave a lasting and
telling imprint on the character of the individual.  Yet how
is it possible that an infant can  "know" so  much  and  so
accurately  about what is going on around him when he cannot
formulate his experiences  and  the  reactions  to  them  in
words, as older children and adults do?  Does a baby reflect
on  his  experiences;  if  so, how?  To answer some of these
questions, Freud reasoned his way to  a  picture  of  mental
functioning  at  birth  and  thereafter that is known as the
psychoanalytic theory of development.

## INFANTS: FACT AND FICTION

        I have many times made the experiment, in  my  classes,
of  asking women psychiatrists in psychoanalytic training to
tell the group about the psychical condition of the  infant.
Just  like  other  well-informed students, they describe the
newborn invented by Freud and his contemporaries as  an  un-
tamed,  asocial,  uncoordinated bundle of needs and unsatisfied passions. Wanting nothing more than to be  left  alone
to  sleep,  the  infant  is  forced  by drive pressures like
hunger and thirst to seek contact with  the  external  world
through primitive reflexes.  Cognitively the baby is a blank
slate,  a tabula rasa upon which society--using the infant's
need for drive gratification  as  leverage--laboriously  and
against  resistance  writes  its  bitter message: conform or
perish.  The helplessness of the infant  usually  guarantees
that  (outwardly, at least) the world will win the struggle,
but for the rest of his life there is rage against the  need
to  adapt.  Under the best of circumstances, the struggle is
eventually sublimated; the anger and disappointment at  having  to  contain one's impulses become transformed into love
for others, into love for abstract goals, and into  enthusi-
asm for work to be accomplished.  More often, the compromise
is  only  superficially  satisfying and the struggle between
drive and the need for adaptation leads to neurotic  pathology of varying degrees.

        When  I  then ask the same candidate if she has a child
and if so to tell us about its infancy,  we hear of a little
human being with a distinct personality,  a  wide  range  of
emotions,  and  an  eagerness for stimulation and companion-
ship--curious about new things and  quick  to  learn.   This

baby is not an angry unhappy blob of unwilling protoplasm, but someone who in his own way converses happily with his mother at great length, has a distinct personality, and re- fuses to behave in the way that psychoanalytic theory would predict. At this point, other members of the class--both male and female--who are parents chime in with their respec- tive experiences, attesting to the individuality and human- ity of infants from the moment they are born. Is a proud mo- ther just deluding herself,reading into her baby the respon- siveness, affection, and ability she wants to see; or does the fault lie with the theory?

It is the belief that "communication and learning de- pend on verbal language" that has been responsible for the misunderstanding of human infancy. However, as the work of the psychoanalyst David Freedman and his colleagues (1971, 1972) has shown, the belief that "communication between in- fant and parent, and the child's subsequent relationship to the world around him, depends primarily on speech" is false.

Ingenious experiments have now demonstrated that in- fants are born with perceptual abilities that are not only sophisticated but selective. The ability to identify the lo- cation of signals, to prefer structure over disorganization, to perceive in depth, to discriminate among colors, and to practice many other skills heretofore thought to be acquired through learning, turn out to be hereditary givens with which the infant meets his environment.

We know today that the newborn infant's goal is not to withdraw into Nirvana, a state of satiated apathy, but, on the contrary, that he is stimulus-hungry from birth, favors novelty (Carpenter 1974, Bower 1971), and, furthermore, is selectively programmed to respond to and elicit human inter- action (Condon & Sander 1974; Carpenter 1974). Throughout life, the function of the brain is to maintain and organize an optimal stimulus input; understimulation is as great if not a greater danger than overstimulation (Basch 1975a).

Equally important for understanding the mental life of the infant is the evidence that, from the first day, the newborn learns from experience very much as do older chil- dren and adults: patterns of expectation are laid down that are reactivated when relevant stimuli are repeated and call forth the appropriate learned behavior. For example, a child of three days can be taught, by being rewarded with

the   sight  of  a  brightly  colored  object,  to  turn  his  head  to
one side when he hears a buzzer and to the other if he hears
a bell (Bower 1971).  Furthermore, like other  animals,  the
infant is alerted to and ready to respond to input important
for his species. Right from birth the human voice, face, and
smile   are   given   preference  over  other stimuli  and greeted
with inborn behavioral responses that remain essentially the
same  throughout  life.   The  normal  infant,  if  given  the
choice,   shows   more  interest  in  unfamiliar  than  in  familiar
ones.   Clearly the evidence favors maternal observation  and
not   professorial   speculation:   the  infant  is  not  a  little
wild animal that needs to be tamed, but an organism  ideally
equipped   to   join   and   participate  in  human  society  (Basch
1977).

## AFFECT AND COMMUNICATION

It came  as a  surprise  to  the   scientific  community  to
learn that infants have the sophisticated perceptual capacities just mentioned.  However, we could excuse our oversight
because  infants lack both speech and large-muscle coordination with which to let us know directly what they  perceive.
But  when  it  comes  to the affective reactions of infancy,
nothing is hidden.  The facial expressions of babies are exactly the same ones associated in adulthood with  such  emotions as joy, interest, anger, fear, surprise, disgust, distress,  and  shame:  however, though open to inspection, the
significance and implications of these reaction patterns  of
early infancy were, for the most part, neglected by psychology.   No  one  really took seriously such obvious questions
as:  Does a baby know he is happy when  he  smiles?   If  he
knows,  how  can he picture it to himself without being able
to reflect, through words, upon his state?   If  he  doesn't
know  he is happy, why is he smiling?  Speculation aside for
the moment, parents have always used the  affective  expressions  and sounds of babies as reliable indicators of an infant's  needs  and  have  responded  accordingly,  implicitly
acknowledging  thereby  that the baby is not just a receiver
but also a sender of messages.

It is a commonplace to point out that human beings  are
born  helpless  compared  to the young of most other animals
who come into the world prepared  to  engage  in  the  basic
life-preserving  activities  of  their kind.  Looking at our
newborn in terms of his inability to  find  and  follow  the

mother, to recognize and avoid danger, etc., this observation makes sense. However, considering that extensive and effective communication is the most important function of a human being and is the basis of this otherwise weak and poorly adapted creature's preservation in and mastery over nature, the infant is superbly equipped. Other animals are limited by their neonatal maturity to a relatively narrow environmental range, one suited to their inherited characteristics and preprogrammed behavioral patterns. Independent of most factors that restrict the development of other animals, man is born with the greatest possibility for adaptation. The only "environment" to which he must be attuned is his mother. His capacity to receive and communicate is ideally suited to that task. How long, I wonder, would even the most primitive race of men have survived if Freud's hypothesis were correct that each new generation had to be forced against its nature to become human?

Darwin (1872) commented on the fact that the facial expression of emotions is similar in adult man, human infants, and animals (especially other primates). Furthermore, it is the same in all cultures, for example, people of other cultures identify the picture of an angry Caucasian baby as being that. Darwin suggested that these facial configurations were originally components of larger behavior patterns essential for survival that gradually came to have a communicative function. They signal attitudes and test the forthcoming response before committing the organism to action, thereby making the total reaction unnecessary in most instances. Substantiating the idea that affective expressions were inborn, stereotyped, and universal, were the experiments of Darwin's contemporary, Duchenne, who demonstrated, using electric stimulation, that the triggering of specific facial muscles automatically created emotional expressions independent of the subject's feeling-state at the time. This is corroborated by neurological evidence now available indicating that affective expressions are under subcortical control. Not much was done with this work or its implications until Silvan Tomkins (1962-63) and his co-workers recognized its significance and carried it further. Tomkins suggests that initially the various affective responses are related to stimulus intensity and to intensity gradients, and are probably mediated by the autonomic nervous system. In other words, affective behavior patterns are—like the infant's perceptual capacity—inherited programs with which the infant meets the world.

Although the affective response, being mediated by the autonomic nervous system, is a total bodily response involving body temperature, the skin, hair, glands, muscles, and viscera, in humans it is the face that has become the prime communicator of affective states. Let me just briefly list the main objective (i.e., recordable and reproducible) signs used by researchers (and by parents) to identify in infants the eight ranges of affective response that form the basis for our emotional life. There is a continuum from surprise to startle, depending on the suddenness and the intensity of stimulus onset, in which we see the infant's eyebrows go up and his eyes blink. The range from interest through excitement is shown on the face when the eyebrows go down, the eyes track the stimulus, and there is an attitude of looking and listening. The scale of mild enjoyment to joy is signaled by a smile; the lips are widened and out, and there is slow deep breathing. On the negative side there is the range from distress to anguish, as indicated by crying, arched eyebrows, corners of the mouth turned down, tears and rhythmic sobbing. Even more dramatic is the behavioral series beginning with anger and culminating in rage. Here the face is in a frown, the jaws are clenched, and the face is red. Contempt or disgust is shown by lifting the upper lip into a sneer. The range from fear through terror is manifested by a cold, sweaty face, with eyes frozen open, facial trembling, and hair erect. And shame through humiliation is demonstrated when both the head and eyes are cast down. (Tomkins 1962, Vol. I:337)

As the mere 26 letters in our alphabet can be combined to generate the wealth and nuance of our verbal language, so the even richer variations to be found in the respective ranges of each of the eight basic affects will in time blend and mix, using not only face and body but the variations of the voice, to communicate our feelings at any given time. Affect is a nonverbal language—sometimes accompanying the spoken word, as the Greek playwrights of old used the chorus to underscore the lines of the principal actors; more often, as Freud taught us, a counterpoint that, for those alert to it, reveals the consciously or unconsciously concealed truth behind the spoken words.

Now, to return to the question I asked at the beginning of this section on affect: Does the baby know how to react when he is as yet in no position to reflect on the nature of the stimuli to which he is exposed? Tomkins (1962-63, 1980) suggests that initially and essentially the primary affect responses are related only to stimulus density (the number of nerve impulses transmitted per unit time) and to the gradient of that stimulation. Traced on a graph, surprise, interest, and fear—and their extensions—are all the result of a continuous rise in stimulus intensity above a particular infant's non-reactive, resting level; which of these three reactions (surprise, interest, or fear) will be elicited is determined by the rate, height, and duration of the stimulus vector. Distress or anger reactions are manifested when stimulus intensity is maintained steadily at a higher than optimal rate for a length of time. Joy and its variant results whenever there is a sudden sharp drop in the stimulus pattern, no matter what the nature of that stimulus may be. (Having playfully tossed the baby in the air; we catch it; and the baby laughs—not because he knows he is safe after having been in danger, but because there has been a sudden decrease in the stimulus input that the act of tossing him up and letting him fall had produced.) Shame is the result of an inadequate stimulus reduction, a tension produced by less than satisfactory resolution of a stimulus gradient that had begun but was not carried to completion. The only exception to the quantitative determination of affect is the qualitatively triggered protective withdrawal reaction of contempt or disgust, a primary affect continuum based on an inherited avoidance response triggered chemically by tastes or odors that register as noxious; for example, the smell of butyric acid invariably produces a disgusted curling of the lip and flattening of the nose in newborn humans.

In the study of affect we find the biological foundation, the primary motivation (Tomkins 1970), that explains the intensity of ideas—i.e., their meaning, which Freud sought in vain in the nineteenth-century conceptualizations of instinct, psychic energy, and drive. How body becomes mind, how tension becomes recognized as need (what Freud termed "the mysterious leap") appears to be much less puzzling given the transformation of quantitative stimulus gradients into the experience of quality that we associate with our affective/emotional life. It seems to involve the same sort of transformation that our brains are able to carry out when they take the purely quantitative stimulation of electrons in varying wavelengths that impinge on our sense organs and transform them into the imagery of sight, hearing, and so on.

We can now reexamine the question of what it is the infant experiences when he signals his affective state. Certainly no reflection is involved. It is the baby's way of
adapting to stimuli in the only way he can,given his inability to move closer to what attracts him or to escape what
hurts or frightens him. His only option is to engage his
caretaker and induce her to do for him what he cannot do for
himself. The highly developed facial musculature of the infant and the automatism of the affective reaction is ideally
suited to setting up an error-correcting feedback system in
the mother-child unit; the child provides the indicator by
which the mother can judge whether her efforts on his behalf
have been successful. We can now answer in the negative the
question of whether a baby knows he is happy when he is
smiling. No, the baby does not know he is happy; but his mother attributes that emotion to him, and that's what is important. "Happy" is a concept that has no meaning as yet
for an infant, nor is there a concept of self that could
serve as a reference for such a subjective judgment. Eventually the affective state, like other experiences, can become
symbolically conceptualized and labeled verbally; only then,
in my opinion, should it be called an "emotion".

The picture of communication in the first eighteen
months of life that I am drawing here has further implications for psychological theory. The wide range of autonomic, inherited experiential patterns described by Tomkins,
which become the basis for an explanation of man's emotional
life, are in keeping with everyday observation and with
clinical experience. As a result we are no longer limited,
as we have been in psychoanalysis, to trying to account for
the quality of experience by categorizing it as being either
love or hate--i.e., the outcome of the discharge of either
sexual or aggressive instincts. Neither do we need to explain emotional life as being due to either the damming up
of a postulated psychic energy with resulting unpleasure, or
to its discharge accompanied by pleasure. I have dealt more
extensively with the issue of affect theory and its significance for psychoanalysis in two earlier publications (1975a,
1976b). But I hope that, though necessarily brief, this review has demonstrated how the variety and quantification of
affect replaces the metaphor of a psychic energy by giving
us a better, as well as a scientific, explanation for what
we experience as the intensity of our thoughts and the force
of our wishes.

Initially linked primarily to stimulus intensity, these affective responses--mediated sub-cortically and transmitted by the autonomic nervous system--become linked to and associated with the cerebrally registered perceptual experiences of the infant. Together they form a cognitive matrix in which the infant's experience of his interaction with the surround can be registered in neurological patterns that bear not only the imprint of what happened but also the affective tone that accompanies the event. I hope I have said enough (although it has been only an outline) to show that the work on perception and affect in infancy provides an alternative to the 19th-century conceptualization of development, and enables one to account for the problem-solving capacity of the non-verbal neonate without resorting to either zoomorphism or adultomorphism (see also Basch 1976b).

## THOUGHT AND COMMUNICATION

If by "thought" is meant the capacity to organize stimuli into meaningful patterns which guide the organism's adaptation, then the process is not a linear one, nor does it progress simply by accretion. The world in which an organism exists depends not only on the kinds of stimuli to which it responds, but what it does with them. Piaget's work (Piaget & Inhelder 1969) seems to show that we humans pass through a series of learning stages each based on its own organizing principles, and that (though there is continuity) the cognitive processes at various ages are different in kind, not just in degree. The infant, the child, and the adult literally live in different worlds.

During the first eighteen months of life, sensory input is organized around the actions that are performed in response to or in connection with stimuli; this is termed the "sensorimotor" stage. These encoded action patterns, the sensorimotor schemata, are the fundamental elements of all thought. They serve as "practical concepts". Once laid down, repetition of stimuli or sensations reproduces the activity previously associated with them; recognition in infancy is in terms of action, and it is through action that the infant gradually creates an intracerebral model of the world, of "reality". Piaget has subdivided the sensorimotor phase of development into six sub-stages progressing from simple habit formation to behavior that anticipates an end result and mobilizes the appropriate means for its achievement. Apparently, recall through imaging plays no part in this process, and therefore reflection (which is such an important aspect of adult thought) is not possible; yet, learning can take place.

Words as symbols representing a particular event or experience do not play a part in sensorimotor learning of infancy. Insofar as isolated words are used in those early years, they do not stand for any concept, they do not name or classify objects, and they have no private significance. As Vygotsky (1934) pointed out, words at that time have the purely social significance of being supplemental gestures, serving simply as a form of reaching or pointing. The word in this case is an action, a part of the sensorimotor schema that the child has learned will lead to a desired result. For example, "milk" will bring food; but it does not, for the child, represent food.

What Piaget does not deal with in his investigations is the motivation for learning behavior and its continued elaboration. Freud postulated a "sexual" instinct striving for organ pleasure and tension discharge as the driving force behind infantile behavior. However, the concept of instinct as a quasi-physiologic push given by the body to the mind is no longer a tenable one (Arnold 1960). But Freud's suggestion that infants, like adults, strive for what is pleasurable and avoid what is painful (the so-called pleasure-unpleasure principle) does seem to apply—though it needs to be expanded and divorced from the concept of instinctual drive.

Affective reaction provides inherent basic controls for adaptation—by placing a qualitative value on percepts, thereby providing the motivation for goal-seeking behavior. The interaction and union of ever more complex sensorimotor schemata, guided by the affective tone of past experience, results in a network of error-correcting feedback cycles which are perfectly adequate for the infant's needs and (most important from the psychoanalytic point of view) explain how it is possible to permanently embody a record of an infant's affective/behavioral transactions before imaging, recall, or reflection are possible.

The affectively colored, permanent sensorimotor record of the infant's experience is the basis for character formation. The parents, as the practically exclusive providers

of external stimulation, play the dominant role in laying
down the basic view of the world that will determine later
attitudes--just as psychoanalytic findings have long suggested was the case. However, the infant does not react to
his mother and father as if they were objects (i.e., identified sources of message input); "mother" and "father",
"thing", "part", and "whole" are concepts made possible by
reflection, and are not part of the infant's perceptual
world. The baby responds to the kind and degree of stimulation provided, and especially to the affective tone of his
parents. The parents' attitudes toward him are conveyed to
the infant through tone of voice, rhythm of action, sureness
of touch, and other signal patterns which are, to a greater
or lesser extent, not in the adult's awareness. The infant,
in responding to the kind and quality of the messages sent
to him, lays down the aforementioned action patterns that
form the basis of his personality and are a response, so to
speak, to what his parents are "telling" him about himself
and the world he has entered. This form of communication remains basic throughout life (though, for the most part, people continue to remain unaware of it). These transactions,
I believe, account for Freud's observation (1915) that the
unconscious of one person can communicate with that of
another without passing through consciousness--or, as we now
should say, affective communication can proceed without the
benefit of discursive verbal description or self awareness
throughout life just as it does in infancy. There is no
mystery about such "non-verbal" communication once it is understood that the basis of thought is not imagery but the
record of affectively triggered interaction with the environment. Such an understanding makes it possible to eliminate the adultomorphization of the infant found heretofore
in psychoanalytic psychology and carried out to its logical
extremes by Melanie Klein and her followers, who in essence
attribute formal reasoning to the infant.

Throughout the sensorimotor period, the infant learns
to adapt to his surround--utilizing his ever-increasing motoric skills guided by the opportunities provided by the environment. Obviously the degree and type of affective stimulation provided, especially by the mother, will determine
the affective axis around which the infant reorganizes his
experience. The broad range of stimulation at reasonable
levels of intensity provided by a self-confident and happy
mother will of course be quite different from that offered
by a depressed mother whose apathy and resentment alternates

with overstimulation generated by fearful concern and guilt.
As the findings of psychoanalysis in adults demonstrate,
such early variations in caretaking lead to significant
differences in character formation. Parents are able to
transmit their unconscious traits and their idiosyncratic
views to their infants by their manner of fostering affective development. In a given situation, one affect may be
stimulated optimally, another overstimulated and perverted,
while a third may be arrested in development for lack of
stimulation. So, already in infancy and long before the
child can reflect on his experiences, he lays down mood
patterns (i.e., broad dispositions to reaction--a readiness,
let us say, for one infant to respond with fear, while
another is generally set for joy).

Piaget notes that dreams, phantasy, and play-acting--
all of which involve deferred imitation--are usually not observed until after the age of 18 months. This behavior ush-
ers in the stage of preoperational development (Piaget &
Inhelder 1969), in which re-presentation of experience
through imagery begins. This means that sensorimotor action
patterns can become transformed by imaging, which makes recall and deferred imitation (that is, imitation in the
absence of external stimuli) a possibility. Imagery is at
first static, idiosyncratic, and syncretic. Everything that
happens is made to fit together on the basis of contiguity
of location, simultaneity of occurence, similarity in appearance, and affective import. What is most important affectively becomes the focus for organization, and events are
distorted to fit the pattern that is personally most significant. Throughout life, our dreams reflect this phase of
thought and the unconscious wishes uncovered by psychoanalysis also exist in this form. It is this organization of
perception that Freud explored and decoded when he investigated neurotic symptoms, and which he called the "primary
process".

The preoperational phase marks the beginning of symbolic control over behavior. Once an experience and its
associated affective components can be re-presented through
imagery, evocative recall can provide motivation for action.
Where previously the infant was dependent on either inner
need or external signals to stimulate behavior, now the
imaging activity of the brain can initiate action by providing motivating stimuli.

Not all experiences are positive, and with the capacity for recall comes the possibility of generating unpleasure through recollection of painful experiences. The need for the psychologic defenses familiar through the analysis of psychoneurotics, but present in everyone, originates in this stage. Once the function of imagination becomes a possibility, ways must be found to avoid imagery experienced as traumatic (i.e., generating disorganizing amounts of negative affect). During the preoperational stage, avoidance through the kind of illogical distortion seen in dreams is still effective. If something is not to one's liking, it is simply changed around until it fits better with one's hopes and expectations. However, in the next transformation of thought—the concrete operational stage—the organizational principles are logical ones, and defense against painful mental experiences will be seen to change accordingly.

During the preoperational phase, words are used to label the egocentric concepts that are being developed. Words are now representative, but they represent the activity of the child more than the labeling or classifying of the object involved. Ask a 3-4 year-old child "What's a bicycle?" and he will begin to pump his legs and hold his hands on the imaginary handlebars in front of him and perhaps say "Something to ride on". Vygotsky (1934) found that, at first, children use words to label the endproduct of their activity (for example, calling some scribbles on a paper a "house" or an "airplane"). But gradually the word is shifted to the beginning of a task, announcing what is going to be accomplished—an image held out as a goal to be fulfilled—and the drawings become much more realistic and accurate.

At this point, words are valuable in planning for activity; they hold experience in focus and promote connections between action and experience. Luria's (1961) experiments have shown that even very young children can master quite complex tasks if they are helped to attach words to the various aspects of the performance. Speech is now employed to gain control over one's own activities, promote goal achievement, and set direction for action. Rather than speech determining what concepts are formed, it is the beginning of concept formation through "preoperational imaging"—what Langer (1942) calls "presentational symbolism" (symbolism which presents the event in its totality rather than subdividing it in linear fashion)—that creates a need for words and makes the child receptive to mastering the vocabulary of his culture.

Conscious reflection—and with it the possibility of objectifying experience, including objectification of the reflexive function itself as "self"—has its origin in the preoperational period. This appears in speech when the child not only describes what he will be doing, but also refers to himself as actor in the third person (e.g., "Johnny go bye-bye"). Then "predicate speech" develops (for example, "go bye-bye"), where the reference to self as actor drops out, indicating that there is an implicit recognition of self as the subject acting.

Logical manipulation of the perceptual world begins at about six years of age. The first step of this development Piaget terms the stage of "concrete operations". "Concrete" because the cognitive manipulations or "operations" are performed on the sensory aspects of percepts. For example, abstractions like "part" and "whole", as well as all those dependent on sensory qualities like color, tangibility, size, etc., become part of thought. When the affective aspect of experience also becomes objectified, "feelings" may be said to exist. The ability to call by name, classify, and mentally manipulate the autonomic reactions that comprise affect permits the associative cortex some measure of control over what had heretofore determined behavior. It marks an important turning point in human development: the organism is no longer totally at the mercy of the old-brain's adaptive mechanisms but instead controls the controller through understanding. During the concrete operational phase, from approximately 6 to 12 years, thought is organized according to what Vygotsky (1934) calls "complexes"—that is, classification based on complex groupings formed through perceptual associations.

About age 11 or 12, a final transformation of thought takes place, which Vygotsky calls "conceptual" and Piaget terms the "stage of formal or propositional operations". This transformation makes possible thought in purely verbal concepts, or propositions—independent of sensory qualities. This means that a common hypothetical element can be abstracted from a perceptual complex, be used as the basis for association, and be subject to further mental manipulation. Concepts can now be freed from physical qualities. For example, it is only now that the concept of "color", as such, can be used as a basis for classification free of particular perceptual input; numbers no longer have to designate something specific, and therefore can be used mathematically and not just arithmetically, etc. Totally abstract constructs (such as "cause and effect") which refer to hypothetical relationships that can never be directly available to the senses become possible at this time.

This development is made possible by a transformation of the speech function. While speech had been used to externalize concepts, to put concepts into words so that they could be communicated and used in the interest of controlling the environment, now what Vygotsky calls "inner speech" makes its appearance. Inner speech does not just communicate thought, it generates it by making possible internal dialogues in the interest of problem-solving. Formal reflection, the manipulation of verbal propositions, is dependent on speech; and it is this final phase of cognitive development that led Freud to assume that speech and thought were interdependent throughout development.

Inner speech raises the concept of self to a more abstract level, so that a self embodied in principles and ideals can become focal for action. Affect too is influenced by this maturation. While moods and feelings existed as concepts before this, emotion—in the proper sense of the word, an affective stance related to the self concept—now makes its appearance.

What one chooses to call "thought" is arbitrary. I favor including all ordering activity of the brain, beginning with the sensorimotor stage (Basch 1976c). I know that others would disagree—some even claiming that only the final development, the manipulation of pure abstractions, deserves to be termed "thinking". Be that as it may, what is important is the recognition that the development of thought involves a series of transformations.

ORDERING, LEARNING, AND ERROR-CORRECTING FEEDBACK

As psychotherapists, our conceptualization of infantile development and mental operations has a significant influence on the way we conduct treatment and the goals we implicitly help patients set. If we think of ourselves and others as basically filled with dangerous urges that seek expres-

sions in forbidden aggressive and genital behavior, then our
emphasis (whether we are aware of it or not) is on control
and on protecting the patient from himself. It is therefore
not just of interest to theoreticians that the studies of
infants alluded to above and recent work in the psychoanalysis of adults (Kohut 1971) lead to very different conclusions. As was pointed out, we are born attuned to a wide variety of stimuli, and with a readiness to organize and respond to the environment. We now know that the function of
the brain is not to process or discharge energy but to selectively organize or order the myriad stimuli that impinge
upon it from the external environment, from the rest of the
body, and, most importantly, those created by its own activity. (See Basch 1976c.)

Perception is fundamental for ordering. Every animal
lives in and responds to a world fashioned by its particular
sense organs and the way it arranges the stimuli it receives. Perception is not simple or passive; neither eye
nor brain operate like a camera obediently registering whatever happens to fall on the film (Brandt 1945, Basch 1976c).
It is always a creative act, involving the exploration of
the surround and--through comparison with past experiences
or through genetic endowment--fashioning a figure-background
configuration that has sufficient familiarity to be recognized. If and when we reflect upon this process, we call it
understanding. As has already been discussed, perception is
selective, making it more likely that certain configurations
and stimuli will be preferred to others. Man, like other
animals, is born with certain perceptual proclivities, but
in addition also has a seemingly unlimited range for developing perceptual preferences or sets based on experience--a
perceptual configuration being sought after or avoided,
depending on the affective experience associated with its
occurrence. I suggest that the increasingly effective application of the perceptual process constitutes learning;
but what does that mean operationally (i.e., How can what
happens be described? How do we learn to do anything better
than we did it before, to say nothing of learning to do
something for the first time?)? Since we psychotherapists
deal with people who seem either to have not learned what
they need to know to function reasonably happily or to have
learned it improperly or incorrectly with self-defeating
results, learning is something we need to know about.

A five-day-old child needs many trials to find its mouth with its thumb; when he is ten days old, he is demonstrably more adept at doing that (Murphy 1973). Clearly, something has been learned over that time; but what? Conventionally the answer has been that trial-and-error has perfected motor coordination and transformed the activity of thumb-sucking into a habit that, to mother's distress, is from now on available and resorted to on a reflex basis in time of tension, boredom, frustration, etc. This is usually explained neurologically by saying that the brain, through trial and error, has learned to control its output more effectively. A whole school of psychology has grown up around the idea that the role of behavior is to create an observable effect in the environment (approximating the thumb to the mouth, for example) and that psychotherapy, when it is needed, should address itself to diagnosing the behavior pattern that is failing the patient and then restructuring it appropriately. In other words: eliminate bad habits, and develop serviceable ones. Would that we could! Then we would not have to concern ourselves any longer with speculations about what goes on inside that unobservable "black box", the brain—what a relief! Practically speaking, however, the application of this philosophy of therapy presents the psychotherapist with some problems.

For every patient who responds to re-education of this sort, there are others who are perfectly well aware both of what they are doing that is wrong and how they could correct it—but they don't do it. Some, no matter how encouraging and persuasive the therapist may be, never do manage to overcome their bad habits; why not? In a way, even more puzzling are those in this group of patients who have not responded to the well-meant and correct advice of friends and family but, when they are told essentially the same thing by the therapist, miraculously pull themselves together and recover. Why should the therapist make such a difference? Another experience that makes it doubtful that behavior modification is the panacea is the number of patients who have had a favorable response in treatment and have seemingly learned to behave effectively, only to relapse into their former difficulties shortly thereafter. If a baby never forgets how to suck its thumb, why do patients "forget" their hard-won gains in therapy when so much is at stake? Instead of bypassing the need to speculate about the nature of thinking, it seems we have added some further questions.

As a matter of fact, the commonsense supposition that "When we learn, we learn behavior" has long been in trouble because of the phenomenon of equifinality. If you restrain the right hand of a baby who has learned to suck the thumb of that hand, he will without great difficulty find his left thumb or, unable to move his thumb to his mouth, he will without further trial and error move his mouth to his thumb. It takes a while for a rat to learn to run a maze and reach the endpoint where the food that it smells has been placed. Having mastered this bit of behavior, the rat, when the maze is flooded, now not only readily enters it but quickly swims along the route he has learned to get his reward. Teaching a child to write his name usually requires a long period of trial-and-error learning; but when he has mastered it, you can ask him to write his name in the sand with his toes, or with his nose, and he will oblige you on the first trial with no difficulty. How does it happen that the baby, the rat, and the child—all of whom needed time to correct error in mastering a particular behavior—can then achieve rapid or even instantaneous success in applying adaptational patterns for which they have not been previously educated? It seems that the achievement of a particular goal, having once been learned, can be managed in many different ways without further training (a phenomenon known as "equifinality"). For behaviorists to explain this by saying that "the learning experience has been generalized" not only begs the question of how that happens but puts them in the position of postulating one of those vexatious, unobservable events inside the head that they swore to eschew.

Equifinality demonstrates that the emphasis in learning should be on the goal, not on the behavioral technique that was used to reach it. The difficulty with that concept is that it raises the specter of what Gilbert Ryle (1949) has called "the ghost in the machine", the little homunculus that sits in the brain and tells it what to do. If you have a goal "in mind", as we say, there must be a little fellow called "I" or "ego" in there who observes the incoming stimuli for the opportune moment and says "O.K., now!" to the brain and gets things working at the right time. And who then controls the controller?

William Powers (1973) has shown that the concept of error-correcting or negative feedback permits us to reintroduce the concept of goal or motive into psychology without reverting to animism. The brain can be pictured as operating akin to an analogue computer--operating by controlling its input, not its output. A thermostat can be used as a model to illustrate this basic concept. The thermostat's world is one-dimensional; it can sense only temperature. It has a dial on which you can choose a temperature which serves as a reference point. In doing so, you are building into the system a perception which serves as the thermostat's goal--let's say 70 degrees Farenheit. As long as the ambient temperature is 70 degrees Farenheit, the thermostat is at peace with itself. From the viewpoint of an outside observer, nothing is happening; but actually the thermostat is continuously active in monitoring stimuli and comparing them to the reference point (70 degrees Farenheit). Perceptual activity is continuous; attention is constant. If the ambient temperature drops below the reference point, a discrepancy between what should be and what is sets up an error signal and activates the only behavior of which the system is capable: the furnace is turned on. As soon as the temperature is raised to 70 degrees Farenheit, the reference pattern and the incoming stimuli match; the error signal stops, and the thermostat's signal for "behavior" on the part of the furnace ceases. The thermostat does not control the behavioral output; how the furnace works is not the concern of the thermostat--all it "cares" about is that the input match the reference signal. So what is controlled is not behavior or output, but input. Nor does the thermostat "care" about the nature of the behavior; it doesn't matter whether the temperature is raised by sunshine, by a fire in the fireplace, or by the furnace. Once the input is controlled at the reference level, the error signal is no longer present; and insofar as consequences are triggered by it, they are ended. As Powers's title indicates, behavior controls perception--not vice-versa, as has always been thought. Goals are perceptual, not behavioral. The baby does not act in order to put his thumb in his mouth (i.e., to generate "thumbsucking" behavior). He is trying to match a particular perceptual pattern (let's say a certain feeling in his lips and oral cavity) and will run through the repertoire of behavior available to him to achieve that end.

Powers (1973) developed a hierarchical model whereby the computing function of the brain subjects the initial sensory input to a series of increasingly complex transformations, each higher level serving to monitor the activity of the lower by providing it with its reference signal. Although one cannot do justice to the concept without studying

the original work in its totality, its significance for my
argument is such that I will attempt to outline the essen-
tials of the model, knowing full well that my explanation
will leave much to be desired.

The model consists of nine levels, each level taking a
number of signals from the level below and generating a new
signal that expresses a different and more complex relationship than the antecedent one from which it derived its input. At the first level, sensory stimuli enter the nervous
system and are transformed into one neural signal which now
stands for that input into the nervous system. This neural
signal and many others like it ascend to the second level,
where intensity is transformed into sensory quality. At the
third level, groups of sensory quality patterns are transformed into signals of static configurations which, in turn,
become evaluated at the fourth level in terms of transition
patterns (i.e., changes in time, space, and motion). The
fifth level of perception transforms patterns of change into
information regarding the sequencing of events. And on the
sixth level, these are interpreted in terms of relationships: for example, spatial relationships like "in", "on",
"above", etc.; logical relationships like cause-and-effect;
abstract associations like "father" and "mother", etc.

The seventh level is that of programs: branching "net-
work(s) of contingencies", with choice points at each bifur-
cation—-each decision in the program being determined by
what went before. The eighth level is the system of principles which determines the choice of programs, and the ninth
level is a supraordinate level which determines the principles that are to be followed.

Each level in the hierarchy of perceptual transformations is dependent on the one below for the signals which it
processes; in other words, the "environment" for the sixth
level is the fifth level, etc. Only the first system is in
touch with extra-neural reality. Control over output is in
the reverse order. Each system controls the output of the
one below it, by providing the reference signal with which
the input of the lower system is compared, influencing that
one below it—-and so on down, until behavior is activated
which generates the input which, in its transformed state,
satisfies the reference signal of the highest system involved.

Though he is careful to present his concept only as a possible model which may or may not correspond to what actually happens in the brain, Powers's conception is elegant and parsimonious. It answers so many questions about the perceptual-cognitive process that I, for one, would certainly accept it until a better model is presented that supersedes it. I would suggest that only through a model such as this can one explain what psychoanalysts subsume under the term "ego functions", meaning thereby those processes which make adaptation and learning possible.

This model, with some refinements that will be mentioned, eliminates the necessity for postulating that instinctual behavior is teleological (that is, that inherited behavior patterns operate according to a design for the future). Relatively simple animals, with a central nervous system whose power of adaptation is limited, need only to inherit the reference signals from the seventh level of the perceptual system in order to insure their survival in the average, expectable environment. Their adaptation will be goal-directed, and may be very sophisticated since the lower six levels of the perceptual system operate in synchrony to match the seventh-level reference signal. Such animals can be responsive to qualities of sensation, transition, configuration, sequence, and relationships--as long as these responses fall within the confines of their particular seventh-level program. In this, even the simplest animal is probably more sophisticated than the most complex computer made by man to date. Where such creatures will run into difficulties are those situations where there is such a radical environmental change that survival depends upon their being able to change the programs for adaptation themselves. However, in the higher forms of animal life, even that capacity is already an accomplished fact.

As Powers points out, flexibility on the level of programs can be assured if there is a reorganizing system independent of, but in communication with, the perceptual system. Such a system might consist of intrinsic physiological states which act as reference signals for the perceptual system. For instance: as long as some range of electrolyte concentration is maintained, the reorganizing system's reference signal is zero and the perceptual system goes about its functions undisturbed. Should the intrinsic physiological state change radically--that is, should the electrolyte balance change very significantly and over a long period of time--the perceptual system would be subjected to continuous error signals, disposing the system toward a change in the programs themselves. Such changes alter the perceptual references and, thereby, the behavior of the animal. For example, an experimentally induced sodium chloride deprivation in rats will alter the program for feeding. Such animals' behavior will be directed toward finding salt and differentiating between minimal variations in the concentration of that substance in a manner not seen ordinarily. More dramatic but less adaptive (at least for the individual) is the amazing behavior of lemmings, who periodically seem to be bent on genocide by marching into the sea and drowning. Ethologists have established that in time of overpopulation, the shortage of food sends the lemmings in search of new land on which to forage. Obedient to the overriding demands of the newly developed program that governs their behavior, they march onward to reach land that seems to appear on the horizon and, oblivious to streams that would ordinarily inhibit their progress, drown. In this case, the intrinsic physiologic state--altered by starvation--called for program changes involving perceptual modifications that resulted in an atypical behavioral output.

The affective response, as was discussed previously, is a total bodily response. Mediated by the autonomic nervous system, it involves precisely the kind of general changes in the electrolyte balance that Powers posits as controlling the programs that govern perception and, thereby, behavior. Or, probably more correctly in phylogenetic terms, affective communication has evolved as the most effective method whereby a complex physiological system can respond as a whole to the demands of an often rapidly changing environment. In man, such changes can be created symbolically. Once the symbolic capacity is developed, often more important than the actual stimulus is the meaning--the affective significance--assigned to what is happening or is anticipated. It is the meaning attributed to the circumstances in which a person finds himself that determines the affective response which, in turn, controls the physiological changes that then control the perceptual programs governing behavior. What we call free will seems to be the ability of human beings to exercise some degree of control over the symbolic meaning that will be given to what is being experienced or imagined (Basch 1978).

We know from geneticists that it is impossible to inherit the large number of behavioral patterns that sophisticated animals display in their adaptation to an often rapidly changing environment and its contingencies. All that can be inherited is the ability to generate certain kinds of informative messages which then influence behavior. Powers provides a mechanistic explanation for the manner in which such a system might actually operate. This model shows that instinctual behavior need not be "blind" and that, through the negative feedback model, it can reach heights of sophistication and complexity that eliminate the need to anthropomorphize the brain as has so often been done in the past.

What I have sought to show in the preceding pages is that Freud's clinical discovery of the method for investigating human motivation can be separated from the explanations he offered for his findings, and can then be placed in the context of a theory derived from observation of a development whose hypotheses are open to experimental confirmation and falsification.

## PSYCHOANALYTIC SELF PSYCHOLOGY

Freud expected his patients to be equal partners with him in the treatment process. The patient was considered to be essentially mature, and he and Freud together would look at the distressing bizarre foreign body of a symptom and make sense out of it. Freud functioned as an interpreter who taught the patient the code with which he could understand the content of dreams, symptoms, and other unconsciously motivated behavior (including that of the transference neurosis).

Although initially the patient's complaint centered on his symptoms, these tended to fade; instead, the unconscious fantasies responsible for precipitating the neurotic symptom came to be directed toward the analyst. In other words, the patient had transferred his neurosis to Freud; the patient now unconsciously expected the analyst to be instrumental in both fulfilling his wishes and punishing him (the patient) for those selfsame wishes. The impulses directed toward the analyst were accompanied by the original, intense affects of childhood appropriate to them; and in the case of patients with symptom neuroses, the presence of these impulses--in their raw form and with their seemingly perverted aims--

aroused the patient's horror, fear, and disgust. Appropriate interpretation led to understanding and acceptance of the underlying experiences (real and imagined) and the resolution of the transference neurosis; this left the patient free to pursue his life as he saw fit, no longer hampered by ununderstood infantile wishes. These cases are rare, but most instructive when they do occur. More common than symptom neurotics, though not abundant by any means either, are the so-called character neurotics or neurotic character disorders. Here too the insufficiently resolved oedipal complex is central; but instead of giving rise to symptoms, it influences the adaptive mechanisms of the individual which collectively are termed his "character". Because of the sexualization and aggressivization of behavior with infantile impulses and fear of punishment, the patient finds himself unable to fulfill himself in some significant way. For example, a gifted individual may find that he always stops short of success in his chosen endeavor by undermining himself. He observes his self-destructive behavior but cannot prevent it. In the transference neuroses, the competitive destructive urges toward the analyst as father emerge; and it becomes clear that, for this patient, to succeed means (on the unconscious level) to have destroyed the father and won the oedipal victory. Therefore, as much as he wants to succeed, he must ensure failure in his endeavors. In these cases the interpretations of the analyst meet with greater resistance, since the patient's objectivity about his conduct is not as great as that of the symptom neurotic. There is more rationalization of his behavior, and a greater readiness to see it motivated by external circumstances rather than by inner expectations. Usually however, these patients too are amenable to the traditional psychoanalytic approach and respond well to it.

There have always been a large number of patients who come for analysis not because they are aware of symptoms or symptomatic behavior which interferes with their lives, but because they are dissatisfied with life itself. They complain of loneliness and boredom; their achievements seem meaningless to them, their goals unattainable; and they voice a dissatisfaction with their relationships past and present. Psychoanalysts have usually approached these patients with the assumption that their immaturities represented a retreat from an oedipal conflict, and that its interpretation would cure them by enabling them to deal with life maturely. Although arrests and conflicts in psychosexual development could be found in these cases and appropriately interpreted, the results were generally disappointing. Much to the analyst's chagrin, these patients transferred to the analyst not unconscious sexual and aggressive fantasies but quite conscious demands for help in living their lives more effectively. They demand advice about their daily problems while, at the same time (illogical as it seems), holding the analyst responsible for those miseries that brought them for treatment. Their attitude seems to be that since they have come to the analyst, it is now his responsibility to relieve them of their unhappiness; and if he fails to do so, it is because he doesn't like the patient, doesn't try hard enough, doesn't know enough, etc. This is a far cry from the partnership envisioned by Freud, and it left the analytic community at a loss for what to do with these cases. It was clear that the approach based on the model of the psychoneuroses was not effective with these patients. However, if the analyst gave up attempts to deal with the complaints as indications of unconscious meaning—instead treating them at face value—and tried to advise, console, and educate the patient, he was no more successful than before.

It was generally agreed that these patients were suffering from preoedipal pathology and problems in early object relations, as opposed to psychoneurotic patients whose problems lay in the area of object love (the oedipal relationship). In practice, there were certain gifted analysts who seemed to work well with such patients and did obtain improvement in these cases. The patients, however, attributed their improvement not to the interpretations of their analyst per se but to their relationship to their analyst somehow having had a beneficial effect in permitting them to mature. However, there was no systematic understanding of these disorders and their treatment until the work of Heinz Kohut (1959, 1971, 1977, 1978) systematically described and classified the forms of transference in the so-called narcissistic personality disorders, as Freud had done in the case of the psychoneuroses.

Kohut's observations of these patients' behavior in the analytic situation led him to the conclusion that they were transferring to him their need for a structured, cohesive, viable, and stable sense of self—a need for validation of their existence and worthwhileness that they could not supply themselves. He found in the analyses of his own patients, in the cases he supervised, and in the published records of narcissistic personality disorders, certain commonalities that permitted clinical generalizations. These patients seemed to relate to the analyst in one of two ways: either behaving as if they were fused with the analyst, who then had no independent existence; or raising him to godlike status and attributing to him all virtue, knowledge, and power. If the analyst does not regard these attitudes as artifacts to be eliminated, and avoids confronting the patient with their unrealistic nature, they develop into the narcissistic (selfobject) transferences on which the analyses of these heretofore unanalyzable patients is based. Kohut has taught us that rather than repeating the conflict of the oedipal period in the transference, these patients repeat the longing and disappointments that accompanied an unsuccessful earlier attempt to establish a viable self concept.

Infants and children are psychologically, not just physically, dependent on their parents. Long before the symbolic abstraction that we call our "self" becomes a possibility, the groundwork for that concept has been laid by the communication between infant and parent. The infant needs his mother to serve, first, literally as an extension of his body as he makes his needs known and, somewhat later, also as a mirror in which he sees and validates his existence. It is the parent or the parent surrogate who reflects back to him whether or not he is worthwhile, admired, and satisfactory. At the same time, he needs his parents to serve as admired models upon which he can pattern himself. Narcissistic character disorders--or, as Kohut later preferred to call them, patients with difficulties in selfobject relationships--have had other than optimal mirroring and/or opportunities for idealization. As a result, it is these aspects of development that are replayed in the analysis in the form of mirror and idealizing transferences.

Kohut's retrospective psychoanalytic investigation of development dovetails with recent findings in experimental psychology. Together they give us a very different picture of the maturational process than we have had heretofore.

In the psychoneuroses, the function of interpretation, according to Freud, is to bring unconscious wishes couched in the primary process across the repression barrier to the level of the secondary process, consciousness and under-

standing.  Freud  attempted  to explain the vicissitudes of
psychopathology on the basis of energy  exchanges  motivated
by instinctual  pressures within the mental apparatus.  Not
only has this explanation been recognized as  scientifically
untenable,  its use as an illustrative metaphor is quite unsatisfactory (Holt  1965,  Peterfreund  1971,  Rosenblatt  &
Thickstun  1970).   The developmental point of view seems to
correct both deficiencies.  In the case of the neurotic conflict, we see a struggle taking place at the time  that  the
child  is passing from a transformation of presentational to
discursive symbolism, from the preoperational  to  the  concrete operational level.  Until then, the lack of discursive
logical  divisions  and their implications had permitted the
child to symbolize his desires  without  fear  of  affective
consequences.   As  Freud  observed,  primary process (i.e.,
preoperational symbolism) does not operate on the  basis  of
mathematical,  logical  niceties.  Contraries can exist side
by side; negation is not present; etc.  As the  next  transformation  is  about to take place there are certain desires
and wishes which, if confronted on the concrete  operational
level,  would  promote  serious  affective  conflicts: overwhelming desire would collide with strong fear of sanctions.
So repression is mobilized to prevent this from occurring.

        As Freud pointed out (1915), repression, the basic  de-
fense in neurosis, is always a defense against affect.  What
we see clinically is that an experience (real or phantasied)
that might generate great fear, distress,or disgust is elim-
inated--not  simply  from consciousness but from thought itself (Freud 1895). This however does not take place as Freud
believed,  by severing word images from pictorial  ones,  but
rather by preventing the concrete operational transformation
of  potentially  anxiety-provoking  preoperational  content.
Successful repression literally does not permit the  offending  event to be thought about discursively, thus preventing
the affect attached to such concepts from emerging. However,
as we know, repression is not always completely  successful;
and  the affective tension set up by the preoperational concept achieves transformation in a disguised  manner  through
symptoms.   The  analyst's task is then, in the case of neurotic symptoms, to help the patient transform the  preoperational  expression  of  a  forbidden wish into a discursive,
concrete operational form--a task which, if  successful,
brings  not only the forbidden desire but the affects stimulated by it into focus and subject to examination and understanding on the level of everyday language. In this way, the

psychoanalytic formulation that the undoing of repression
involves "the replacement of the primary by the secondary
process" is given roots in terms of the developmental pro-
cess--a move which eliminates the necessity for both reify-
ing "mind" in the form of a mental apparatus and then power-
ing it with an imaginary psychic energy (hypotheses which
persist in psychoanalytic psychology for want of an acceptable substitute theory). (See Basch 1977.)

From the developmental point of view, the traumatic
effects leading to narcissistic personality conflicts have
taken place at the sensorimotor level, before symbolization
and evocative recall have become possible. These patients
present a very different picture for the analyst, since the
subsequent maturation of the symbolic function is used not
to advance the patient's maturation but to protect the infantile sense of entitlement at all costs. What becomes
symbolized is the need to maintain integrity by controlling
others.

Here the analyst does not have the task of connecting
presentational symbols of the preoperational period with the
more discrete symbols of words and of concrete operations,
but rather the job of helping the patient in turning what
has not been symbolized into symbolic form. As Kohut (1959)
has shown, the analyst's capacity for empathy and introspection made possible by his own analysis permits him to become
aware of what the patient is experiencing, though the patient is completely unable to become conscious of or express
his need. Not what the patient says, but how he says it and
in what context, enables the analyst to make him aware of
what might be going on within him. The exquisite sensitivity of these patients makes them respond to any variations in
the familiar routine of the analysis as if these were traumatic threats. The analyst voices for the patient the needs
and the disappointments his behavior indicates he might be
experiencing. Gradually the patient himself can become
aware of these vicissitudes both in and away from the analysis. This in itself is already very helpful to the patient; the ability to translate his sensitivity into words
promotes understanding and offsets his feeling of helplessness.

Extant metapsychological models in psychoanalysis are
incapable of explaining either narcissistic pathology or its
cure. In developmental terms, however, it is possible to

describe what has taken place. The failure of empathic communication during the sensorimotor period fails to stimulate affective development and/or perverts it to the point that its symbolic transformation in the preoperative stage is experienced as disorganizing and threatening. The concept of self that then develops is one which continuously anticipates danger and misunderstanding. The analyst uses his empathy to experience the patient's world preoperatively, transforms these experiences symbolically and conveys his insight to the patient in an appropriate manner. The patient's development proceeds in reverse fashion, as he hears the empathic formulations of the analyst he gradually trusts himself to transform his sensorimotor experiences symbolically so that they may gain entrance to logical thought.

In both narcissistic and psychoneurotic disorders, once the arrest of development has been overcome and concrete operational thought in the area of the conflict is possible, the working through of the interpretation proceeds. The analyst has succeeded in grasping the nature of the concepts underlying the symptoms and has translated them into words with and for the patient. This is therapeutic for the patient in that it gives him a degree of mastery through understanding the content of his conflict and the opportunity to relive its attendant affect in the nondestructive setting of the analysis. However, even though the patient may feel well and his symptoms are relieved, the analytic task is not yet finished. The patient must now be helped to transform what the analyst has conveyed to him into inner speech. It is only with this final transformation or "working through" of the interpretations to the level of propositional thought that the patient is enabled to become reflective about his conflicts. This transformation corresponds to what clinicians have called "insight"--the ability to recognize not just how conflict originated and developed, but how it influences the ideals held out as goals for the self in the present.

Psychoanalysis as a method of investigation of arrested development can make significant contributions to general psychology. Actual human functioning is determined in great part by those unconscious dispositions and tendencies that covertly set goals which the individual is usually not aware of and would deny if they were called to his attention. Because they could not be explained by experimental psychology, teleological concerns have been disregarded, treated as

if they were meaningless concepts, or subjected to reductionism. Yet it is the capacity to form and manipulate symbolic goals that makes human life unique.  Once psychoanalysts are able to express their findings in the language of
development, psychoanalysis can potentially restore to psychology the understanding of motive, intention, and goal
that is necessary for the formation of a general theory of
the self.

### INDIVIDUAL DEVELOPMENT AND THE SOCIAL SYSTEM

Freud's instinct theory led him to the conclusion that
the individual was by nature asocial and had to be forced by
his caretakers--against his will and through fear--to conform to the demands of civilization. The developmental process previously outlined, demonstrating that the infant is
geared toward human communication from the outset, disproves
that contention.

Systems, once generated and made viable, tend to develop their own goals and the feedback patterns to attain them.
Though it may be that human society at first evolved as a
means to further biological interests (the purpose society
seems to serve in non-human species), it appears that human
social systems exist on a higher hierarchical level where,
though not independent, they exercise control over the biological systems which they exploit for their own symbolic
ends.

By linking its patterns and goals symbolically to the
biological needs for survival and reproduction, society motivates the individual organisms to perpetuate the social
system.  The goal of the social system however is not to
preserve the individual or the species, but to perpetuate
itself and its manner of communication. Not just biological
existence, but the manner in which individual elements in
the system are related, becomes paramount.

Considered from the biological viewpoint, the human
infant achieves individual existence after nine months of
union with the mother. However, thought of as part of the
social system and in terms of communication theory, the biological birth of the human infant does not establish him as
a separate individual. Biological birth is a maturational
stage in the infant-mother system, a unity established at

conception and no less a functional unity after parturition. The change at birth lies not in the physical separation but in the manner of intrasystemic communication. The processing of signals had been primarily on a biological level during the embryo's intrauterine existence; at birth, an exchange of behavioral signals between the two parts of the mother-infant system is activated.

If society is a system based on symbolic exchange, then the first two years of life are the initial stages in preparing the organism for symbolic activity. In the sensorimotor stage of development, the infant brain lays down permanent and repeatable patterns which form the basis for later symbolization. The sensorimotor patterns of the neocortex are established through transactions in the mother-infant system as these stimulate the affective inherited behavior patterns that are encoded in the autonomic nervous system and controlled by the old brain.

With the advent of symbolic communication, the child still does not reach independent existence as a communicating system but again reaches a new and different level of communicative transaction. It should be emphasized that the message interchange within the mother-infant system is transactional, and not simply interactional; each input modifies the substrate and influences the nature of the system and subsequently its output (an interaction would imply only that each part of the system responded to the other's output without being permanently influenced by that activity).

The advent of symbolic capacity makes it possible for the infant brain, a subsystem of the mother-infant or mothering-infant system, to both recreate—and (eventually) manipulate—past communicative transactions. It is this activity that psychoanalysts refer to as "internalization", "introjection", and "identification".

The evolutionary acquisition of symbolic capacity led social system to become independent of the biological one through making it possible for the former to set its own goals, which are symbolic rather than biological. Once symbolic values can be transmitted, the social macrocosm can be reproduced in the microcosm of individual brains—thereby establishing a program for that subsystem's future activity. How successful that encoding will be depends primarily on the efficacy of the mother-infant system. Insofar as the mother responds consistently to the infant's behavior, the infant brain gains increasing security in its ability to predict the response of the environment and the consequences of its own activity. This forms the basis for later adaptation on the symbolic level. For too long, child-rearing practices have been seen in terms of biological survival and the communicative transactions have been neglected, thereby setting the stage for future maladjustment. Psychoanalysis recognized the significance of the early years of development and the need for healthy mother-infant interaction. Benedek (1973) initiated the concept that not only the baby but the mother too matured through her experience with pregnancy and motherhood. We would want to go even further, postulating that it is the maturation of a single system that is involved, and that the spatial separation of mother and infant does not imply separateness on the level of communication. Indeed, in the sense of communication, the mother-infant unity is not ever ended; it is only expanded, modified, and transferred.

Once the capacity for symbolic abstraction has been established (during the period that Piaget terms the pre-operative stage), the child, now 5-6 years old, enters the first stage of learning to manipulate symbols (Piaget's period of concrete operations). He is now put into the school system, a subsystem of society, where increasingly complex commmunications are transmitted and the tools for receiving and handling these patterns are imparted. What enables the child to participate in this next step is not just his cognitive development in the Piagetian sense, but the ability to selectively extend what was the mother-infant and parent-child system of communication to individuals in society as a whole.[2] His ability to appropriately form transactional systems of greater or lesser duration determines how his symbolic armamentarium expands and how efficaciously it is used. The previous mother-infant, parent-child experiences must be of a sort that make this expansion possible. This is what psychoanalysis calls "transference". What is transferred is the readiness to form affectively toned analogies between new patterns of experience and previous ones. If the groundwork has been laid by steady, predictable, consistent experience, adaptation to new situations will be easily achieved; if the past has also been reasonably gratifying in the sense that the balance was in favor of positive experiences, then the child will enter new situations with similar expectations. Insofar as the mother-infant communication has been inconsistent, frustrating, and/or inappropriate, or the infant has for any reason been unable to utilize what was offered, later transferences will fail to form or to be effective.

We tend to blame children's school failure on a learning disability, but it is not (except in cases of organic damage) his cognitive skills that are impaired. The child's failure represents his inability to establish the transactional communication which would permit him to succeed in the new relationships of the classroom. The parent-child system having been improperly programmed, the child fails in its transference attempts. The child cannot communicate in the accepted social code and is rejected by the school subsystem in the same manner and for the same reason that a foreign body is rejected by the physical organism: the messages it sends are not properly coded and either are not received or indicate danger (generate anxiety) for the recipient. For example, intelligence tests measure the ability to adapt oneself to a particular situation with cognitive skills, not the possession of the skills themselves; failure in these tests may not be a learning deficit but a lack of capacity to communicate effectively what has been or could be learned. The systems viewpoint of society clarifies that the entrance into the school situation is not the beginning of the process of socialization but already represents a test of the suitability of the subsystem we call a child to be a functioning element in the structure of society. "School failure" occurs in the first six years of life, not after school starts for the child.

The existence of a psychological "self" or "individuality" is an erroneous concept when tied to the idea that a person is basically a biological unity. We tend to believe that society is made up of individual selves, combinations of bodies and minds. For centuries, this notion would have been not only heretical but incomprehensible. Man could not even conceptualize himself apart from his society and its goals. The motive for existence lay in socially oriented concepts--whether these were symbolized as the will of the gods, the destiny of the nation, or the continuity of tribe or family. It is only in the recent past that the individual, as such, has become focal. A myth has arisen that it is society that somehow shackles the individual self and prevents its development. Society is seen as a disease that must be altered, possibly eliminated (at least controlled), in order for the suppressed self to emerge. However, society is not a collection of many individual systems called selves.

The structuralist viewpoint clarifies that reality lies in relationships, not in the elements that make the relationship possible by serving as its substrate (Basch 1976c). Man, the element, does not give rise to the totality, Society; rather, the relationships of symbolic communications collectively identified as "society" generate the concept of "man". Man apart from society is a structural contradiction.

Just as society evolved from biology and subordinated it, so the self created in a social matrix is reaching for that endpoint where society will be in its service. Ultimately, the self could represent an essentially new and controlling system. Just as the evolution of society was not based on a struggle with biology but on cooperation with the realities of physical existence, so the evolution of the self will not come about through a struggle with society (with the aim of eliminating social existence), but through understanding and using the social realities which are a necessity for human existence. Kohut (1971, 1977, 1978) has described the evolution of the self system in individuals, tracing the development of selfobject relationships from their origin in the initial mother-infant communication system to their controlling position on the level of ideals. He describes how, in the potential final stage of individual psychological development, mature ideals govern the fate of the individual--not by rebelling against society but through grasping both the limits and the possibilities open to the subsystem we call the self. For our purposes, Kohut's work enables us to see that on the larger scale the discontent with human existence that is evident all around us is not the result of social pressure or demands. Quite the contrary, it is the misunderstanding of the significance of the earliest years of life that leads to the development of individuals who, being unable to communicate effectively on the social level, cannot continue to progress and achieve maturity.

Only a systems viewpoint can liberate us from a nonproductive atomism that has marked the study of man. Man cannot be studied as either a biological, social, or psychological entity. Man is best studied as an activity, one delineated at any given time by the relationships in which he is active.

## AN OPERATIONAL DEFINITION OF SELF

The self is not a thing or an entity; it is a concept; a symbolic abstraction from the developmental process. The self refers to the uniqueness that separates the experiences of an individual from those of all others while at the same time conferring a sense of cohesion and continuity on the disparate experiences of that individual throughout his life. The self is the symbolic transformation of experience into an overall goal-directed construct.

The infant is born with a proclivity for human communication; the vicissitudes of that development, based on the infant's inherited predispositions and his experiences, weave the tapestry of the complex transactions that collectively are called the self. The matrix of the self lies in the affectively toned sensorimotor patterns of infancy, which remain the reference points for the ordering of all later experiences. Therein lies the basis for the person's aforementioned uniqueness and sense of individuality. In spite of all later complexities, all the combinations and permutations generated by subsequent experiences, the interpretation and significance of future events is always colored and indelibly stamped by the manner in which the world was first experienced by the infant.

Operationally, when we speak of studying or investigating the self (our own or another's), what we are actually referring to is the exercise of observing behavior and inferring from it--through introspection and empathy--what the motivation for that behavior, actual or intended, might be. As Freud's discovery of psychoanalysis demonstrated, our motivation is never totally available to what we mistakenly call self consciousness. Consciousness of self is limited at best, and in its pre-verbal sensorimotor roots is available only through inference. However, Kohut's discoveries enable us to penetrate more deeply and more accurately into the vicissitudes of that motivational network than was heretofore possible.

Although our observation of behavior can usually readily establish what the external goal of an activity might be,

that knowledge is almost always trivial. What we really
want and need to know, as students of human nature, are the
affectively toned goals that determine that behavior--what
the meaning of that behavior might be. Meaning, affect, and
motivation are essentially equivalent and are related to the
concept of self. The everyday usage of these words tends to
obscure what they signify or should signify. Affect does not
refer to a special set of experiences separate from something called cognition, antithetical to reason and therefore
not understandable and closed to scientific investigation.
Affect refers to the autonomic reaction patterns that are
basic for communication, and therefore for tension-control,
from the beginning of and throughout life. The dispositional effect of an experience is its meaning. What will it
move us to do? How is it interpreted by the goal-seeking
reference patterns that constitute the self? Motivation
always involves the management of autonomic tension or
affect. In all our behavior, we are either seeking to re-
capitulate in some way qualities and levels of tension experienced as optimal or trying to avoid experiences that
threaten to overwhelm the brain's capacity for ordering
stimuli and thus give rise to anxiety.

The subjective experience of self is not coextensive
with the self. It is a late development, both in the individual and in the race. It stems from symbolizing symbolic
activity, objectifying the brain's function of arranging
stimuli in terms of figure and background. As psychoanalysts have established, our awareness of self is always incomplete and often in error--usually in those areas that are
most significant for tension-control. The psychoanalyst,
when successful, is able to correlate the experience of
failures in the management of tension with what can reasonably be presumed to be its antecedents in earlier failures
of communication. It should be noted that such correlation
is not necessarily causal. What brought about the communicative deficit (whether it is transactional, circumstantial,
based on inherited deficits, or based on the misinterpreta-
tion of experience) often cannot be established retrospec-
tively, nor is it always essential to know that. To know
that such failure occurred, and in what manner it affects
the goal-seeking behavior of the present day, fulfills the
dictum "know thy self".

ENDNOTES

1. Psychotherapists may object that questions of scientific philosopy are not germane to practicing clinical
   psychology, but this is not correct. The above-mentioned confusion in the fields of science is an off-
   shoot of a much larger misunderstanding about the
   nature of reality which is part and parcel of the up-
   bringing of people in Western civilization. The distortion and confusion of the nature of reality begins with
   infancy and eventually plays a very significant part in
   predisposing the individual to psychological difficulties which are much more readily understood and
   treated if the psychotherapist has himself come to
   grips with and corrected them.

2. My use of the term "mother-infant relationship" should
   not be construed as a playing down of the importance of
   the father and other care-giving individuals in the
   newborn's environment. These individuals influence the
   child directly as well as through their transactions
   with the infant's mother, whose response to her baby is
   significantly affected by her relationships past and
   present. I use the term "parent-child" relationship to
   indicate that the child, once it is able to do so, actively turns to others--especially the father--to compensate for those inevitable shortcomings in the mother-infant, mother-child transactions that leave his
   selfobject needs significantly unrequited (Kohut 1971,
   1977).

REFERENCES

Arnold, M.B. 1960. Emotion and Personality, Vols. I and
        II. New York: Columbia University Press.
Basch, M.F. 1975a. "Toward a Theory That Encompasses De-
        pression: A Revision of Existing Causal Hypothe-
        ses in Psychoanalysis". In Depression and Human
        Existence (ed. by E.J. Anthony & T. Benedek).
        Boston: Little, Brown, & Co., pp. 483-534.
Basch, M.F. 1975b. "Perception, Consciousness and Freud's
        'Project'". The Annual of Psychoanalysis, III:
        3-19.

Basch, M.F.   1976a.   "Theory Formation in Chapter VII: A
        Critique".   Journal of the American Psychoana-
        lytic Association, XXIV:61-100.
Basch, M.F.   1976b.   "The Concept of Affect: A Reexamina-
        tion".   Journal of the American Psychoanalytic
        Association, XXIV:759-77.
Basch, M.F.   1976c.   "Psychoanalysis and Communication Sci-
        ence".   The Annual of Psychoanalysis, IV:385-421.
Basch, M.F.   1977.   "Developmental Psychology and Explana-
        tory Theory in Psychoanalysis".   The Annual of
        Psychoanalysis, V:229-63.
Basch, M.F.   1978.   "Psychic Determinism and Freedom of
        Will".   The International Review of Psychoanal-
        ysis, V:257-64.
Benedek, T.   1973.   Psychoanalytic Investigations: Selected
        Papers.   New York: Quadrangle/New York Times Book
        Co.
Bower, T.G.R.   1971.   "The Object in the World of the
        Infant".   Scientific American, CCXXV:30-8.
Brandt, Herman F.   1945.   The Psychology of Seeing.   New
        York: The Philosophical Library.
Carpenter, G.   1974.   "Mother's Face and the Newborn".   New
        Scientist, 21 March:742-44.
Condon, W.S. & Sandor, L.W.   1974.   "Neonate Movement as
        Synchronized With Adult Speech: International
        Participation and Language Acquisition".   Sci-
        ence, CLXXXIII:99-101.
Darwin, C.   1872.   The Expression of the Emotions in Man and
        Animals.   1965 ed., Chicago: University of Chi-
        cago Press.
Freedman, D.A.   1972.   "Relation of Language Development to
        Problem Solving Activity".   Bulletin of the Men-
        ninger Clinic, XXXVI:583-95.
Freedman, D.A., Cannaday, C., & Robinson, J.S.   1971.
        "Speech and Psychic Structure: A Reconsideration
        of Their Relation".   Journal of the American
        Psychoanalytic Association, XIX:765-79.
Freud, S.   1888.   "Preface to the Translation of Bernheim's
        Suggestion".   In Standard Edition, I 1966, Lon-
        don: Hogarth Press, pp. 75-87.
Freud, S.   1891.   On Aphasia.   1953 trans. by E. Stengel,
        New York: International Universities Press.
Freud, S.   1895.   "Project for a Scientific Psychology".   In
        Standard Edition, I 1966, London: Hogarth Press,
        pp. 281-397.

Freud, S.   1900.   "The Interpretation of Dreams".   Standard
        Edition, IV and V 1953, London: Hogarth Press.
Freud, S.   1915.   "The Unconscious".   In Standard Edition,
        XIV 1963, London: Hogarth Press, pp. 159–204.
Holt, R.R.   1965.   "A Review of Some of Freud's Biological
        Assumptions and Their Influence on His Theories".
        In Psychoanalysis and Current Biological Thought
        ed. by N.S. Greenfield & W.C. Lewis).   Madison:
        University of Wisconsin Press, pp. 93–124.
Jones, E.   1953, 1955, 1957.   The Life and Work of Sigmund
        Freud, Vols. I, II, and III.   New York: Basic
        Books.
Kohut, H.   1959.   "Introspection, Empathy and Psychoanaly-
        sis".   Journal of the American Psychoanalytic As-
        sociation, VII:459–83.
Kohut, H.   1971.   The Analysis of the Self.   New York: In-
        ternational Universities Press.
Kohut, H.   1977.   The Restoration of the Self.   New York:
        International Universities Press.
Kohut, H.   1978.   The Search for the Self: Selected Writings
        (1950–1978).   Ed. by P.H. Ornstein, New York: In-
        ternational Universities Press.
Korzybski, A.   1933.   Science and Sanity.   1948 ed., Lake-
        ville, Connecticut: International Non-Aristote-
        lian Library Publishing Co., Institute of General
        Semantics.
Langer, S.K.   1942.   Philosophy in a New Key.   1951 ed.,
        Cambridge: Harvard University Press.
Luria, A.R.   1961.   The Role of Speech in the Regulation of
        Normal and Abnormal Behavior.   New York: Pergamon
        Press.
Murphy, L.B.   1973.   "Some Mutual Contributions of Psycho-
        analysis and Child Development".   In Psychoanaly-
        sis and Contemporary Science, Vol. II (ed. by
        B. Rubinstein).   New York: MacMillan, pp. 99–123.
Peterfreund, E.   1971.   "Information, Systems and Psycho-
        analysis".   Psychological Issues, VII, 1/2, Mono-
        graph 25/26.   New York: International Universi-
        ties Press.
Piaget, J. & Inhelder, B.   1969.   The Psychology of the
        Child.   New York: Basic Books.
Powers, T.   1973.   Behavior: The Control of Perception.
        Chicago: Aldine Publishing Co.
Rosenblatt, A.D. & Thickstun, J.T.   1970.   "A Study of the
        Concept of Psychic Energy".   International Jour-
        nal of Psychoanalysis, LI:265–78.

Ryle, G.  1949.  The Concept of Mind.  1968 ed., New York:
        Barnes and Noble.
Tomkins, S.S.  1962–63.  Affect, Imagery, Consciousness,
        Vols. I and II.  New York: Springer Publishing
        Co.
Tomkins, S.S.  1970.  "Affect as the Primary Motivational
        System".  In Feelings and Emotions (ed. by M.B.
        Arnold).  New York Academic Press, pp. 101–10.
Tomkins, S.S.  (1980).  "The Quest for Primary Motives:
        Biography and Autobiography of an Idea".  Journal
        of Personality and Social Psychology, (to be pub-
        lished).
Vygotsky, L.S.  1934.  Thought and Language.  1962 ed.,
        Boston: M.I.T. Press.

STEPS TOWARD A MODEL OF THE SELF

Gil G. Noam, Lawrence Kohlbert, and John Snarey

Harvard University and
McLean Hospital

A day will come when the psychology of cognitive functions and psychoanalysis will have  to
fuse in a  general  theory which will improve
both through mutual  correction,  and  starting
right  now we should be preparing for that pro-
spect by showing the relations which could  exist between them.

J. Piaget, 1973

## INTRODUCTION: PIAGET AND FREUD

At a time when many Piagetian social-cognitive psychologists are applying their theories to real-life problems beyond  the  school setting (e.g., Gilligan 1982, Noam & Kegan
1982, Selman  1980),  applied  developmental  psychology  is
gaining  increased  importance.  Application  to  clinical
phenomena is, in turn, renewing an interest in psychoanalytic and developmental integrations and in an inclusive theory
of the self.  But the integration of Freudian and  Piagetian
psychologies,  which we call clinical-developmental psychology, is a difficult task, involving integration of the study
of social-cognitive, developmental, clinical, and  affective
phenomena.  Through  this chapter, we hope to advance a few
steps in the understanding of  ego  development,[1]  psychopathology,  and clinical intervention by reflecting on steps
taken previously.

We are particularly interested  in describing  how  the
central  concerns  of clinicians are addressed when developmental psychology shifts  its focus to "person-objects", in-

59

terpersonal relationships, the process of developmental transition, and the organizational principles of the self which mediate both cognition and affect. But psychoanalytic psychology and cognitive developmental psychology have separate theoretical traditions, research orientations, and methods of intervention. Although both share a developmental viewpoint, there are profound differences between a psychology that focuses on affective development and psychopathology, and a psychology that focuses on cognitive development and in human development. While the different questions and methods have contributed to the separateness of these two psychologies, other researchers, educators, and clinicians have recently emphasized the importance of theoretical integrations and empirical explorations in order to further our understanding of normal and abnormal development (e.g., Greenspan 1979, Selman & Yando 1980, Noam & Kegan 1982). Piaget (1973) himself has suggested the need for a general psychology that would integrate psychoanalysis and genetic epistemology, and provide the basis for a "developmental psychopathology".

Previous attempts to bridge the gap between psychoanalysis and developmental theory have been based on Piaget's work rather than on a more explicitly social-cognitive neo-Piagetian theory. Peter Wolff (1960) was one of the first to systematically compare the theories of Freud with Piaget's theory of six sensorimotor substages. He concluded that the two theories are essentially compatible, especially through the modification of psychoanalytic ego psychology.

James Anthony (1976) describes both Freud and Piaget as essentially epistemologists and developmentalists. He demonstrates a number of similarly held positions, e.g., (1) moral sense is not innate, but acquired; (2) acquisition takes place in the course of early development; (3) moral acquisition begins with external sources (parental injunctions) and is subsequently internalized; and (4) the subsequent shift from an external to an internal locus of control. While conceptual discussions of psychoanalysis and Piagetian developmental psychology have not been common, comparative or integrative research has been even less frequent. An important exception is the work of Decarie (1978), who systematically studied the relationship between the Piagetian construct of object constancy and the psychoanalytic notion of object relations. She specifically considered how the timing of the development of internal representations parallels the devel-

opment of the affective tie between the child  and  the  mother.  Recently, Basch (1980) has presented a model in which
he  attempts to reinterpret or modify the psychoanalytic notions of regression and repression in the  analytic  process
from  a Piagetian perspective.  Basch reworked the early developmental stages of psychoanalysis by describing cognitive
development in the sensorimotor,  pre-operational  and  con-
crete-operational stages postulated by Piaget.

Greenspan  also attempts to integrate the Piagetian and
psychoanalytic model and applies his synthesis  to  clinical
work  with  infants and  young  children  (1979, 1981).  Two
aspects of his work are especially relevant to  our  discussion.  First, he was able to show that to a large extent the
structures that represent cognitive development are the same
or  at  least parallel and are highly interactive with those
structures that represent socio-affective development.   He
describes  structural  principles regulating interaction between the boundaries, principles which are  also  useful  in
understanding  a  number  of  psychoanalytic and psychiatric
concepts in cognitive-developmental terms:  regression,  latent  psychosis,  fixation and repression, and encapsulation
of neuroses.

Malerstein and Ahern (1979) hypothesize three character
structures based on characteristics of the  thought  pattern
described by Piaget.  These three structures—the operational,  the intuitive, and the symbolic—each represent the individual's most basic concerns and systems of processing interpersonal relations.  Santostefano's (1978)  bio-develop-
mental  perspective  is a very important synthesis of George
Klein's ego psychology concerning cognitive controls and ego
functions and Piaget's stages of  intellectual  development.
The  outcome of this integration is a Piaget-informed method
of clinical intervention, namely cognitive controls therapy.

This comparative work has generated considerable interest but,  with few exceptions, has  not led to  a systematic
integration of the two theories. The research, for instance,
does not address the important paradigmatic differences  between  a theory focusing on early childhood experiences, unconscious motivations, affective development, and regression
and a psychological model that is primarily  concerned  with
cognitive organization and transformation. Furthermore, Piaget's work was mainly dedicated to the uncovering of the development  of  intelligence  as  it encompassed the changing

meaning and understanding of physical properties. Most previous attempts have used Piaget's work alone rather than
neo-Piagetian social-cognitive theories, which are much more
suited for such integrations. Social-cognitive theorists,
building on Piaget's cognitive theory and expanding his original work on moral development, address psychological phenomena directly relevant to the psychoanalytic psychologist
by providing an ego and object relations view on social interaction. This implies a theoretical and empirical orientation that traces the way children, adolescents, and adults
understand and experience themselves (self), important others (person-objects) and their interaction (internalization,
identification, communicative acts). This new phase in the
history of develomental psychology may represent the missing
link in the previous attempts to integrate Piaget and Freud.
Social cognition can serve as a bridge between a psychoanalytic understanding of the self and social interaction and a
Piagetian cognitive-developmental approach.

This rapprochement is also aided by other developments.
The field of "emotional development" is one of the fastest
growing subdisciplines in developmental psychology. Now that
cognitive theory has come of age, interest has refocused on
the age-old question regarding the relationship between cognition and emotions. Cognitive theorists in the Piagetian
tradition are well advised to "reinvent the wheel" in an in-
formed fashion by reworking the knowledge and formulations
of psychoanalysis. The study of cognition and emotions is,
in our view, one of the principle rationales for a theory of
the self, since we view the self as the mediating structure
and process encompassing these two psychological subsystems.
It is this mediational function that gives the self an organization separate from but encompassing intellectual, moral, and social subdomains (see also Edelstein & Noam, in
press). No psychoanalytic-Piagetian integration can occur
without a concept of the self as the agent of mediation and
a concept of personal history (biography and identity). This
approach readdresses the life-cycle concepts of Erik Erikson
in a manner compatible with social-cognitive theory through
a focus on stages and crises of development, and on
adolescence and adulthood as life phases. Although many
constructs remain incompatible (which we will detail in this
chapter), a bridge between clinical psychoanalysis and academic developmental psychology can be built on the foundation laid by Erikson. Previous attempts at integration have

not traced structural ego development as the central dimension of the life cycle.

In our attempt to compare and contrast the theories in the Piagetian and Freudian traditions--by clarifying their underlying assumptions, discussing implications for the life span, and giving special attention to social-cognitive or Piagetian reinterpretations of psychoanalytic theory--it also becomes necessary to distinguish terms, concepts, and assumptions. In fact, many terms, e.g., stages and phases, functions and structures, have been used interchangeably in the literature and have thus perpetuated confusion and ambiguity. We will distinguish between two psychological models: functional and structural. After introducing these two basic approaches, we apply them to the work of psychoanalysts and developmental psychologists. We attempt to explicate how this distinction between models helps to clarify differences between psychoanalytic and structural developmental theories and to place "mixed models" in context. Specifically, we will critically review psychoanalytic-functional, structural-developmental, and structural-functional models of ego development and then draw a set of conclusions which may provide a foundation for future steps toward an integrated developmental theory of the self.

## THE PARADIGMS: FUNCTIONALISM AND STRUCTURALISM

### Functionalism

Functionalism can be traced within at least three separate social-science traditions: psychological, psychoanalytic, and anthropological or sociological. We will not describe in detail here the differences between these disciplines; rather, we will limit ourselves to delineating the underlying similarities. The functionalist school of American psychology goes back to William James and John Dewey. The functionalists, building on E.B. Titchener, focused on the function of consciousness, viewing consciousness as being task related and having the role of solving tasks. Psychoanalysts discussed ego functions--defenses and adaptive ego organizations that mediate between internal and external reality, between the id and the superego demands. Erikson's explication of psychosocial stages in the life cycle wed the psychoanalytic functional perspective to a more sociological and anthropological functional perspective.

Anthropological and social functionalism has its roots in the work of Emile Durkheim, although he did not coin the term. Bronislaw Malinowski labeled as functionalism an approach in which individual behavior and social phenomena are interpreted by the way they relate to each other and the function they fulfill within a sociocultural system. Malinowski's description (although general and paradigmatic) pointed to an overall functional model of explanation which was elaborated by anthropologist A.R. Radcliffe-Brown (1935) and sociologist R.K. Merton (1947), who were less oriented toward the individual and more toward society. These two theorists, of course, also differ in important ways: Radcliffe-Brown has an organismic analogy for society, in which the social system is seen as a unitary whole, while Merton's position is that functions must be demonstrated for certain elements and subgroups, but not necessarily for the whole social system. Merton also introduced the distinction between latent and manifest functions to distinguish between intentional and unintentional actions and their anticipated and unanticipated consequences.

The great functionalist integration of psychology and sociology, representing the peak of functionalist analysis in the social sciences, came from Talcott Parsons, who applied a functional framework to the study of family systems and to the relationship between personality development and social structure (1964). Parsons described institutionalized cultural organization and internalized norms and roles in individual development as a means of understanding the social system and its functional consistency in the face of divergent interests. The connection between individual need systems, cultural organization, and overall social systems provides the basis for the fulfillment of role expectations and the experience of them as legitimate. Parsons (1964) relied heavily on Freud's personality theory to link the social and individual motivational structures.

What common theoretical configuration do these different functional traditions share?
1. Functionalist thought addresses not only the functional content of a thought, a feeling, or an activity, but asks what purpose is being fulfilled. Thus, from a functional point of view, the life cycle refers to cyclical patterns of acquisition and performance of social roles through occupation, marriage, and parenthood, "and in the later years through retirement, death of marital

partner, the launching of children along adult paths".
(Elder 1975, p. 6).
2.  The parts of a system are functionally interrelated: one
    part is required in order for the other parts and the
    system as a whole to function. Disfunction in one part
    places the entire system into disequilibrium; the parts
    of a system are geared towards the overall and continued
    functioning of the system. "It is their emphasis upon
    contributions which are defined as functions, that
    earned functionalism its name" (Abrahamson 1978, p. 5).
3.  Most systems contribute to the functioning of other sys-
    tems, thus subsystems are part of an overall system, a
    concept very important for a functional sociological
    analysis of institutions and society.
4.  Systems do not change easily; they are integrated and
    stable. Basic relationships among elements change little
    over time. In general, fundamental changes are resisted
    by systems and occur only infrequently.

Some systemic structuralistic orientations may also
sound quite similar; systems analysis is not confined to
functionalism, but is the basis of functionalism. There is
no functional analysis without a systemic point of view because phenomena are always interpreted in relation to their
function within a certain form of psychological, biological,
or sociological organization.

To conclude, while the theoretical differences and
practical implications of the diverse functional theories
are considerable, we are here concerned with the overall
functionalist orientation. This orientation views small
groups and interacting persons as social systems that are
regulated by action (elements) and serve a function in the
larger unit. Every idea, cultural phenomenon and custom fulfills an interpretable function and has a task to fulfill.
Elements in a system are interpreted in terms of their relationship to the purpose for which the system was created.


## Structuralism

Like functionalism, structuralism is an overarching
theoretical orientation influencing all natural and social
sciences. In this sense both structuralism and functionalism
are truly interdisciplinary. Again, we cannot detail the
differences of structuralist thought in each discipline (for

general introductions, see Piaget 1968 and Gardner 1972). For linguists, structuralism of the DeSausurrean sort represented a departure from the isolated study of linguistic phenomena, such as tracing words historically to their origins. A structural approach to linguistics focuses instead on overall language systems. This trend culminated in Noam Chomsky's formalization of transformational (generative) grammar. Structure derives its wholeness not through a static underlying gestalt but, similar to the work of Piaget, from laws of transformation. Although Chomsky leaves room for psychological interpretations of language development through the study of individual speech acts (creative aspects of language), the "heart" of his theory--his laws of "generative grammar"--is considered biologically determined or "innate". For sociologists, structuralism has drawn on and reinterpreted aspects of Marxist thought, e.g., distinction between real economically based social relations and ideological superstructures. Althusser (1965) analyzes Marx's model of contradiction between the relations of production and forces of production from a structuralist perspective and formalizes an inherent social transformational system. This structuralist interpretation of the development of social systems also leads to structuralism in the study of history. But the most important contributor to structuralism outside of psychology proper is Claude Levi-Strauss. His anthropological studies (1962, 1963, 1969) of cultural kinship systems, rituals, and myths build on the assumption that all social relations are guided by "conceptual structures", and that all social practices and cognitive products lend themselves to a structural analysis.

Psychological structuralism, of central interest in this chapter, preceded Piaget and was an early response to the associationist tradition that reduced complex behaviors, perceptions, and thoughts to their basic elements, losing sight of the overall unified whole, e.g., behaviorism the gestalt psychologists, for instance, were the first group in psychology to build on phenomenological traditions in philosophy (Kohler 1947, Wertheimer 1912, and extended by Lewin 1951). Essential to gestalt psychology is the concept of wholeness, in which the perceptual elements never exist as separate entities, but always as part of the whole within their context or, as Lewin later put it, "within their field". The theories of perception in gestalt psychology are not developmental in nature, the data is derived mainly from adult subjects. Thus, gestalt "structuralism", like the structuralism of Levi-Straus, is ahistorical and lacks a concept of transformation or development.

Piaget's great contribution was to present a developmental structuralist model involving an integration of the concepts of (1) wholeness, (2) internal regulation, and (3) developmental transformation. The basic orientation of Piaget's structural-developmental theory and all subsequent cognitive-developmental formulations can be summarized as follows:

1. Development of cognitive structure is the result of processes of interaction between the structure of the organism and the structure of the environment, rather than being the direct result of maturation or of learning. This presupposes a distinction between behavior changes or learning in general and changes in thought structure. Structure refers to the general characteristics of shape, pattern and organization of response rather than to the rate or intensity of response or its pairing with particular stimuli. Cognitive structure refers to rules for processing information and for connecting experience events. Cognition (as most clearly reflected in thinking) means relating events by an active process, rather than a passive connecting of events through external association and repetition. In part this means that connections are formed by selective and active processes of attention, information-gathering strategies, motivated thinking, and so forth.

2. Transformation, the temporal idea of structure, replaces a concern with static forms. Development involves underlying transformations of cognitive structures which are understood as the organizational wholes or as systems of internal relations, rather than by elements of association (e.g., stimulus-response).

3. Development of cognitive structure always means moving toward greater equilibrium in the organism-environment interaction. This balance--between external objects and internal assimilating and accommodating structures-- represents knowledge, truth, logic and adaptation.

4. The balance of equilibration is more than a momentary act, but represents periods of underlying stability, thus the concept of stage. The structures are whole entities or systems rather than aggregates of elements that are independent of the whole. Elements of structured wholes are defined by the laws governing the roles that they play. The elements derive their power as phe-

nomena only as parts of a whole. The structure that orders the elements raises them to a form that they do not
possess in isolation.

These four general principles apply to all cognitive development, as well as to physical and social objects. The basic processes involved in "physical" cognitions and in stimulating developmental changes in these
cognitions are also basic to social development. But
since our topic here is the development of the self and
social cognition, additional assumptions about social-
emotional development are important.

5.  Social cognition always involves role-taking, that is,
    awareness that the other is in some way both like and
    unlike the self, and that the other knows or is respon-
    sive to the self in a system of complementary expecta-
    tions. Accordingly, developmental changes in the social
    self reflect parallel changes in conceptions of the so-
    cial world.

6.  Affective development and cognitive development and
    their functioning are not in distinct realms, but rather
    they represent different aspects in defining the same
    structural transformations.

7.  There is a fundamental unity of personality organization
    and development termed the ego or the self. While there
    are various strands of social development (psychosexual
    development, moral development, etc.), these strands are
    united by their common reference to a single concept of
    self in a single social world. Social development is,
    in essence, the restructuring of the concept of self in
    its relationship to other persons in a common social
    world with social standards. In addition to the unity of
    social development stages due to general cognitive de-
    velopment, there is a further unity of development due
    to a common holistic factor of ego maturity.

8.  Ego development is also directed toward an equilibrium
    or reciprocity between the self's actions and those of
    others toward the self. In its generalized form, this
    equilibrium is the end point or definer of morality,
    conceived of as principles of justice, i.e., of reci-
    procity or equality. In its individualized form, for
    instance, it defines relationships of "love", i.e., of
    mutuality and reciprocal intimacy. The social analogy
    to logical and physical conservation is the maintenance
    of an ego-identity throughout the transformations of
    various role relationships.

TABLE 1

COMPARISONS OF THEORIES OF THE SELF

| | Erikson | Loevinger | Kegan | Baldwin |
|---|---|---|---|---|
| Principle of Development | Epigenetic-embryological; invariant sequence | Hierarchical-transformational invariant sequence | Hierarchical-transformational invariant sequence | Inter-functional, differentiation and reintegration |
| Logic of Stages | Psychosocial crisis; institutions tasks | Reference of self and world. In domains of interpersonal, moral, etc. | Equilibration of self-other relations | Social role taking. Process of imitation and identification |
| Concept of Regression | Regression in service of ego & as fragmentation | None | None | None |
| Theory of "ego strength" | Developmental model of ego virtues | None | None | Concept of will and choice |
| Age/Stage Relationship | Age and stage synchrony | No age – stage synchrony | No age – stage synchrony | Age – stage synchrony-early years |
| Developmental Pathology | Stages different pathologies, & resurfacing of earlier symptoms | None | Stage-related symptoms. Transition as source of problems | None |
| Measurement | None | Washington sentence completion test | Structural clinical interview and coding in progress | None |

These general structural principles, as they apply to social and ego development, serve as the basis for our renewed attempt to expand a theory of the self and to explore similarities and differences between different ego models

## PSYCHOANALYTIC-FUNCTIONAL APPROACHES TO EGO DEVELOPMENT

The most important psychoanalytic contribution to our understanding of ego development comes from the systematic study of ego functions. According to psychoanalytic theory these functions serve to mediate external requirements and internal wishes, to transform id impulses into socially acceptable behaviors, and to create a buffer against unreasonable superego demands. The defense mechanisms are the most original and important categorizations of ego functions, although other formulations are also of special interest from both functional and structural perspectives--ego synthesis, stress tolerance, and impulse control.

David Rapaport (1967) has traced the development of the concept of ego functioning. According to Rapaport, the first phase of ego psychology arises from Freud's early writings, in which he equated the terms self and consciousness with a rudimentary, limited concept of ego. He developed the first theory of defenses at a time when the role of reality occupied a major place in his thinking. Defenses ward off the affect associated with real memories and with real past and present experiences. The second phase of ego psychology, however, was marked by Freud's lesser emphasis on the reality experience. He believed that infantile sexual experiences, as reported by patients in analysis, were not necessarily based on reality but on fantasy. His center of interest shifted to the agent that creates the fantasies and the processes by which this agent works. The discovery of the instinctual drive followed and its exploration dominated the second phase. The third phase of ego psychology concerns the Freudian description of the ego as a coherent organization of mental processes. Although the ego is structured around the "conscious-preconscious" systems, it includes unconscious forces similar to those of the id. In addition, although the ego can transform id energies into ego energies, it is also made up of neutral energies. Rapaport describes the limitations of the ego concept of the third phase: "...the ego still appears as a result of the promptings of id, superego, and reality..., the ego is still the helpless rider of the id horse" (1967, p. 749). Rapaport's fourth phase deals with the theoretical formulations of modern ego psychology, the starting point for our discussion of functional theories of the ego. We will limit our discussion to central contributors of the functional-ego school.

## Anna Freud and the Theory of Defense

In her book, The Ego and the Mechanisms of Defense, Anna Freud (1936) provides an important bridge between psychoanalysis and developmental psychology both theoretically and empirically (a fact we will later address by describing a recent contribution to the study of defense mechanisms that potentially brings these two disciplines closer). The understanding that defense mechanisms serve adaptive as well as pathological functions has been one of the more fruitful contributions of psychoanalysis to both psychotherapy and developmental research. The list of defenses has been augmented and reorganized by many theorists (Bibring, et al. 1961, Laughlin 1979, Vaillant 1977, Hauser, et al. 1979). Though specific definitions have varied, initial work on the systematizing process originated with Anna Freud. She writes: "The view held was that the term psychoanalysis should be reserved for the new discoveries relating to the unconscious psychic life, i.e., the study of repressed instinctual impulses, affects, and fantasies" (1936, p. 31). Anna Freud argues that adjustment to the outside world had been underemphasized in psychoanalytic theory. She posits a more comprehensive definition, giving the ego and its defense-functions a central place in both the theory and practice of psychoanalysis. She defines the task of analysis as follows:

> To acquire the fullest possible knowledge of all three institutions of which we believe psychic personality to be constituted and to learn what are their relations to one another and to the outside world. That is to say: in relation to the ego, to explore its contents, its boundaries, and its functions, and to trace the history of its dependence on the outside world, the id, and the superego; and in relation to the id, to give an account of the instincts, i.e., of the id contents, and to follow them through the transformations which they undergo. (ibid., p. 4f.)

Anna Freud describes the analysis of the ego as more difficult than the traditional psychoanalytic focus on the id because the ego is the source of resistance. She classifies the ego's defenses and adds to the list Sigmund Freud had postulated. New mechanisms include regression, reaction formation, isolation, undoing, projection, turning against the self, displacement, and reversal. Anna Freud links different neuroses to characteristic defenses, e.g., undoing and isolation are linked with obsessional neurosis. She underlines the developmental aspect of defenses and psychopathology within the framework of psychoanalytic thought and she traces defenses to their developmental origins:

It is meaningless to speak of repression where the ego is still merged with the id. Similarly, we might suppose that projection and introjection were methods that depended on the differentiation of the ego from the outside world. The expulsion of ideas or affects from ego and their relegation to the outside world would be a relief to the ego only when it had learned to distinguish itself from that world. (ibid., p. 51)

Two more recent contributions on defenses have integrated psychoanalytic concepts further into developmental psychology and subjected them to more empirical scrutiny.

George Vaillant and Norma Haan: Defenses and Coping

Vaillant's recent descriptions of adaptive processes of the ego expand on Anna Freud's notion that defenses are hierarchically organized. Much psychoanalytic work had been built on the conceptualization of developmental levels of defenses and their relationship to psychopathology; for example, neurotic defenses like repression and reaction formation were attributed to hysteria, and isolation and intellectualization were attributed to obsessive-compulsive disorders. In more severe mental disorders, such as paranoia, defenses of projection and denial were implied, while in character disorders acting out and more impulse-oriented mechanisms were observed. Psychoanalytic work with patients at all levels of psychological functioning had convinced Vaillant (and Semrad) that in the process of psychological decompensation and reorganization patients demonstrate important shifts in their use of defenses.

Vaillant advanced these insights by systematically describing a developmental hierarchy of four levels of defense (adaptational) processes: psychotic, immature, neurotic, and mature (see Table 2). But even more importantly, he systematically applied the analysis of these defense analyses to a longitudinal study of healthy male adults from an elite sample, the Grant study of Harvard College students, who were followed over a period of forty years. His data suggest that even immature and neurotic defenses are connected with normal adaptation to challenges and crises over the life span, not only with psychopathology. Vaillant found that important developments occur in adulthood which he attributes to adaptational style. With increasing age, the mean use of low-level defenses decreased. These findings are also linked to better life adaptation in areas of work, love, health, recreation, and self-esteem. Thus Vaillant elaborates Erikson's life cycle perspective as well as Anna Freud's question about the hierarchical organization of defenses and takes a step towards describing lines of defenses. For example, altruism is part of the family of mature adaptational styles. Its neurotic precursor is reaction formation, a mechanism through which feelings (especially anger) are turned to their opposite (often positive regard and caring). On the immature level, Vaillant describes projection as the precursor to these mechanisms. Each one of these mechanisms shares an orientation toward others, while they differ in their shape regarding knowledge about psychological processes, emotions, and interpersonal relationships. Similarly, Vaillant interprets a developmental line from fantasy to intellectualization to sublimation, all of which function to ward off uncomfortable feelings and interpersonal closeness.


TABLE 2

A Glossary of Vaillant's Defense Hierarchy*

Level I--Psychotic Mechanisms
For the user, these mechanisms alter reality.
To the beholder, they appear "crazy".
1. Delusional projection. Delusions about external reality, usually of a persecutory type.
2. Denial. Denial of external reality.
3. Distortion. Grossly reshaping external reality to suit inner needs.

Level II--Immature Mechanisms
    For the <u>user</u> these mechanisms most often alter distress
        engendered either by the threat of interpersonal in-
        timacy or the threat of experiencing its loss.
    To the <u>beholder</u> they appear socially undesirable.
1.  Projection. Attributing one's own unacknowledged feel-
        ings to others.
2.  Schizoid fantasy. Tendency to use fantasy and to indulge
        in autistic retreat for the purpose of conflict reso-
        lution and gratification.
3.  Hypochondriasis. The transformation of reproach towards
        others arising from bereavement, loneliness, or unac-
        ceptable aggressive impulses into first self-reproach
        and then complaints of pain, somatic illness, and
        neurasthenia.
4.  Passive-agressive behavior. Aggression towards others
        expressed indirectly and ineffectively through passiv-
        ity or directed against the self.
5.  Acting out. Direct expression of an unconscious wish or
        impulse in order to avoid being conscious of the
        affect that accompanies it.

Level III--Neurotic Mechanisms
    For the <u>user</u> these mechanisms alter private feelings or
        instinctual expression.
    To the <u>beholder</u>, they appear as individual quirks or "neu-
        rotic hang-ups".
1.  Intellectualization. Thinking about instinctual wishes
        in formal, affectively bland terms and <u>not</u> acting on
        them. The idea is in consciousness, but the feeling
        is missing.
2.  Repression. Seemingly inexplicable naivete, memory
        lapse, or failure to acknowledge input from a selected
        sense organ. The feeling is in consciousness, but the
        idea is missing.
3.  Displacement. The redirection of feelings toward a rel-
        atively less cared for (less cathected) object rather
        than toward the person or situation arousing the feel-
        ings.
4.  Reaction formation. Behavior in a fashion diametrically
        opposed to an unacceptable instinctual impulse.
5.  Dissociation. Temporary but drastic modification of
        one's character or of one's sense of personal identity
        to avoid emotional distress. Synonymous with neurotic
        denial.

Level IV--Mature Mechanisms
   For the <u>user</u> these mechanisms integrate reality, interper-
     sonal relationships, and private feelings.
   To the <u>beholder</u> they appear as convenient virtues.
1. Altruism. Vicarious but  constructive and instinctually
     gratifying service to others.
2. Humor.  Overt  expression of ideas and feelings without
     individual discomfort or  immobilization  and  without
     unpleasant effect on others.
3. Suppression.  The conscious or semiconscious decision to
     postpone  paying  attention  to a conscious impulse of
     conflict.
4. Anticipation.  Realistic anticipation of or planning for
     future inner discomfort.
5. Sublimation.  Indirect or  attenuated  expression of in-
     stincts without either adverse consequences or  marked
     loss of pleasure.

*Adapted from: Vaillant, 1977.

Vaillant's hierarchy is very useful for any developmental  understanding of clinical  processes and organizes many
psychoanalytic constructs in creative ways.   In  fact,  his
hierarchy, functional in orientation, can be integrated into
a structural understanding of development.  This integration
requires  an additional analysis: what are the cognitive and
cognitive-affective transformations that  underlie  the  defense levels?  What is the structural basis for altruism being  placed "higher" than reaction formation?  Vaillant himself is interested in this question and discussed  Piaget's,
Loevinger's, and Kohlberg's work in the final chapter of his
book;  Vaillant,  of course, does not go beyond a comparison
of parallels but he does point the way for future   research.
His  adaptational  styles require a structural reinterpretation, because he addresses overall cognitive  and  emotional
<u>organization</u> and  <u>meaning constitution</u> throughout the life
cycle.   Such an analysis integrates defense process and systematic distortions (left out of most  structural   accounts)
but cannot be based on those mechanisms, because they do not
represent the <u>basis of meaning</u>.

In our attempt to build a bridge between structural ego
theory  and psychoanalytic ego process we make a distinction
between stages of ego development  and  personality  styles.

Defenses represent mediational processes between these stages of cognitive-affective self organizations. Thus intellectualization, fantasy, and sublimation are connected to cognitive development as are altruism, reaction formation, etc. But the preferred use and the overall orientation (towards people and relationships vs. towards things or ideas) are determined by the style, traditionally referred to as character. The dynamic between style, stage, and the mediational defense processes are central to an understanding of ego development.

Norma Haan (1977) introduced another way of understanding adaptational processes. She applies psychoanalytic defense concepts more closely aligned with Piaget's and Kohlberg's developmental theories of structure and cognitive transformation than with Vaillant's theory. Haan's view of the ego is one of a process of assimilation and accommodation to environmental demands restructuring thinking, feeling, and action. These ego processes are not themselves organized structurally but are expressions of a person's changeable forms of adaptation. Haan's taxonomy (as shown in Table 3), includes a hierarchy of ten generic processes which are subdivided into defense, coping, and fragmentation.

The three modes are distinguished by formal properties: most generally coping is implied when the ego activity involves logic, choice, flexibility, affective expression, and an interpersonally defined reality. Defense occurs when the ego activity is rigid, irrational, reality distorted, anxiety guided, affect and conflict denying, impulsive, and not under the guidance of logic. Fragmentation is ritualistic and it violates intersubjective reality by being guided by the idiosyncracies of inner thought and feeling. The processes are defined as a utilitarian hierarchy: "the person will cope if he can, defend if he must and fragment if he is forced, but whichever mode he uses, it is still in the service of his attempt to maintain organization" (Haan 1977, p. 4). This definition is clearly going beyond the ego psychology tradition in psychoanalysis (while building on many of their assumptions), because it explains the function not

TABLE 3

Taxonomy of Ego Processes*

| Generic Processes | Modes | | |
|---|---|---|---|
| | Coping | Defense | Fragmentation |
| *Cognitive Functions* | | | |
| 1. Discrimination | Objectivity | Isolation | Concretism |
| 2. Detachment | Intellectuality | Intellectualizing | Word salads, neologisms |
| 3. Means-end symbolization | Logical analysis | Rationalization | Confabulation |
| *Reflexive-intraceptive functions* | | | |
| 4. Delayed response | Tolerance of ambiguity | Doubt | Immobilization |
| 5. Sensitivity | Empathy | Projection | Delusional |
| 6. Time reversion | Regression-ego | Regression | Decompensation |
| *Attention-focusing functions* | | | |
| 7. Selective awareness | Concentration | Denial | Distraction, fixation |
| *Affective-impulse regulations* | | | |
| 8. Diversion | Sublimation | Displacement | Affective preoccupation |
| 9. Transformation | Substitution | Reaction formation | Unstable alternation |
| 10. Restraint | Suppression | Repression | Depersonalization, amnesic |

*Norma Haan, Coping and Defending, 1977, p. 35.

as mechanisms in the struggle between id, drive, and physiological processes, and the ego, but purely as tools in the
interaction between person and environment.

An example might help to illustrate the distinction between coping, defense, and fragmentation. Indeed, Haan's
point that psychoanalytic formulations do not stress enough
of the adaptive side of cognitive processes is well taken.
Her introduction of a generic process of detachment (which
has a coping side called "intellectuality", a defense side
called "intellectualization", and fragmentation is observed
as both confabulations and neologisms) is a useful attempt
to distinguish adaptive and maladaptive forms of ego processes. Although psychoanalysts have interpreted defenses as
having the potential for adaptive and maladaptive direction-
ality, the lack of defining markers between coping and defense and the traditional view of the intellect in conflict
with the id easily led to cognitive performance being interpreted as defensive intellectualization.

Haan posits, and we agree with her, that ego processes
are more than an operational expression of a given cognitive
structure. "Ego functioning is characterized by mobility,
inventiveness, and changeability, while structures being
based on consolidated and irreversible principles of knowing
are stable, integral and certain" (Haan 1977, p. 47). But
Haan gives support to our position of a necessary but not
sufficient relationship between cognitive stages and defenses (Noam, in preparation). Her example of a Piagetian
stage of concrete operations characterized by objectivity,
decentering, and classificatory ability establishes the
first basis for the discrimination of generic ego processes
(coping: objectivity; defense: isolation; fragmentation:
concretism). Embedded in this view is also the idea that
regression does not necessarily mean a return to earlier
functioning, but a fragmentation within the constraints of a
present cognitive organization (i.e., concrete operations).

What is still missing is a model that integrates both
the functional aspects of ego processes and the structural
definitions of ego stages. Such an analysis would logically
reconstruct the defense processes and their hierarchy from a
constructive-developmental approach and places it into a
personality framework. Yet, the work of Vaillant and Haan
has taken the first important steps in that direction.

### Erik Erikson's Epigenetic Model

Whereas Freud and Hartmann introduced an expanded vision of the ego's functions within the context of Freud's metapsychology, Erikson developed new and far-reaching principles that partly transformed psychoanalytic thought itself. In their excellent philosophical interpretation of the evolution of psychoanalytic thought, Yankelovich and Barrett (1972) elaborate the fact that Erikson presented a new viewpoint in psychoanalysis (and the social sciences), stopping short of introducing a new paradigm.

When we move from Hartmann to Erikson, we move from the world of forces and energies to the world of meanings, possibilities, genuine development and ethical realities. In making this move we have truly made a decisive leap; for we have not only changed our unit of analysis, we have begun to change our philosophy. (p. 154)

Erik Erikson's approach is based on a conception of the ego that is difficult to integrate with structural-developmental theory, although we will show some overlap that could lead to fruitful integrations and important advances in psychology. Erikson does not assume an ego originating from autonomous motivation for competence and adaptation to the world (a step White and the British object relations school took), but rather retains the Freudian and Hartmannian view of the ego emerging from libidinal drives and eventually mediating between the id, the superego, and reality. By applying biological drive to psychosocial development, Erikson remains in a field of contradictions.

Yankelovich and Barrett support our view:

He (Erikson) implicitly depends upon a philosophy at odds with the scientific materialism that underlies the classical metapsychology, but he has not developed such a philosophy explicitly to the point where he can lean on it for support. (ibid., p. 153)

But Erikson rarely thinks in the Freudian dynamic, topographic, or economic model. Rather he puts forward his own developmental principles and interprets all case material from that vantage point. His attempt to use an expanded metapsychological construct within traditional psychoanalysis has led to a concept of an ego rooted in an original ego-id matrix, which gives his theory a strong biological and psychosexual connotation. In describing the ego and "its

counterplayers", Erikson reiterates his indebtedness to a
traditional view of personality.

But who or what is the counterplayer of the ego?   First,
of course,   the id and the superego, and then, so theory
says, the "environment"... The ego's overall task is, in
the simplest terms, to turn passive into active, that  is
to screen the impositions of its counterplayers in such a
way that they become volitions. This is true on the inner
frontier where  what  is experienced as "id" must become
familiar, even tame and yet  maximally  enjoyable;  where
what  feels like a crushing burden of conscience must become a bearable, even a "good"  conscience.   (1968,  pp.
218-19)

The  British  object  relations  school,  represented by
Fairbairn, broke with this biological aspect of Freud's work
at about the time Erikson introduced his theory in Childhood
and Society in 1950.  Guntrip (1971), the eloquent  theorist
who  attributes an important place in the history of psychoanalysis to Erikson, criticizes:

In his handling of infantile sexuality, Erikson  actually
takes  up  the  same  position [as Freud] without clearly
stating it.  He treats it as a complex of ego  reactions,
not as an id drive. In what he writes about the id, Erikson  clings  illogically to a theory he has in fact abandoned "a view of human personality as constructed of layers: primitive and biological at the beginning; cultural,
social, sophisticated and psychological at  the  top,  id
and ego.  This I believe to be a false view, which needs
to be superseded by a view of the psyche--soma as a whole
that does not have primitive survivals inside it,  but  a
whole  in  which  everything  that is taken up into it is
transformed in a way that makes it appropriate to its being part of this whole".  (p. 84)

Erikson devoted a chapter of his original  1950  expose
Childhood  and Society on the life cycle to the relationship
of infantile sexuality into psychosocial stages.  He distinguishes between  animal  instincts  and  human  "inborn  instincts",  which are "drive fragments to be assembled, given
meaning and organized during a prolonged childhood" (p. 95).
But the theory he criticizes is  still  the  frame  of  reference; sexual maturation remains a focus in the development
of  meaning, virtues, and ethical commitments over the life-
span.

To a considerable extent, Erikson derives age and stage-typical content of concern from Freudian psychosexual stage theory. This theory specifies what is on the child's mind in terms of oral, anal, and phallic interests and drives (together with anxieties and reactions to these). Erikson extends these concerns into related interpersonal modalities. The links between psychosexual and psychosocial development are the organ modes which dominate the psychosexual zones of the human organism. These modes of incorporation, retention, elimination, inclusion, and intrusion, are central in development in connection to libidinal zone, e.g., oral-incorporation and anal-retention and elimination. The psycho-social aspect of zones and modes is their organization of relationships to things and persons as well as responses by significant others. An example might clarify this point: at the oral stage, sucking emerges as the first modality. But it is, Erikson convinces us, not only the physiological process of food intake, nor the sexual pleasure of the mouth that fully explains what occurs with the infant. "The 'sucking' mode is the first social modality learned in life, and it is learned in relation to the maternal person, the 'primal other' of first 'narcissistic' mirroring and of loving attachment" (in press, p. 28). This first encounter involves the mutual recognition of infant and caregiver, the eye contact, the touching, and the sound.

Erikson expands on this idea of ego strengths and virtues by describing the interpersonal aspect of this earliest mode; by getting from the parent, the infant learns how to get someone to give and to establish the foundations of a future giver. When moving from the oral-incorporation stage to the anal-muscular stage, and later to the infantile genital stage, the child is involved in expanding space-time experiences and his radius of social interaction. This radius is partly determined by the body's ability to move in time and space, whether crawling or upright, alone or with a helping hand.

In Erikson's descriptions of these early stages cf childhood, the link between psychosexual zones and psychosocial modes is over-generalized. The radius of time and space and interpersonal encounters is not only defined by psychosexuality but by other organs upon which Erikson focuses, especially the eyes and other sensing organs, the legs and their relationship to body and movement. Although Erikson argues convincingly for a strong correspondence between an

expanded biological version of maturation and psychosocial
development, following the child into adulthood, the relationship between sexuality and social development becomes
more difficult to justify, since it is not primarily biological change or psychosexual maturation that accounts for
change in adult psychosocial adaptation.

Let us now turn our attention to the model Erikson uses
to describe development over the life cycle. We have said
that Erikson's ego stages develop out of the ego's efforts
to cope with all the forces implied by the three agencies of
id, superego, and ego. The id, or the libido, was thought to
develop as a maturational function by Freud, and since the
account is accepted by Erikson, an embryological model is
appropriate.

Embryology now understands epigenetic development, the
step-by-step growth of the foetal organs. I think that
the Freudian laws of psychosexual growth in infancy can
best be understood through an analogy with physiological
development in utero. In this sequence of development
each organ has its time of origin. This time factor is as
important as the place of origin. If the eye, for example, does not arise at the appointed time, it will never
be able to express itself fully, since the moment for the
rapid outgrowth of some other part will have arrived, and
this will tend to dominate the less active region and
suppress the belated tendency for eye expression. (1950,
p. 65)

We have earlier described some basic differences between a structural and a functional model. We will explore
what the functional requirements and changes of each stage
of Erikson entail. Erikson's functional position is summed
up in a statement about the relationship between psychological stages and social institutions.

Each successive stage and crisis has a special relation
to one of the basic elements of society, and this for the
simple reason that the human life cycle and man's
institutions have evolved together. (1950, p. 250)
Erikson's epigenetic chart formalizes a progression through
time, the differentiation of parts in which each item that
develops is related to all other items and changes the relationship of all other items, and in which each item exists
before its emergence as a critical period: "But they all
must exist from the beginning in some form for every act
calls for an integration of all" (1950, p. 271). Erikson

(1968, p. 104) describes six components of psychosocial
stages:
1. The expanding libidinal needs and the new satisfactions
   and frustrations that occur due to these changes.
2. The widening social radius involving the people one is
   in contact with and the changing importance as partners
   of interaction.
3. The differentiating capacities that become important
   at each stage. Capacities to move, to orient oneself in
   the world, to use one's hands and one's intellectual
   abilities.
4. Results of new relations with institutions and the need
   for new forms of mastery and synthesis.
5. New sense of estrangement, forms of separations from important ideas, people, and institutions of the past.
6. Psychosocial strengths and weaknesses that come out of
   the resolution of the crisis and create the foundation
   for all future strengths.

This epigenetic principle represents a universal law of
development of a hierarchy of stages of growth.  Developmental crises are at the source of these developmental transitions and create potentials for new relationships with important others, institutions, and the self.
    Personality, therefore, can be said to develop according
    to steps predetermined in the human organism's readiness
    to be driven toward, to be aware of, and to interact with
    a widening radius of significant individuals and institu-
    tions. (ibid., p. 93)
Each resolution of the stage specific crisis leads to ego
strength, to a new form of ego synthesis, which Erikson
calls virtue.  Erikson's epigenetic model presents us with
the first life span approach to ego resilience and self-esteem, concepts that have only recently drawn attention within the structural-developmental framework.  The epigenetic
stages extend far beyond the childhood stages that Freud uncovered.  In fact, perhaps the most lasting and important
contribution that Erikson made within the field of psychoanalysis is the articulation of a "postoedipal" theory of
adolescence, adulthood, and old age. Erikson, the psychoanalyst, agreed with the basic pschoanalytic premise that the
recovery of early childhood experiences in therapy could be
a means to break with the repetition of compulsions in later
life. Oedipus, the man and the myth, remained for him a symbol for the playing out of early wishes merging with later
actions: "The fact that human conscience remains partially

infantile throughout life is the core of human tragedy"
(Erikson 1950, p. 257). But Erikson, the innovator with an
anthropological interest in the study of the social and moral world, has introduced a different model: the person "in
search of meaning". The person, that is, facing crises at
each phase of life which are solved through interaction with
the social world "on their own ground", rather than a function of early childhood experience. Erikson states:

In passing, however, we must note another essential
attribute of all developmental unfolding as the radius of
counter-players increases, graduating the growing being
into ever new roles within wider group formations, certain basic configurations such as the original dyad or
triad demand to find a new representation within later
context. This does not give us the right without very
special proof to consider such reincarnations as a mere
sign of fixation or regression to the earliest symbiosis.
They may well be, instead, an epigenetic recapitulation
on a higher developmental level and, possibly, attuned to
that level governing principles in psychosocial needs.
(in press, p. 48)

This position has brought Erikson closer to a Piagetian
model of interactionism and has led to numerous controversies within the psychoanalytic movement. Jacobson, for example, criticizes Erikson for failing to place enough emphasis
in the process of identity formation in early childhood.

Originally Erikson did not overlook the genetic approach,
but he seems increasingly to remove himself from it. His
studies on identity certainly placed the focus mainly upon the adolescent and pre-adolescent period. This is reflected in his use of the terms ego identity and identity
formation... No doubt the term ego identity in this sense
will lend itself readily to psychosociological study
which relates "individual identity" to "group identity".
I find it very difficult to distinguish personal identity
from ego identity, all the more since Erikson links the
the latter with "realistic self-esteem" and relates the
individual super-ego to the value systems of the society
in which the individual is reared. (1964, pp. 25-26)

Jacobson is only one of many critical voices who disagree
with Erikson's attempt to trace identity as a psychosocial
phenomena over the entire life cycle. But are we dealing
with a paradigmatic question--is ego development mainly determined by the early period of development--or with a crit-

icism that Erikson's concepts integrate too many disciplines and traditions? The question has not been answered fully. Putting Erikson into the honorable company of both social psychologists and social anthropologists appears to be a disassociation from his psychoanalytic roots. In fact, Erikson's questions will continue to pose themselves for psychoanalysts in ever new ways (cf. Vaillant, Kohut, etc.). Table 4, based on an extensive review of Erikson's writings, stresses his original contribution explicating the ego's task to master social relationships, to resolve crises, and to build strengths in the process. But the conflicts can also lead to regressions and "ego weaknesses".

We will now trace Erikson's stages of ego development along four dimensions which might help clarify a number of theoretical questions:

1. radius of social interaction,
2. ego tasks,
3. ego strength/self-esteem,
4. ego weakness/pathology.

Erikson's model is not a transformational model based on Piaget (see Table 5), which we will describe in greater detail when we turn to Loevinger and Kegan. Although his model shares the Piagetian principle of invariant universal sequence of development, it differs on a number of counts: 1) the definition of the stages of ego development; 2) that each stage is linked to social phases of development, and that each crisis is connected to an age-specific functional requirement that makes new adaptational modes necessary; 3) that each of the crises exists from the beginning but has its focal crisis phase and its key potential for resolution at a given time in the life cycle. This theoretical position allows Erikson to relate old crises to new ones as well as retaining the important concept of regression and multiple functioning that psychoanalysis has made meaningful. If old crises have not been resolved they will re-emerge at a later stage. It is not necessary to understand normal personality change but psychopathology as well, within the confines of a given stage. Thus, it is not surprising that Erikson's theory has an explicit concept of regression. It generates the possibility of tracing thought and action to its precursors for an understanding of clinical and everyday phenomena, and gives us new avenues to a scientific study of regression.

TABLE 4

STAGE 1 – BASIC TRUST vs. MISTRUST

| Radius of Social Interaction: "Dyadic-Maternal" | Ego Strength/Self Esteem: "Hope" |
|---|---|
| Mutuality of recognition, by face and by name between infant and caretaker. Need for consistency, continuity and sameness of experience. Images firmly correlated with familiar and predictable things and people. Food and love from mother/father, combination of sensitive care of baby's individual needs and a firm sense of personal trustworthiness, direct and fulfillment. Guidance by prohibiting and permitting, and representing deep and almost somatic conviction of meaning to actions. | Trust and confidence in self and others. The necessary groundwork to become the giver, "Because one was given." First crawling leads to exploratory interest (curiosity). Relatedness to others and through others to self. Ability to develop greater capacity for renunciation and transfer of disappointed hopes to better prospects. A sense of rootedness. Ability to perceive enduring quality of things and people. |
| **Ego Tasks: "To Let Take Care of"** | **Ego Weakness/Pathology: "Withdrawal"** |
| Comprehension of existence of primary other. Ability to move and crawl, contributing to later competent steps. Maternal care, the basis for the resolution of the first tasks of the ego. Infant's needs to let caretaker out of sight without undue anxiety or rage. Need to rely on sameness and continuity of providers and resultant knowledge that one can trust one's organs to cope with urges. Testing and differentiation of inside and outside. | Feelings of having been deprived, divided and abandoned. Becoming closed up, refusing food and comfort. Autisim, lack of eye contact and facial responsiveness, stereotyped gestures. Masochistic tendency of hurt whenever unable to prevent significant loss. Infantile schizophrenia, withdrawal into schizoid and depressive states, psychotic breaks, addictive. |

STAGE 2 - AUTONOMY vs. SHAME, DOUBT

Radius of Social Interaction: "Dyadic-Parental"
Outer control at this stage must be firmly reassuring.
Firmness of parents, must protect against potential anarchy of untrained sense of discrimination.
Parents must support independence and will, must protect from shame and doubt.

Ego Strength/Self Esteem: "Will"
Cooperation, willfullness, self expression.
Good will and pride, awareness, attention, manipulation.
Keeping rivals out fosters autonomy.
To learn how to hold, to own and to let be.
To gain the power of increased judgement and decision in the application of drive.
Verbalization and locomotion.
Exercise of free choice and self-restraint.
Sustaining the emerging split between good and bad, self and others.

Ego Tasks: "Holding and Letting Go"
Wishes for choice.
To stand on own feet.
Learning how to hold and let go.
Appropriately demanding.
Cooperation.
Learning to say "no."
Stubborn elimination.
Guided experience of autonomy and free choice.
Learning how to master shame and doubt.

Ego Weakness/Pathology: "Compulsion"
Cruel retainment or restraining.
Sadistic destruction, letting loose of destructive forces, hate, rage.
Turning against self, over-manipulation of self.
Power orientation and precocious super ego, inability to share.
Compulsion neurosis, feeling small and over-defiant.
Too much self-doubt which finds extreme adult expression in paranoid fears.

(continued)

## STAGE 3 – INITIATIVE vs. GUILT

**Radius of Social Interaction: "Familial"**
Within the family, the caring for siblings.
Ideal adults recognizable by their uniforms and functions.
Parental ideal figures become more important.
From the parents to a larger set of ideas and ideals.
Insight into institutions, functions and roles "which will permit a responsible participation".
First interactions with teachers.
The family, protecting the child from unmanageable conflict, hunger and danger, for creation of realm of play.

**Ego Tasks: "Taking on Parental Role"**
Often embittered attempts at demarcating a sphere of unquestioned privilege.
Discovery of eroticized genitals (and the possibility for them to be harmed).
From pre-genital attachment to parent, to becoming the carrier of tradition.
Self-guidance, self-observation and self-punishment.
To cooperate with peers for the purpose of constructing, planning and playing.
Looking for opportunities for work identifications that promise field of initiative.

**Ego Strength/Self Esteem: "Purpose"**
Making and being on the make.
Suggesting pleasure in attack, conquest, self direction and cooperation.
Learn where fantasy is out of place and learned reality becomes all demanding.
Envisioning of goals beyond the family.
Development of a strong conscience as a consistent inner voice.
Active participant in language which verifies the shared actuality.
Courage to envisage and pursue valued goals uninhibited by earlier defeat and fear of punishment.

**Ego Weakness/Pathology: "Inhibition"**
Failure of demarcating a sphere on one's own leads to resignation, guilt and anxiety.
Inhibition, over-control and over-constriction, to the point of obliteration.
Over-obedience and hysterical denial.
Development of lasting resentments and hate because parents do not live up to new conscience.
Hopes and fantasies repressed.
Suspiciousness and evasiveness due to an all-or-nothing super ego.

STAGE 4 - INDUSTRY vs. INFERIORITY

Radius of Social Interaction: "Family/School Neighborhood"

School, "whether school is field or jungle or classroom," and systematic instruction.
Wider society becomes significant in its ways of admitting the child to an understanding of meaningful roles and its technology and economy.
The neighborhood becomes a social sphere for friendships, playgroups, and social exchanges (shops, etc.).

Ego Tasks: "Mastery of Tool World"
Ability to push away fantasy and obey the laws of impersonal things.
Mastery of the three "R"s.
To become a worker, a potential provider.
Learning to win recognition by producing things.
Mastery of the organ modes.
Applying self to skills and tasks and adjustment to the inorganic laws of the tool world.
Learning fundamentals of technology.
Becoming literate if culture permits.
Understanding of first division of labor, differential opportunity.

Ego Strength/Self Esteem: "Competence"
Productivity, planning ahead.
Development of work principle.
Steady attention and persevering, concentration.
Setting priorities, punctuality and order.
Becoming ready for a variety of specialization in learning.
Participation in ethos of production.
Free exercise of dexterity and intelligence in work and play.
Understanding and accepting logic of tools and skills.

Ego Weakness/Pathology: "Inertia"
Sense of inadequacy and inferiority.
Discouragement of identification with tool partners and tool world.
Despair over equipment and consideration of horizons to include only work.
To see work as the only criterion of worthwhileness.
Danger of becoming a conformist and thoughtless slave of his technology and of those who are in a position to exploit it.

(continued)

STAGE 5 - IDENTITY vs. IDENTITY CONFUSION

Ego Strength/Self Esteem: "Fidelity"
Form of ego identity - the accrued confidence of inner sameness and continuity.
Taking the chance to fall in love and become intimate.
Ideological mind: the love for ideas, ideological seeking after inner coherence.
Ability to sustain loyalties although contradictions of value systems perceived.
High sense of duty, accuracy, veracity, conviction, authenticity.

Ego Weakness/Pathology: "Repudiation"
Role confusion based on doubt on one's sexual identity; delinquent and psychotic episodes can occur.
Inability to settle on an occupational identity, totalizing of world image.
Temporary over-identification to the point of loss or confusion of identity.
Clanishness and cruelty in exclusion of others who are "different."
Friendships and affairs are attempts to delineating fuzzy outlines of identity, narcissistic mirroring.
Cynical, apathetic, lost, role repudiation.

Radius of Social Interaction: "Peer Group"
Becoming part of a peer group or larger crowd.
Intimate relationships with important others.
Creation of a moratorium - a time of non-commitment in a place protecting adolescent from decision making.
Role opportunities for selection and commitment.
Apprenticeship - professional and school.

Ego Tasks: "Commitments"
Confronting adult tasks ahead; how to connect the roles and skills cultivated.
To sustain a relationship with a person of another sex, sex role experimentation.
Creation of beginning of adult morality, of ethical standards and of ideological commitment.
Development of small group rituals.
Participation in public events, sports, political and religious arenas.

STAGE 6 - INTIMACY vs. ISOLATION

| Radius of Social Interaction: "Partnership" | Ego Strength/Self Esteem: "Love" |
|---|---|
| Fusing identities in mutual love and counterpointing them to an individual who in work, sexuality and friendship promises to prove complimentary.<br>Willing to jointly regulate the cycles of work, procreation and recreation.<br>Preparation for stages of satisfactorily securing of off-spring and their development.<br>Commitment to concrete affiliations in partnership cultivating styles of in-group shared patterns of living.<br>Greater participation in social institutions. | True mutuality, ethical sense, morality.<br>From the resolution of the synthesis between intimacy and isolation emerges love.<br>Loyalties and ethical strengths due to commitments to concrete affiliations.<br>Exclusivity as a counterforce to love is important for the boundary of the relationships.<br>The ability to choose and actively love a mate creating a shared identity.<br>Ability to affiliate with others poten-tially ready to care for off-spring. |
| **Ego Tasks: "Intimacy"** | **Ego Weakness/Pathology: "Exclusivity"** |
| Intimacy: the capacity to commit to con-crete affiliations and partnerships.<br>Persons of very different backgrounds must fuse their habitual ways to form a new milieu for themselves and their offspring.<br>Ability to share mutual trust.<br>Ability to sacrifice and compromise for sake of being with other; adult duty.<br>Body and ego must now be masters of organ modes to face fear of ego loss when close. | The counterpoint to love is exclusivity: elitism cultivating cliques marked by snobbery.<br>Avoidance of intimacy, isolation and consequent self absorption.<br>Prejudices or destruction of others who are not part of the familiar.<br>Isolation as regressive and hostile re-living of the identity conflict.<br>Partnerships in which the intimacy leads to an "Isolation a Deux."<br>Cruel and combative relationships. |

(continued)

STAGE 7 - GENERATIVITY vs. STAGNATION

Radius of Social Interaction: "Work and New Family"

Concern in generativity of the establish-ment and guiding of the next generation, including productivity and creativity.
Children, family and institutions which reinforce generativity and that codify the ethics of generative succession.
A generation of new beings, new products and new ideas, including self-generation.
The ritualization of parental, didactic, productive and curative in the family and at work.

Ego Strength/Self Esteem: "Care"

Care and altruism for the "creatures of this world" - universal care.
Widening commitment to take care of the persons, the product and ideas one has learned to take care of.
The ability to become a model in the next generation's eye and to act as judge of evil and transmitter of ideal values.
Generativity includes a measure of true authority rather than authoritarianism.
Ability to be challenged by that which has been generated.

Ego Tasks: "To Raise"

Need to be needed.
Guidance and encouragement from what has been produced and must be taken care of.
Interest in that which is being generated.
Altruism of taking pride in fostering.
Securing productivity and creativity of self and others.

Ego Weakness/Pathology: "Rejectivity"

Where procreativity and generativity fails, stage-wide regression may occur (pseudo-intimacy or kind of preoccupation with identity issues).
Indulging as if people were their own children; excessive self love.
Early invalidism (physical or psychologi-cal) as a vehicle of self concern.
Lack of faith and belief in the species.
Physical and moral cruelty against one's children.
Moralistic prejudice against family or com-munity; authoritarian use of power.

STAGE 8 - INTEGRITY vs. DESPAIR

## Radius of Social Interaction: "Mankind"

Cosmic world order

Comradeship with the ordering ways of different times and different pursuits

The knowledge of one life cycle as being only one segment of history.

Ordering the world by generations.

Emotional integration permitting participation and responsibility of leadership.

## Ego Strength/Self Esteem: "Wisdom"

The post-narcissistic love for humankind.

The new meaning for life in face of death.

A timeless love for those others who have become the main partners in life's most significant contexts.

Existential identity.

Wisdom, a kind of detached concern with life itself in the face of death itself.

Part of wisdom is disdain. Wisdom can well contain disdain as a refusal to be fooled in regard to man's antithetical nature.

## Ego Tasks: "Closure"

Integrity, "ego's accrued assurance of its proclivity for order and meaning."

Acceptance of one, and only one life cycle that had to be a new and different level of one's parents.

Preparation for death.

Ego integrity therefore implies an emotional integration which permits participation by fellowship as well as acceptance of the responsibility of leadership.

Commitment to world and history.

Dealing with bodily decline.

## Ego Weakness/Pathology: "Disdain"

The over-anxiety of death, the terror of time being short.

Disgust hiding despair.

Feeling finished, passed by and helpless.

Despair of the knowledge that a limited life is coming to a conclusion.

The philosophical ritualization can become dogmatism and cohesive orthodoxy as well.

Boring repetitiveness of early reminiscence.

Rigidity.

TABLE 5

Structural and Functional Stage Approaches

| Piaget's Structuralism | Erikson's Functionalism |
|---|---|
| 1. Stages are different for a single function, e.g., moral judgment or logical reasoning. Later stages replace earlier stages. | 1. Stages are choices or uses of new <u>functions</u> by an ego --earlier functions or choices remain as background to the new stage. |
| 2. Each higher stage integrates all former forms of thinking into a new structure. | 2. Each higher stage brings new relationship between new crisis and its resolutions and former crises, but does not integrate former stages. |
| 3. Experience leading to development is cognitive experience, especially experiences of cognitive conflict and match. | 3. Experience leading to development is personal experience, especially experiences and choice of personal conflict. |
| 4. The developmental change is primarily a changed perception in the physical, social, and moral world. | 4. The developmental change involved is primarily a self-chosen identification with goals in a choice or a commitment. |
| 5. Later stages are more cognitively adequate than earlier stages:<br>1) including the earlier stage pattern;<br>2) resolving the same problems better; and<br>3) in being more universally applicable or justifiable, i.e., in the universality and inclusiveness of their ordering of experience. | 5. Later stages are more adequate than earlier stages, not in cognitive inclusiveness, but in virtue of ego strength, i.e., in their ability to order personal experience in a form that is stable, positive, and purposive. Attainment of a stage and adequacy of stage use are distinct however. |

The epigenetic principle states that <u>all</u> crises exist before and after the time of their primacy. While basic trust over basic mistrust is the psychosocial crisis of the earliest stage of life, components of the second crisis, autonomy versus shame/doubt, are already existent. Erikson gives the example of the infant who may show autonomy striving from the beginning by moving away or reacting angrily. The crisis of autonomy versus shame in which the two-year old "begins to experience the whole critical opposition of being an autonomous creature and being a dependent one" (Erikson 1950, p. 271), only emerges on the second stage, though. Each component "comes to its ascendents, meets its crisis and finds its lasting solution during the stage indicated. But they all must exist from the beginning in some form for every act calls for an integration of all" (ibid., p. 271).

When asked during a recent interview (Hulsizer et al. 1981) how adult crises pre-exist in childhood or how, for example, generativity conflicts exist at the first horizontal line of Erikson's chart, Erikson answers in terms of an early preparation for a later task. He states that the learning, the experiencing of the infant, of being given, will one day create the giver, the generative adult. Actually the only example in which Erikson has specified both horizontal and vertical lines is the stage of identity crisis in adolescence. Identity formation neither begins nor ends in adolescence:
    It is a lifelong development largely unconscious to the
    individual and to his society. Its roots go back all the
    way to the first recognition; and the baby's earliest ex-
    change of smiles, there's something of a self-realization
    coupled with a mutual recognition. (1968, pp. 133-34)
And the resolution or non-resolution of the identity crisis will have implications on all the later stages of the life cycle. The conflict of the identity crisis might re-emerge at later times, and thus
    many of our patients break down at an age which is
    properly considered more pre-adult than post-adolescent,
    is explained by the fact that often only an attempt to
    engage in intimate fellowship and competition or in
    sexual intimacy fully reveals the latent weakness of
    identity....For when an assured sense of identity is
    missing, even friendships and affairs become desperate
    attempts at delineating the fuzzy outlines of identity by

mutual narcissistic mirroring; to fall in love then often
means to fall into one's mirror image, hurting oneself
and damaging the mirror. (1968, p. 167)


Even the most regressed patients have to deal with
their age-specific crises, though. The young adult facing
Stage 6 (Intimacy versus Isolation) without having developed
a firm identity might regress to early functioning. Here we
find another difference between the Piaget-Kohlberg tradition and Erikson. Regression is possible and earlier forms
of functioning can still regulate thinking, feeling and act-
ing--"Indeed, from the point of view of development, former
environments are forever in us" (ibid., p. 24). Although
Erikson is aware of the fragmentation evolved in the recourse to these former environments to solve present conflicts, he uses the term regression mostly in the sense of
regression in the service of the ego or regression in the
service of development. Erikson states that the ego contin-
ually subsumes in fewer and fewer images and gestalts, "the
fragments and loose ends of all infantile identifications".
This work is partly achieved through regression to unresolved conflicts of the past.

Regression, in this sense, is not a general pull to
some primitive mental functioning, but the "early environments that are within us" are part of the general logic of
development, implicit in the theory. The unresolved vulnerabilities to which we return in the face of new tasks are,
at the same time, part of the forces that can hinder developmental progressions. Their resolutions make the successful handling of the present psychosocial crisis possible.
In this respect, Erikson and his psychoanalytic heritage
have fostered the hope and introduced the therapeutic technique to describe and influence the working through of early
experiences. While psychoanalysis had contributed to this
understanding for almost fifty years, Erikson implied that
regression always has to be understood in relation to the
present-day functional, psychosocial requirements that he
described in a developmental sequence. Within the adolescent and adult stages of development he opened the possibility to study regression from two points of view: that of
regressing to, and that of regressing from. Erikson only
briefly outlines how such developmental precursors or developmental lines of unresolved problems in the past could
look.

A more elaborated theory of regression builds on Erikson's sketches and reinterprets his concepts from a structural-developmental perspective. Such a theory builds on the notion that regression never exists per se, but is always a process and a dynamic between the highest attained psychological stage and the earlier "internalized and unresolved environments". The functional requirements of the earlier stages are not being worked out in the process of regression, but the forms of earlier constructions of the self and the social world in relation to new tasks (Noam 1982b). This brings us back to our attempt to understand Erikson structurally and to indicate how one could rethink another important focus in Erikson's work: ego strength. Again, Piagetian psychology has by and large neglected this important area of self psychology.

Erikson views his virtues as representing certain types of active competence. This meaning is clarified by considering their parallels in Erikson's theory, the forms of self-esteem (and their opposites). Each term in the Erikson hierarchy is a form of self-esteem. If this is not completely clear in the positive forms, it is in the negative forms (mistrust, shame, guilt, inferiority), which are all forms of negative evaluation of the self whether by others or by the self. If Erikson's attitudes are forms of self-esteem, his parallel virtues seem to be forms of "will". Erikson specifically labels virtues of the second stage "will", using it to refer to the sense of freedom in willing or acting. "Purpose", is clearly another example of ego strength, as is "competence", the free exercise of intelligence in the completion of tasks. With regard to cognitive-structural contributions to psychosocial or ego stage, Erikson is generally accepting of the Piagetian account. It is possible to abstract a simplified picture of Erikson's developmental phases and their related accruals of ego-strength, a picture which stresses the universals of cognitive, ego, and moral levels. From this point of view, the psychosexual and interpersonal content of an Eriksonian stage is less important than the form of the developmental task and its resulting strength (or weakness). Furthermore, the stages can be seen as having an inner logical sequence independent of earlier virtue in the sequence.

From a structural-developmental point of view, then, each of Erikson's virtues is a component of the "attention-will" cluster of cognitive-style traits which we suggest de-

fined  the thread of ego strength in childhood and in adulthood.  At its intuitive roots, the concept of  ego  strength
seems  to  denote  qualities  of will.  As higher stages are
reached, the Eriksonian virtues become more and more matters
of "moral will" as  opposed  to  ego-willing  as  such. Adolescent  "fidelity",  young-adulthood  "love", and adulthood
"caring" represent successive capacities for ethicality, for
socio-moral commitments that are freely made  but  that  are
ethically  binding.  Erikson's childhood virtues have a less
moral flavor, except in their negative  poles  which  derive
from  the negative moral structures (shame, guilt, inferiority).

It is important to note that the virtues  as  forms  of
will  and  forms  of  self-esteem can be said to represent a
logical sequence, in the sense that each higher form presupposes the lower and  represents  a  differentiation  of  it.
Trust  and hope are positives that do not differentiate what
the self can do or what is positive in the  self  from  what
the parent or the world can do.  A sense of autonomy or will
makes  this distinction but does not distinguish the goal of
a free choice or activity from the activity itself.  A sense
of purpose or initiative is  a  positive  valuing  of  goals
which  presupposes this distinction.  Competence or industry
implies or takes for granted a positive goal and focuses  on
confidence  in  the  self's  generalized  means  (skills and
tools) to these goals.  Erikson himself is not  very  interested  in  the fact that he defines an abstract logical succession of forms  of  competence  and  self-esteem  somewhat
apart from his detailed psychosocial theory of stages.  From
our point of view, however, a focus on this logical sequence
helps  us  to  relate  his  developmental description of ego
strengths to the universals of  cognitive,  moral,  and  ego
structural  development.  Such a linkage can be made without
commitment to his more complex theory of the  way  in  which
psychosexual,  psychomotor,  and  psychosocial  developments
synchronize into age-specific developmental crises.

To conclude, Erikson addresses four concepts  important
in  the exploration of personality theory from a developmental perspective: such developmental  concepts  of  the  life
cycle  are  also  used as a foundation for clinical-developmental psychology, which focuses on adaptation and  maladap-
tation,  coping  and defending, and normality and pathology.
These developmental concepts are:
1.  Erikson's theory allows for the tracing of  later  stage

development to early development. This opens the possibility for an understanding of different developmental lines, which Erikson brought together in one single concept of the ego. George Vaillant's work, for example, has built on Erikson, describing one such line: the hierarchy and processes of transformations of defense mechanisms in adulthood.
2. Erikson's concepts allow for the study of ego strength. Structural-developmental psychology has not introduced any convincing theory of ego strength nor the illuminated relationship between ego stage and ego strength.
3. The developmental theory of the ego and ego strength for Erikson leads also to a theory of regression. Structural-developmental theory has not solved the theoretical problems of a transformational model and the phenomena of recourse to earlier forms of mental functioning.
4. The clinician Erikson has presented us with a developmental theory of psychopathology. Although this theory has not had much impact on psychiatry, with the exception of his clinical formulations of the adolescent identity crisis, his conceptualization can help guide the work to attempt to uncover a Piagetian life-span perspective for an understanding of the self and developmental aspects of psychopathology.

PIAGETIAN-STRUCTURAL APPROACHES TO EGO DEVELOPMENT

Jean Piaget, the most important and far-reaching structural psychologist, never put the ego or the self into center stage of his theoretical and empirical concerns. On the other hand, however, it was his starting point. As he states, for instance, "in estimating the child's conception of the world, the first question, obviously is...can the child distinguish the self from the external world?" (1926, p. 33). His lifelong preoccupation centered around the "epistemic self", which is based on the generalized universal structures of cognitive development. From such a position, ego development was of secondary importance. Similarly, the domain of emotional development so central to an ego theory, never was a focus in Piaget's work, although he did address emotions in their relationship to cognitive structures (e.g., Piaget 1973, and most elaborately in 1981). Piaget posits a parallelism in the development of cognitive structure and affective energies and argues that emotions are the phenomena that indicate disequilibrium and stage

transition.    But  Piaget's position, even with the detailed
elaboration of emotional development at each stage of cognitive reorganization has left a central question  unanswered.
What  is  the  mediation between the cognitive and affective
domains?  Does this mediation follow structural lines?  What
are the consequences for a self theory and its  implications
for  an  understanding of clinical phenomena?  Neo-Piagetian
theorists, however, have begun to focus on  these  questions
in recent years.  Two of the most notable theorists to bring
a  structural perspective to the self are Jane Loevinger and
Robert Kegan.  We will begin this  discussion with  Kegan's
concepts, because they follow more directly from a Piagetian
model.

### Robert Kegan's "Evolving Self"

Adapting the frameworks of Piaget and Kohlberg to study
the self from a structural perspective, Robert Kegan has addressed  processes  of  "meaning-making" over the life span.
For Kegan, meaning-making is the core of ego development. He
seeks an understanding of the ego by defining an  underlying
structure  of the self-other relationships that were originally outlined in more general terms by Kohlberg (1969, 1971)
and later by Selman (1980).

> This subject-object relation is  the  common  ground,  or
> deep structure for all Piagetian theories.  That develop-
> ment,  which each Piagetian theorist studies...is, I con-
> tend, a direct consequence of developments in  this  more
> basic  activity.   It  comprises  as  well the underlying
> structure and process missing  from Loevinger's  theory,
> the  closest  work  yet to a Piagetian conception of "ego
> stages".  (Kegan 1979, p. 9)

The self-other differentiation refers not only to the  relationship between a person and other persons, but to ways the
self  understands and deals with itself as well as the meaning real relationships have in the social world.   Thus  the
theory addresses:

1.  A perspective of a self "on itself";
2.  A  perspective  of  important  others that have been internalized, and are part of an "inner dialogue";
3.  A perspective of real others with whom the person is interacting and the understanding of  those  others  in  relationship to the self.

One way of understanding Kegan's work is that he has
translated Kohlberg's stages of moral development into
stages of the self. Kohlberg's emphasis on people's understanding of rules, norms, and social systems has become a
structural orientation of the self and an understanding of
relationships. Kegan's framework addresses the process of
identity formation through the stages and shares with Selman's model of interpersonal understanding and role-taking a
focus on the process of social perspective taking. Much as
Kohlberg and Selman describe the underlying social perspective for each stage of moral development, Kegan describes
the perspectives underlying the self at different developmental positions. Whereas psychoanalysis has looked at how
people internalize "objects", the process of taking important others into the self and relating this to childhood,
Kegan describes a number of internalizations over the life
span by focusing on the structure of balance in the relationship between the self and "person objects". In this
model of subject-object development, the object refers to
those feelings, thoughts, constructs, and relationships that
we can step out of, observe, and thus manipulate. In other
words, what has become object can also be objectified and
becomes potentially conscious.

The subject side of the balance refers to those aspects
of the self which the person is "embedded in" and has no
distance from and thus has no awareness of. Kegan's distinction between subject and object is paralleled by the distinction between "being" and "having", two concepts which
have long been a philosophical and psychological preoccupation (e.g., Fromm 1976). "Being" refers to that part of the
self (subject) that is lived out without being reflected
upon. For instance, at the interpersonal stage my identity
derives from being the dyad that I am part of--"I am my relationships". As soon as I have disembedded myself from
identifying self with relationships, and can reflect upon
them, I have moved beyond that stage. What was the subject
(being relationship) moves to the object side. Now I have
relationships and my identity is no longer derived from my
dyadic context. The move from subject to object is a process
of disembedding and of internalization; what once needed the
dyadic embeddedness to define the self, now has internalized
the dyads which allows for having relationships without
making them the central definers of the self. A new self
emerges in the transition which now is its role system and

coordinates in relationship to self. This example will become clearer when we describe Kegan's five stages of ego development. Since he develops his model in detail in this volume, we will only summarize the central features of subject and object at each level in Table 6.


TABLE 6

Kegan's Stages of the Self*


Stage 0 and the 0-1 Transition

Subject: reflexes--Object: none

Stage 0 is defined by the absence of a self-other boundary. All experiences of infants are extensions of the self. They are embedded in their reflexes and senses and do not possess these reflexes as object yet, but in Kegan's language, "are" these reflexes. During the course of the first transition to Stage 1, the person separates from "being" reflexes to "having" them. This internalization of reflexes integrates the elements of a more complex system, leading to the ability to retain an image and a gradual ability to "hold" one's own experiences through memory. Kegan interprets the acquisition of object permanence and object constancy as a consequence of this first subject-object differentiation.


Stage 1 and the 1-2 Transition

Subject: impulses/perceptions--Object: reflexes

Between ages 2 and 5 the child is embedded in his perceptions and impulses. Again, the child "is" his perceptions and impulses, rather than "having" them, i.e., when the perceptions of an object change, the object itself has changed. Kegan explains tantrums as a result of the child's not being able to express a given impulse, because his very organization, rather than simply an element of his organization, is frustrated. In other words, impulse control and delay are internal. This balance expresses itself in the affective realm through a limit that two feelings cannot be held si-

*Sources: Kegan 1982; Kegan, Noam, & Rogers (in press).

multaneously (inability to experience ambivalence). A shift
in the psychological organization occurs in the child between the ages of 5 and 7. Fantasy now becomes more real-
ity-bound. During this well researched transition in the
Oedipal years the child is gradually disembedding from his
impulses and perceptions ("having" them, rather than "being"
them). The significant manifestations of this transition in
the cognitive and the affective domain are separation of
appearance from reality.

## Stage 2 and the 2-3 Transition

### Subject: needs—Object: impulses/perceptions

There is the realization that persons are distinct and cannot be perfectly attuned to their own experience, needs, interests, wishes, impulses/perceptions. When the impulses become "object", Kegan states, the new system can now coordinate impulses over time. The "enduring disposition" emerges,
a way the child feels over time and the overcoming of the
"moment-to-moment-lability" of the earlier stages. The child
seems to "seal up" by setting boundaries, and trying to manage tasks alone. No longer can the parents read the feelings
and know all inner secrets. The limits of this organization
is its embeddedness in the "class", the self cannot coordinate two points of view; conflicting needs (enduring dispositions) cannot be integrated into a bigger whole. The self,
in Kegan's language, is its needs, rather than "having"
needs. In the development to Stage 3 the early adolescent
becomes an "interpersonal self" that coordinates interpersonally and intrapsychically between need-perspectives.

## Stage 3 and the 3-4 Transition

### Subject: interpersonal mutuality—Object: needs

At Stage 3, the self is "interpersonal", embedded in the
shared reality created through the third person perspective.
This stage brings construction of mutually reciprocal relations of co-equal obligation and expectations. Its strength
is the capacity to create an intimate interpersonal reality;
the developmental ceiling is the inability to objectify and
step out of the shared reality. The ability to experience
the world through the eyes of another person leads the self

to account for experiences beyond narrow parameters. But
any thought and emotion that threatens the interpersonal
fabric goes to the core of self's anxiety of abandonment and
"loss of self". Thus anger, a differentiating emotion, is
rationalized or remains unexperienced because it could lead
to a lasting separation too threatening to the "inclusive"
self. The transition to Stage 4, which usually occurs in
adolescence or adulthood, eventually leads to a new psychological independence or internal authority which corresponds
most closely to Erikson's identity stage.

## Stage 4 and the 4-5 Transition

### Subject: authorship, identity--Object: interpersonal mutuality

Kegan shows that in moving from "I am my relationships" to
"I have relationships" there is now somebody who is doing
this having, the new I, that creates a "psychic institution". This new self coordinates mutuality and the different
interpersonal contexts of Stage 3. Emotions are now more
internally controlled. "The immediacy of interpersonalist
feeling is replaced by the mediacy of regulating the interpersonal. In this sense, ego Stage 4 is inevitably ideological". But the limitations of Stage 4 is its boundary, the
orientation towards the "here I stand" without an ability to
create an intimate shared reality--two or more selves creating a new joint reality that goes beyond a "partnership".
Only Stage 5 brings a true integration of identity and intimacy.

## Stage 5

### Subject: interindividuality, interpenetrability--Object: authorship, identity

The new embeddedness of the self at Stage 5 coordinates the
"institutional selves" and creates a new sharing of the self
which "permits the emotions and impulses to live in the
intersection of systems, to be "resolved" between one self-
system and another". The self can sacrifice without being
masochistic, can be close without losing a sense of purpose,
and can delay needs, wishes, yearnings for the sake of
principles that integrate individuality, and social purpose
in a larger whole.

Each stage of ego development includes an underlying social perspective. Each higher stage of ego development consists of a more differentiated and more complex social perspective. For example, a systemic social perspective, which in moral development leads to an understanding of a social system that is regulated by laws and social obligations, in ego development underlies a systematic understanding of the self, as a self-regulating "agency" that has an orientation toward self-authorship, understanding of self as role with obligations toward others. Kegan's underlying ego structures are harder to capture than Selman's social-cognitive perspective, however. His orientation towards a hierarchy of biological and social organizing principles, such as reflexes (0), impulses (1), needs (2), and interpersonality (3) raises an important question. Are these constructs structural-developmental? Do impulses transform into needs, for example, or are they parallel processes? Also, Kegan seems to give up the important Piagetian distinction between judgment and action. The ego incorporates thinking, feeling, and acting. These structures are thus determined or characterized by the ways a person lives (e.g., observation of childish impulsive behavior) as well as by the framework of meaning a person gives to the self and life situations (e.g., responses to interview questions).

The logic and the process of self-other relations in Kegan's theory suggest that ego development might be best represented by a helix in which personality swings back and forth, between two poles of development. This model interprets growth and adaptation as a process of differentiation and integration. It sees each balance and organization (each stage) as a new tension between the wishes and needs for inclusion or affiliation and the wishes and needs for autonomy or differentiation (see Kegan in this volume).

Kegan turns to phenomenological and existential theories to describe the psychological activity and meaning-making of the self. Kegan proposes an integration of Piaget and phenomenological tradition, by applying cognitive theory to the processes of counseling. Focusing on the transition points in development, as periods of disequilibrium, he postulates that psychopathology is strongly related to those times in the life cycle in which an old balance of knowing has fallen apart and a new one has not yet been established. Kegan functionally emphasizes the role of the holding environment. He describes problems arising from a mismatch

of the self and the holding environment within the family,
school, and interpersonal relationships at any given stage.
For him, therapeutic endeavors include more than traditional
patient-therapist relationships. He also defines "natural
therapists" as those people and institutions that support
development.

More recently Noam and Kegan (1982) have explored the
process of externalization as it relates to ego boundaries.
These two concepts are inherent to basic psychological functioning and clinical treatment: the determination of where
"self" ends and "other" begins (boundary) and the process
through which intrapsychic phenomena become interpersonal
(specifically, externalization, or projective identification). An important limitation of the present use and understanding of concepts like "boundary" and externalization in
psychoanalysis is that individual differences tend to be
looked at almost exclusively around a "quantitative" continuum. In other words, the question becomes—how strongly is
the boundary between self and other maintained? What con-
stellation of defenses does the person have in order to
maintain the boundary? How flexible or rigid are the boundaries upheld? The question is never, which boundary is being maintained? Instead, orientation of defended differentiation implies that there really is only one boundary and
the only important questions have to do with whether or not
it exists and how strongly it can be defended. This is partly due to the fact that questions of boundary and externalization are usually understood in terms of early childhood
experiences, regardless of where the person is in the life
cycle.

Noam and Kegan argue that both of these limitations—
the reduction of all boundary phenomena to early experience,
and the reduction of all boundary analysis to degree of defended differentiaton—derive from the same source: the lack
of a life-span developmental perspective to the understanding of boundaries between self and other and their relationship. From a constructive-developmental view there is a
life history of qualitatively different subject-object balances and a full picture of a person will encompass not only
that person's early subject-object experience, but his present subject-object position, as well; not only that
person's degree of defense, but the particular subject-object boundary he is defending. The analysis of boundaries
and their relationship to processes of externalization is

just one example of tracing the ego and its transformations across the life span. It also exemplifies the research potential of this framework.[2]

The attempt to expand a theory of the self into a method of understanding and potentially altering clinical phenomena poses a number of interesting questions. Kegan's transformational model of the self, building on Piaget, does not account for unintegrated earlier vulnerabilities, forms of "regressions", and ego weaknesses stemming from earlier unresolved conflicts. This is not an oversight or a lack of focus, but is inherent in the model itself. Thus symptoms and conflict dynamics are interpreted solely as part of a process captured through a developmental process, captured by stage or transition. Indeed, Erikson's orientation we described earlier derives from clinical work and a tradition of clinical theorizing, thus taking such phenomena as regression and ego strength very seriously and trying to integrate them into a developmental life cycle model. This is also true for the processes of distortion, especially defense mechanisms, which Erikson uses and develops in his conceptualization. Kegan also shares with Erikson a more conceptual-theoretical approach, a fact which Loevinger criticizes. Kegan's work, however, is in progress. The development of a clinical and research methodology will be needed in order to judge the full implications of his model.

## JANE LOEVINGER AND EGO DEVELOPMENT

Jane Loevinger's concept and construct of ego development builds on a number of theoretical traditions of interpersonal, cognitive, and character psychology. Most importantly it draws on neo-psychoanalytic theory, earlier attempts at studying character development (Sullivan 1953, Grant & Grant 1959, Peck & Havighurst 1960, Isaacs 1956), and Piagetian model of structural stage hierarchy.

It is this sequence of differentiated, hierarchically-ordered stages which are defined independently from age, that Loevinger has termed ego development. Dan Candee (1974) explains:

In general, Loevinger's approach of ego development is marked by a more differentiated perception of one's self, of the social world, and of the relations of one's feelings and thoughts to those of others. (p. 621)

The  essence of ego development is the meaning of experience
and the developmental process of "meaning coherence". Loevinger's conception of character development (also  a  dimension  of individual differences) differs from the psychoanalytic view in which the term ego development had  traditionally  been  restricted to the early ego and its differentiation from an id-ego matrix (a concept she rejects). In Loevinger's theory, ego development refers to  a  collection  of
functions  skills  (cf. Haan,  Stroud, & Holstein 1977; Bel-
lack, Hurvich, & Gediman 1973).  Stuart Hauser, in a  review
of Loevinger's work, summarizes the difference as follows:

> In psychoanalytic theory, the orientation is often toward
> ...the  conception of  ego...in terms of it as a collec-
> tion, an "inventory" of related processes whose  overall
> function  is  "task  solving"  or "attempted solution" as
> contrasted with instinctual expression...we see a  strik-
> ing shift of meaning when we turn to Loevinger's descrip-
> tion  of  ego development....Loevinger's approach is best
> characterized as one which takes account of the individu-
> al's integrative processes and overall  frame  of  refer-
> ence.  (1976, p. 928)

Loevinger (1976)  states that there are at least four meanings given to ego development in  psychoanalysis,  of  which
only one, Erikson's, is compatible with her use. Loevinger's
model  differs  from  Erikson's, however.  Loevinger assumes
that the ego is a unitary system of functions; Erikson  does
not  necessarily  assume this unity.  Erikson cites age-specific tasks, like  entering  kindergarten,  courtship,  marriage,  and  childbearing in his stages of ego development;
Loevinger does not.  The meaning attributed to  these  tasks
goes along stage lines, thus Loevinger can account for individual  differences  within an age cohort.  Loevinger (1969)
states: "The striving to master, to integrate, to make sense
of experience is not one ego function among  many,  but  the
essence  of the ego" (p. 85).  Her stages differ on a number
of dimensions: impulse control,  interpersonal  style,  conscious preoccupation, and cognitive style.

The  descriptions of Loevinger's ego stages that follow
also clarify the point that the stages are  not  defined  by
phase-specific tasks and their resolutions but rather by the
constitution and complexity of the self in relationship with
the world.  Loevinger calls the first level (I-0) the preso-
cial stage where the  child  experiences the world totally
egocentrically and does not distinguish the  self  from  the
non-self.  Animate and inanimate  objects  are also not dis-

tinguished. The first step of development is to differentiate the self from the surroundings. The child who remains
at this stage past infancy is called autistic. Loevinger
refers to this stage as I-0 because one cannot really call
this "oceanic" experience an ego processed one.

In the symbiotic stage (I-1) the self-mother unit (or
other caretaker) is differentiated from the external world.
A first differentiation of self from mother occurs leaving
the boundaries between self and non-self blurred.

In the impulsive stage (I-2) the child shows a basic
awareness of the external world, but manifests a simplicity
of cognition through stereotyping, preoccupation with bodily
feelings, and impulsivity. The child asserts separateness
through the exercise of will, and his impulses help him to
affirm a separate existence. Older children and adolescents
at this stage tend to characterize their emotions in limited
terms such as "mad", "upset", "sick", "high", "turned on",
and "hot": they remain impulsive and tend to act out. They
consider actions bad because they are punished (mainly by
authority figures). Interpersonal relations are exploita-
tive, dependent, and aggressive in tone. Adults tend to
characterize a child who remains at this stage too long as
incorrigible or uncontrollable.

The self-protective stage (delta) brings the first
ability for self-control and impulse regulation. The child
anticipates immediate rewards and punishments, understands
and accepts the existence of rules, but considers them mainly in terms of immediate advantages. A guiding principle is
to avoid getting caught. Older children and adults at this
stage tend to be opportunistic, preoccupied with control and
advantage in interpersonal relations. Relationships are ex-
ploitative, but are not marked by strong overt dependency.
The adolescent or adult focuses on self-reliance with a
sense that others are "not needed, anyway".

In the conformist stage (I-3), the child or adult partly internalizes social rules and moves to an understanding
of self in a group with which the self identifies. The person at this stage belongs and helps, tries to conform, and
attempts to be nice and acceptable. The worries about
appearance and acceptability are paramount. Morality is expressed as guilt for breaking rules that are understood in a
singular manner. Interpersonal role-taking, putting oneself

into the shoes of another person or seeing the world through
another person's eyes, becomes possible. Mutual trust is the
result of this psychological achievement of the Golden Rule,
but the role-taking is frequently not extended beyond a
close circle of family and friends, the "in" group. Conformity to the group can include harboring stereotypes of "out"
groups. Preoccupation with reputation, status, and superfi-
cial descriptions of inner states also marks this stage. A
person in transition following the conformist stage (I-3/4)
recognizes individual differences and allows for not meeting
conformist standards and agreeing with generally held stereotypes. This recognition marks the onset of individualized
inner standards.

In the conscientious stage (I-4), understanding of behavior patterns and traits emerge. Inner moral standards
rather than group norms guide choices. Not living up to
one's inner norms leads to a pervasive sense of guilt. Relationships at this stage are rich in complexity, potential
depth, and meaning. People at this stage are oriented toward
achievement, toward fulfilling obligations, and toward
internally-set goals.

The person in the I-4/5 transition, called individualistic, develops a concern for emotional independence and a
sense of respect for individuality.
Where the conscientious person has an understanding of
individual differences in traits, the person at this level is concerned about tolerance. (Loevinger 1978a, p. 19)

The last stage with clear empirical support is the
autonomous (I-5). Now the person can acknowledge and deal
with inner conflict and conflicting responsibilities in a
way that contributes to self-fulfillment, self-respect, and
tolerance for others who choose differently. Interpersonal
relationships integrate the need for feeling separate with a
recognition of interdependence.

The highest stage (I-6) Loevinger calls integrated.
Here every aspect of I-5 holds true only if the person adds
a sense of integration that moves beyond autonomy, intimacy,
and toleration of conflict.

In sum, then, Loevinger charts ego development from
pre-conformist (self-interest, fear, impulsivity, stereotyping, exploitation, and anticipation) to conformist (genuine

interest in others, awareness of social standards, awareness of self in relation to others, helpfulness, multiplicity, mutuality) to post-conformist stages (which add intrapsychic insight to coping, creative resolution of conflicts, respect for autonomy, inter-dependence, conceptual complexity, and integration of identity).

An important difference that distinguishes Loevinger's model from the theories of Erikson, Kegan, and Baldwin is the rigorous empirical tradition she is aligned to. She has dedicated twenty years to the development of an instrument that measures ego development. This research instrument is a sentence completion test consisting of thirty-six incomplete sentence stems; there are slightly different versions for adults and children, males and females. Subjects are directed to "Complete the following sentences". Typical sentence stems include, "When I am nervous, I...", "A woman's body...", and "Education..." (see Table 7). With the use of a complex scoring manual (Loevinger et al. 1979), the responses to these stems are rated for level of ego development. Validity and reliability studies are reviewed by Hauser (1978) and Loevinger (1979). They indicate that the instrument has a high degree of reliability and construct validity. Research in Curacao, Israel, Germany, French-speaking Quebec, and Japan also suggests that her model and method has a significant degree of cross-cultural validity (see Snarey 1982).

We will now further compare views of Loevinger's ego stages with other structural models. We will begin with an examination of their shared assumptions, the most basic of which is the concept of ego. Both Loevinger and those with a purely structural perspective believe in a relative unity to personality—the ego—which reasons, judges, evaluates, and generally functions to make sense of the world. Second, both accept the applicability of Piaget's hierarchical stage model to the characterization of ego development. These stages form: (a) an invariant sequence of (b) hierarchical transformations, which are (c) structured wholes. In addition, both accept the idea that moral judgment and character are major aspects or dimensions of ego development, relating to a more general ego stage. A third area of agreement concerns test construction and test scoring. Both Loevinger, who comes from a psychometric background, and other structuralists, who have been governed by Piagetian assumptions, move away from traditional psychometric procedures and con-

TABLE 7

Stage Related Responses to "Education..."
Sentence Stem*

| Ego Stages | Education... (summaries) |
|---|---|
| I-2<br>Impulsive | ...is fun; is no fun; is hard<br>...is good for you<br>...helps you learn, helps in school |
| Delta<br>Self-Protective | ...is good, nice, find; is good to have<br>...is good for getting a job<br>...is boring; is for the birds, is not good<br>...is needed but I don't like it |
| I-3<br>Conformist | ...is essential, necessary, important<br>...is necessary, good for everyone<br>...is something everyone should have<br>...is wonderful, desirable, great, etc. |
| I-3/4<br>Conscientious/<br>  Conformist | ...is an essential, important part of life<br>...is a valuable possession, asset, goal<br>...is worthwhile; is worth it<br>...can't be taken away<br>...everyone should have as much as possible |
| I-4<br>Conscientious | ...is an opportunity, must be worked for<br>...should be improved<br>...is important for social goals<br>...should be available to all<br>...valuable but difficult<br>...leads to growth, improvement, prepara-<br>    tion for life<br>...is important for women as well as men<br>...is a way of solving problems |
| I-4/5<br>Individualistic | ...should never end; continues throughout<br>    life<br>...is essential for a full life; for enjoy-<br>    ment of life<br>...is not always what it seems to be |
| I-5<br>Autonomous | ...is intrinsically valuable, admirable<br>...is important socially, individually<br>...helps you cope with life |

*Adapted from: Loevinger et al., 1970, Vol. II, pp. 97-107.

struct tests that attempt to uncover underlying structures. Loevinger is consistent with structuralism in that she agrees that the test constructor finds developmental structures, not by the inductive method, but by an "abductive" method, a sort of mutual bootstrapping, which involves a working back and forth between theoretical reflections and the responses subjects actually give. Fourth and finally, there are striking conceptual parallels in the stage descriptions that emerge from a comparison of Loevinger's and other stage descriptions, such as Kohlberg's (Erickson 1977c).

The parallelism between the stages of Loevinger and Kohlberg is empirically supported, even when researchers control for age and IQ. A study that presents one of the first major comparisons of Loevinger's theory of ego development and Kohlberg's theory of moral development is that of Howard Lambert (1972). His 107 subjects ranged in age from 11 to 60 and included both men and women. The results of a cross-tabulation of his data on ego stages and moral stages yielded an overall correlation coefficient of .80 which decreased to .60 when controlled for age. The "Lambert effect" was true for both men and women in his sample and provides general empirical support for our theoretical position that moral development is necessary but not sufficient for the parallel level of ego functioning in that an individual's moral development score was generally the same as or higher than his ego development score. Lambert's research must be interpreted cautiously, however, since there have been significant changes and refinements in Kohlberg's scoring since his research was completed. There have been three recent studies, however, which have made use of the revised moral development scoring manual, and found similarly high correlations (Snarey 1982, Erickson 1977a, 1977b, 1980).

Leaving the shared theoretical assumptions and the empirical parallels, we will now turn to divergences between Loevinger's model and more consistent structural approaches. To begin, Loevinger's orientation draws from neo-psychoanalytic conceptions of the ego, even though she has consciously departed from her earlier psychoanalytic moorings and has clearly drawn her conception of hierarchical stages from Piaget (Loevinger 1978a). In contradistinction, Kohlberg's, Selman's, and Kegan's orientation is not at all oriented to psychodynamic theory, but to the cognitive-developmental

theories of Piaget (1965) and of the American forerunners of
Piaget (J.M. Baldwin 1902, J. Dewey 1911, G.H. Mead 1934).
This difference in theoretical orientation leads to at least
four differences in the nature of the development that Loevinger and structuralists seek to define and measure.

First, there are differences of how to articulate the
inner logic of ego stages. Loevinger defines her stages
partly in terms of structures, but also partly in terms of
functions and motives pertaining to self-enhancement and defense. The self-protective stage, for instance, is characterized primarily by an interpersonal style that functions
to defend the self and less by the structures that stage
uses to make sense of the world. Thus Loevinger has been
criticized for a lack of an underlying logic guiding her developmental sequence (Habermas 1975, Broughton & Zahaykevich
1977). What makes a higher stage better? What is the structure underlying each stage? In contrast to Loevinger, who
has not spelled out the theoretical inner structural logic
of each stage or the logic of the sequence from one stage to
the next, Piaget, Kegan, Kohlberg, and other structuralists
have defined stages solely in terms of cognitive structures
or ways of thinking. For example Kegan and Loevinger have
an "impulsive" stage, for instance, but Kegan defines, as we
have seen, a basic logical relationship between self and
other, theoretically derived evolutionary balance which
Loevinger addresses descriptively. In fact, the descriptions
are almost identical. Both agree that the child at that
stage is preoccupied with bodily feelings, is at the mercy
of his impulses, and has a limited repertoire of communicated feelings, usually expressed in most simple language,
like "mad", "sick", and "bad". Kegan gives a theoretical
rationale for these phenomena by hypothesizing that the self
is embedded in impulses, having objectified the earlier self
that was "made up" of reflexes. Only with the moving over
of the impulse system to the object side and the new self
now being ruled by "impulse over time", namely needs, can
the child begin to set limits, delay gratification of immediate wants and take over the role of impulse control for
which parents were needed so desperately before. This transition usually occurs in the "5-7 shift", and is, of course,
one might add, connected with the child's move into the institution of learning which necessitates concentration and
delay and the child's new travels between two worlds and two
demand systems—school and home, which put the burden on the
child to sustain separations and conflicting rule systems.

By describing the equilibration between self and other, Kegan can focus on the process of ego development. Loevinger,
on the other hand, presents more of an ideal type characterization, in which a profile of traits which have some  logic
for cohering or going together are described by constructing
an ideal or extreme case. Specific cases are seldom found
to fit such pure case descriptions, but are the  conceptual
anchors for concrete, empirically based descriptions and
classifications. Piagetian stages, in contrast, do not lend
themselves to ideal-type character portrayals. The stage of
concrete operations, for instance, is not characterized by a
typological characterization of an adult character fixated
at the concrete operational level or by a portrait of a typical child at that stage.

     Second, there are differences in dividing the domain of
ego development. Loevinger posits an indivisible ego simultaneously engaged in what she calls impulse control, interpersonal style, conscious preoccupations, and cognitive
functioning. From Loevinger's point of view, there is no
need to divide the ego domain. From our perspective, the
unitary ego includes separation. Whereas Loevinger seeks to
capture within her stages the interpenetration of cognitive
style, self-concerns, and moral character, we consider these
to be governed by separate substructures within the inclusive ego. What Loevinger calls the "cognitive style" facet
of ego development, for us, points to the subdomain defined
by Piaget's stages of cognitive or logical operations. "Interpersonal style and self-concerns" indicates for us the
subdomain studied by Selman (1980) in his attempt to delineate stages of social cognition; "impulse control and character" indicates domain of structure called moral judgment. In
sum, this difference has led Loevinger to hypothesize an ego
system of which subdomains are actually artificial distinctions or indistinguishable. It has led us to hypothesize
separate, and in some cases necessary-but-insufficient relationships between substructures within a more overarching
unifying ego.

     Third, building on the first two differences, there are
contrasts between what Loevinger and other structuralists
are trying to measure. Although her test has been used for
educational and experimental change studies, this was not
her goal in constructing the test. Kohlberg, on the other
hand, wanted to construct a test that would not only assess
the current stage of moral functioning but would also re-

flect his concern with educational goals and his belief that
a higher stage is a better stage. This normative concern has
led Kohlberg to rely upon philosophical as well as psychological theory in defining what is being studied and to give
a philosophical rationale for why a higher stage is a better
stage. This philosophical conception of moral judgment is
based on principles of justice and is dependent upon the
theories of Kant and Rawls. Loevinger has not attempted a
similar normative justification as to why a higher ego stage
is a better stage or why ego development might be an aim of
education. She makes no claim that a higher ego stage is a
healthier stage. However, Loevinger's negative-sounding
early stages, even if she is reluctant to acknowledge it,
may actually be symbolic for social norms or standards that
define the sense in which a higher Loevinger ego stage is a
better stage. The underlying norm is the norm of psychology
itself, that is, the norm of <u>adequacy</u> of the implicit psychological theory used by individuals. Behind the perceptions of self and of others by an individual at Loevinger's
self-protective stage (Delta) is an implicit psychological
theory. If this theory were formalized it would sound something like Skinner's theory of operant learning. It would
say that behavior is instrumentally motivated to get rewards
or payoffs and that ordinarily each individual is oriented
to his own set of payoffs. The theory of operant learning
is, of course, much more complex and sophisticated but, like
Skinner's theory, the psychology of the self-protective
stage must reduce shared sociomoral interests to individual
instrumental interests. Loevinger's next I-3 conformist
stage, while naive in its stereotyping, is a more adequate
psychological theory than the self-protective's psychology
because it postulates trust and socially altruistic motives
as givens in explaining social behavior. By the standards
of adequacy of psychology, rather than of moral philosophy,
then, one might claim that a higher Loevinger stage is a
better stage.

The fourth and final area of disagreement is that Loevinger diverges from a structural approach to test scoring.
Briefly: (a) Loevinger does not clearly differentiate between content and structure; (b) Loevinger constructs and
scores test items so as to be able to infer to a hypothetical entity, a kind of underlying structure akin to the psychoanalytic ego, rather than to the ego's structure, form,
or quality; (c) Loevinger makes use of a psychometric "sign
approach" which combines empirical probabilities with theo-

retical considerations in a process that she calls "saving circularity". Piagetian structuralists have rejected the sign approach and have required each item in a scoring manual to reflect clearly the structure of the stage to which it is keyed. Loevinger's scoring manual, in contrast, does not include interpretive statements that explicate the structure of manual items, tying them to the overall structure of the stages they represent. A comparison of an item in Loevinger's scoring manual (e.g., Delta stage) with the parallel level from Kohlberg's scoring manual (e.g., stage 2) illustrates this contrast. For instance, with reference to the sentence stem, "My conscience bothers me if...", the response "...I steal" is classified as a self-protective Delta stage response. The Loevinger manual notes that stealing is a more concrete content of moral valuing than lying or cheating. But in our terms, it is clearly content, a statement of what is wrong or right, not a direct expression of structure or form of reasoning about why something is wrong or right. The response "...I steal" has no clear face validity as reflecting a self-protective stage. Loevinger has placed it as an example of the self-protective stage based on statistical item analysis which indicated it is associated with the clinical assignment of the total protocol to that stage. Clinical inference of that stem alone, however, could just as easily place it at the conformist stage. Because Loevinger sees her test items as signs that are probabilistic indicators of an underlying personality organization, she sees little need to distinguish clearly between content and structure or for her stage definitions or scoring categories to be structurally defined.

Loevinger, of course, does see the ideal approach as one that is both empirically verifiable and logically satisfying but, since she also sees empirical data as the only court of last resort, she has focused on it almost exclusively. Her approach has been a tradeoff between significant empirical gains and obvious philosophical shortcomings. In contrast, a more consistent structural approach attempts to focus on a method that "unites philosophic and psychological considerations" by "being based on the psychological and sociological facts" of development and by "being based on a philosophically defensible concept" of development (Kohlberg 1971, p. 25). In sum, Loevinger's work is a quasi-structural model that can be understood as having significant functional elements. We will now consider a more balanced approach to ego development--that of James Baldwin.

## JAMES MARK BALDWIN AND THE SOCIAL SELF

Baldwin's insights into social and self development have to this day not been fully explored. Our interest in the self from structural and functional perspectives brings us back to one of the great thinkers of this century. Baldwin had been ignored by American psychologists. The Swiss psychologist Edouard Claparede, however, with whom Baldwin kept a close intellectual relationship after leaving the United States, recognized the importance of his thinking. Claparede later turned his institute over to his student, Jean Piaget, who developed this orientation into a general science with great logical and empirical vigor. One of the areas in which Piaget developed the basic insights of Dewey and Baldwin was in the area of moral judgment (Piaget 1965), an area which Piaget never fully explored because of his subsequent focus on the problem of the development of intelligence. Baldwin addressed a larger domain in that he attempted to provide a unified account of all mental development, the development of emotion, imagination, morality, religion, and the self. The fundamental distinction between Baldwin and Piaget is that Piaget's psychology has no self. Piaget starts with an ego knowing objects, knowing them first egocentrically with a progressive movement in development toward objectivity. In contrast, for Baldwin all experience is experience of a self, not just of a bodily and cognitive ego. Central to the self is not cognition, but will. From the beginning of life, experience is social and reflective. The child derives his sense of self from other selves and ejects it back into them. The sense of self is a sense of will and capacity in the relation of self to others. The individual is fundamentally a potentially moral and ethical being, not because of social authority and rules (as Durkheim and Piaget thought) but because his ends, his will, his self is that of a shared social self.

While basically cognitive, the processes on which social development are based are different from those responsible for development of physical concepts because they require role-taking. In fact, persons and social institutions are only "known" through role-taking. Social-structural influences on social development may be best conceived in terms of variations in the kind and shape of role-taking. In general, Baldwin's "symbolic-interactional" theory of society shares the structural position, in which the primary meaning of "social" is the structuring of action and thought by role-taking, by the tendency to react to the other as someone like the self and by the tendency to react to the self's behavior in the role of the other (Baldwin 1906, Mead 1934). Thus the structure of society and morality is a structure of interaction between the self and other selves who are like the self, but who are not the self.

There are two meanings of "social", the first being affectional attachment, the second, imitation. Both human love and human identification, however, presuppose the more general sociality of symbolic communication and role-taking. Before one can love the other or can model his attitudes, one must take the other's role through communicative processes (Mead 1934).

The basis of any analysis of self and growth of social knowledge, then, implies the ability to share, or to take the viewpoint of another self or group of selves. This fact is paralleled on the active side by the fact that all social bonds, ties, or relationships involve components of sharing.

The central claim made by Baldwin and also by Mead (1934) is that the child's self-concept and concept of other selves necessarily grow in one-to-one correspondence. The child cannot observationally learn the behavior pattern of another without putting it in the manifold of possibilities of acting open to the self. Once it becomes something the self might do, when others do it, they are ascribed the subjective attitudes connected with the self's performance of the act. As stated by Baldwin,
> What the person thinks of himself is a pole or terminus and the other pole is the thought he has of the other person, the "alter." What he calls himself now is in large measure an incorporation of elements that at another period he called another. (1906, pp. 13-18)

Baldwin (1906) and Mead (1934) engaged in extensive debate as to the relative priority of the two mechanisms in social development. Baldwin viewed the similarity of self and other as striven for through imitation, whereas Mead viewed it as the indirect result of role-taking involved in communicative acts. According to Baldwin, the basic unit of the self is a bipolar self-other relationship, with a resulting tendency to play out the role of the other, i.e., the child either has an "imitative" or an "ejective" attitude toward another person.

Baldwin finds in imitation not only the origins of accommodative cognition but of self-control or will. It is this process which follows structural lines of reasoning and has influenced Piaget's cognitive-developmental theory. Role-taking is a broader term than imitation. (All role-taking has imitative components or roots.)

Baldwin's view that the child's knowledge of society initially develops through imitation is similar to the social-learning truism that the child's knowledge of society grows through the observational learning of the behavior of others (learning, which translated into performance, is termed "imitation"). The more distinctive feature of Baldwin's view is that such imitation provides the structure of the child's social relationships, i.e., of his self as it relates to other selves. In the social-learning view of imitation, it matters little whether a response is learned by imitation or by reinforced trial and error since it functions in the same way once learned. In contrast, Baldwin argues that imitation is important because it determines the structure of the child's self-concept, and of his concepts of others, a structure, in turn, determining the use of the behavior pattern learned through imitation.

According to Baldwin:
...the growth of the individual's self-thought, upon which his social development depends, is secured all the way through by a two-fold exercise of the imitative function. He reaches his subjective understanding of the social copy by imitation, and then he confirms his interpretations by another imitative act by which he ejectively leads his self-thought into the persons of others. (ibid., p. 527)
When the child is imitating or learning from the other, his attitude is one of "accommodation", i.e., his behavior is being structured by the structure of the behavior of the other. Behavior modeling is taken as an implicit command, and an explicit command can always be modeled (i.e., "do it this way", accompanied by a demonstration). In either case, the structure of the activity belongs to both parties, but it is being passed on from the active to the passive one. The central focus is upon a novel structure which the active agent has and the passive agent does not have.

In contrast, the active or assimilative self is one that knows what it is doing, and that ejects its own past attitudes into the other. Whether the child is active or passive, there is a focus upon an activity of one person with an attitude of accommodation to it in the other. The attitude of the child practicing something he has already learned through imitation or compliance, then, is always different than the attitude he held in the process of learning it. In learning an activity associated with the superior power or competence of the adult, the child's attitude is accommodative and, in that sense, inconsistent with the prestige of the activity being learned. Accordingly, the child tends to turn around and practice the activity on, or before the eyes of, some other person whom he can impress, into whom he can "eject" the admiration or submissiveness he felt when learning the act.

The actual "ejective" phase of self-development arises initially during the second year of life. At that age, the child seems to attribute feelings, show things to, and communicate with others. At the end of the second year, a "negativistic crisis" (Ausubel et al. 1980) typically occurs. This is a phase in which self and other are sharply differentiated and the difference between copying the self and copying (or obeying) the other are distinguished. An example of the counter-imitation of this era is a 2-1/2 year old's consistent response of "not goodbye", when someone says "goodbye" to him.

Interestingly, however, it is just at this age of "independence" that the child acquires a need for an audience, a need reflected in "look at me" or attention-seeking behavior. Indeed, it is striking to notice that "look at me" behavior often is a phase of imitative acts at this developmental level. The father takes a big jump, which the same 2-1/2-year-old imitates, demanding that the father look at him just as he looked at the father in imitating. Imitation immediately places the child in the model's role, and leads him to eject into the adult his own capacity for admiration.

Thus Baldwin would account for the "show-and-tell" behavior of the 3- to 4-year-old as the reverse form of the sharing involved in the initial imitative act. Imitation, then, generates social sharing at both the learning and the practicing phases. Seeking to act competently almost always requires another person for the imitative young child. First, the child needs another person as a model for what to do. Second, because he has learned from a model, he needs to

practice what he has learned on another person, e.g., to be
a model to another.

The major step in ego development in the years 4 to 8
in the Baldwin scheme is the growth of the differentiation
of the self as mental from the self as physical. According
to Baldwin, this differentiation arises largely in play:
  Play makes possible the determination of the great dualism of Mind and Body, a dualism developing out of that of
  Inner and Outer, but not possible in the mode of fantasy
  and memory. (ibid., p. 16)

Baldwin was impressed by the role of imagination and
the "let's pretend" attitude in all adult experience, not
only in aesthetic creation and experience, but in the construction of scientific hypotheses and of socio-moral ideals. In referring to childhood play as the origin of later
imaginative activity, Baldwin points to the paradoxical
quality of childhood play, the paradox that in play the
child both does and does not believe in the reality of what
he is doing, making, or saying.

According to Baldwin, the "let's pretend" attitude also
gives the child a sense of himself and his mind as a source
of inner freedom distinct from his body and the external
material world. Before the play experience, the child is at
the first stage in his self-experience, the stage in which
the self's physical and mental processes are either copies
of external realities or are fantasies with no validity except as bodily impulses. Play provides a new set of experiences of mind and self for the child. Play ideas and purposes are not literal copies of reality, they are the
child's own inventions and ideas. At the same time, play inventions are not mere fantasies, they are freely held. The
child assimilates his experience into a self by constructing
the experience in play as something the mind or self chooses
and controls. At an earlier point the child has memories and
wishes which, when reproduced in external reality, cease to
exist as mental. In contrast, in the play attitude the
child sets up his intentions in a separate realm, a play
sphere, and while in one sense he is realizing his intention
in external reality, in another sense he is keeping his intentions and plans as his own. In this sense, then, he is
building up a sphere of mind and self distinct from his
body, using his body to realize this sphere, but keeping it
as subjective and as his own.

Baldwin claims the problem of the genesis of the moral self is not solved by the internalization of prescribed habits in the sense of the learning theorists or the associationists of Baldwin's day. Even even if the child accepts certain conforming habits or rules as part of himself, this does not guarantee the existence of a concept of a moral or altruistic self as determining action. There need be no difference between a child's feeling of self-assertion and dominance in carrying out a conforming habit and his feeling in carrying out a deviant or an asocial one.

How does such a concept of a self that wants to be good and to conform to rules arise? Imitation and the bipolar self do not themselves provide the answer. Baldwin says that the bipolar self involves an active, assertive, controlling self and a passive, submissive, imitative self. The child may be either as occasion arises. With a younger child or a parent in a permissive mood, the child himself defines what is to be done and the other is merely an object to be manipulated, an agent in terms of whom the action may be carried out. With an older person or in a novel situation, the child expects to be the object in terms of which action determined by the other is carried out.

Whether the self determining the child's action is the other (the adult) or the child himself that is "selfish," it is a bipolar self, not a shared or sharing self. The act of adjusting to or obeying the adult need not imply an experience of self-control and unselfishness by the child since the self controlling the child and demanding sacrifice of his wishes is not his own self. Though the child's action may be determined by the dominating self of the other, still that other self is conceived by the child in its own image as a basically impulsive or need-gratifying self (insofar as motives are assigned to it at all). The experience of unselfish obligation requires that the two selves be identified or unified with one another, an integration which is not achieved by motives to imitate or obey in themselves.

How does such a concept of a shared self that wants to be good and to conform to rules arise? The experience required, says Baldwin, is one in which the child perceives the parent as putting pressure on the child to conform to something outside the parent. Such an experience is not bipolar since the parent wants the child to be like himself vis-a-vis his attitude toward the rule. The parent's self is seen as simultaneously commanding (the child) and obeying (the rule). Thus imitating action and conforming are seen as both parts of the same self, a self-controlling self.

The difficulty in developing an ideal self is that of getting a situation in which the child conforms, but does not see such conformity as either enhancing his own impulses or due to his own weakness relative to the other who is enhancing himself. The child must come to see the denial of his own wishes as somehow self-enhancing. This is a form of reciprocity, but a different form of reciprocity than that stressed by Piaget.

Such conformity to a third force might simply be perceived by the child as indicating that a third person dominates the parent as the parent dominates the child. However, the fact that such pressure to conform goes on in the absence of the third person or authority tends to give rise to the concept of a generally conforming self. In addition, the fact that the conformity is shared in the family or group gives rise to a sense of a common self which the child is to become.

Originally such a general or ideal self is largely in the image of the parents. It is ideal to the child, it is what he is to become, but it is largely realized in the parents. This does not mean that there is no differentiaton of parents from the rule; the parents are seen as obeying the rule. It does mean that the image of a good, conforming self that obeys the rules is in the parent's image. Baldwin's account suggests that much of the need for approval is born from the fact that most of the child's accomplishments are imitative. Almost everything the young child strives to do or accomplish is something he sees another person do first and which he learns, in part, imitatively. The young child's accomplishments, his talking, walking, dressing himself, toileting, etc., are all activities that he sees others do and knows they can do. Because they are his models for activity, their approval of his performance counts. This focus on age-specific accomplishments through which the child becomes a member of the group places him in a structural-functional tradition, rather than in a purely structural one. Baldwin's theory of human attachment and mastery takes functional aspects of motivation, self-esteem, and shared social activity into consideration.

If, in contrast to physical theories, one takes the desire for a social bond with another <u>social self</u> as the primary "motive" for attachment, then this desire derives from
the same motivational sources as that involved in the
child's own strivings for stimulation, for activity, for
mastery, and for self-esteem. Social motivation is motivation for shared stimulation, for shared activity, and shared
competence and self-esteem. Social dependency implies dependency upon another person as a source for such activity,
and for the self's competence or esteem. The basic nature
of competence motivation, however, is the same whether self
or the other is perceived as the primary agent producing the
desired stimulation, activity, or competence—i.e., whether
the goal is "independent mastery", social mastery (dominance), or social dependence.

Human attachments, even in the first two years of life,
reflect the fact that they are attachments to another self
or center of consciousness and activity like the self. This
fact of human attachment implies the following characteristics:
1. <u>Attachment involves similarity to the other</u>. Attachment
is only to another person, not toward physical objects.
2. <u>Attachment involves love or altruism toward the other</u>,
an attitude not felt toward bottles or cloth mothers.
Altruism, of course, presupposes the "ejective" consciousness of the feelings and wishes of the other, i.e., empathy
or sympathy.
3. <u>Attachment and altruism presuppose self-love</u>. Striving
to satisfy another self presupposes the capacity or disposition to satisfy one's own self. Common sense assumes
that the self (as body and center of activity) is loved intrinsically, not instrumentally (i.e., not because the body
or the body's activities are followed by reinforcement or
drive reduction). This nucleus of self-love also is involved
in organizing attachment to others.
4. <u>Attachment presupposes the desire for esteem in the eyes
of the other or for reciprocal attachment</u>. In other words,
it presupposes self-esteem motivation and the need for social approval, again presupposing ejective consciousness.

We end our chapter with Baldwin, because his theory integrates a number of important concepts into an overarching
model of the developing self. In contrast to Erikson, Baldwin explicates the underlying <u>principle</u> that unifies the
stages of the self: social role-taking. Erikson's epigenetic

model has, we believe, an underlying structural core, the
development of mutuality across the life cycle. But Erikson
has not organized his stages around a single process that
transforms each stage. Although Loevinger comes closer to
Baldwin's notion of a single process of self (or ego) coherence, she has not evolved a theory of process, which explains the changes she theoretically and empirically describes. Baldwin's theory of imitation introduces a process
orientation of self-other relationships and the mechanisms
of "internalization" and "externalization" (projection and
ejection). This brings Baldwin close to Kegan's neo-Kohl-
bergian point of view. Both share an orientation toward underlying self-other relationships in their process analysis
of development. However, Baldwin's theory encompasses more
functions of the self: e.g., play, idealization, representation, and aesthetics. His approach shows how these different domains are functionally related to each other. The organization of the relationship of these psychological functions is Baldwin's conception of structure. He proposes
that these structures of function change in a developmental
sequence of differentiation and integration. Baldwin's
theory reconstructs the self as a social self, while Kegan
mixes biological and social metaphors. Baldwin falls behind
all three theorists in one area--he does not present an explicit theory of life-long developmental change. In fact,
Loevinger (1969) contends that his principle of imitation
stands in the way of addressing ego development past the
conventional or conformist level, since "in principle,
conscience remains an internalized version of social
judgment" (p. 101), although the individual can develop
ethical standards which conflict with the prevailing opinion
of society.


CONCLUSION: TEN PRINCIPLES OF THE DEVELOPING SELF

We began this chapter by pointing to important developments in social-cognitive theory that have led to new and
provocative theories and research in the field of ego development. The advance of academic developmental psychologists
into a traditional psychoanalytic domain has created exciting possibilities for theoretical synthesis and rigorous empirical investigation of clinical constructs. But the cros-
sing of disciplinary boundaries has also led to confusion
due to imprecise applications of terms and concepts. In
this chapter we have attempted to distinguish functional and

structural approaches to the development of the self. Within the framework of these approaches we have reviewed four theoretical traditions, psychoanalytic ego psychology, neo-Piagetian developmental psychology, Loevinger's model and measure of ego development, and Baldwin's theory of the self. Their assumptions are compared with a focus on the distinctions of functional and structural paradigms. The study of the self presupposes a _psychological_ process that mediates between external and internal reality. Functional models do not emphasize such constructs and focus more on the social organization of the life cycle. Rites of passage, for example, have always had an important place in functional theories because they represent modal points in individual and cohort development seen from the point of view of culture and society.

The emergence of ego psychology as an academic psychological discipline presents us with the opportunity far greater than comparing psychological paradigms anew. Integrations and theoretical synthesis between psychoanalytic ego psychology, self and object relations theory, and Piagetian ego models have come into closer reach. Piaget's vision of a general psychology that would one day integrate psychoanalysis and genetic epistemology is still far from realization, though by our focus on the self and the ego, in combination with our cumulative knowledge in social cognition, a greater synthesis with contemporary psychoanalytic thought will be attained. Such new directions include the metapsychological reorientation in psychoanalysis of Kohut and his colleagues (see this volume). The re-emerging of theoretical prominence that Erikson's work is experiencing is demonstrated through the present interest in adult development.

We cannot systematically propose a theoretical and applied synthesis here. We will close this chapter, however, by sketching directions for such an integration. The following ten points will address key issues for a systematic and encompassing study of the developing self:

1. _The self is unitary and integrates developmental subdomains._ While most of the theorists presented in this chapter share a unitary view on the developing ego or self, they do not effectively address the relationship between this overarching process and developmental subdomains, such as moral development, interpersonal perspective-taking, in-

tellectual and affective development. From our perspective, the unitary self includes subdomains, each involving a distinct substructure and each capable of empirical separation. Interpersonal style is captured by social perspective-taking and levels of interpersonal understanding, cognitive styles are partly defined by Piaget's stages of logical operations, and impulse control and character development are partly governed by the domain of moral development. Self theory has to focus simultaneously on the overall system and the subdomains with the propensity of decalage. In fact, this relationship could account for an important "self dynamic", in which the striving for integration into an overall unity is counterbalanced by conflicting developmental patterns of the subdomains. The empirical evidence points toward a necessary, but not sufficient relationship between substructures (e.g., intellectual, social-cognitive, moral). The structural relationships between subdomains and the developing self, however, are only in the early phases of empirical exploration.

2. The self structures cognitively. Cognitive processes—their developmental patterns and the structuring of knowledge about the self, others, and the physical world—are not peripheral but rather central to the developing self. Psychoanalytic ego functions, or Erikson's identity concepts, which have cognitive components (e.g., ideological orientation of the adolescent) do not make use of a genetic epistemology in a Piagetian sense, that is, in which all knowledge is a construction growing out of the interaction between structures in the environment and structures of the organism. The discovery that these structures follow a sequence of hierarchically organized transformations is one of Piaget's great contributions to psychology. These insights can establish the core of innovative post-Eriksonian self theory and bring fruitful possibilities for integrations. The advances in the cognitive-developmental tradition ground such attempts in an empirical tradition and place great emphasis on cognition and reasoning as previously neglected topics of study in personality and clinical theory.

3. The self mediates cognition and affect. While psychoanalytic psychology explored the psyche without a convincing cognitive model, cognitive-developmental psychology lacks an understanding of affective development and a model of how feelings and thinking relate. Despite Piaget's almost exclusionary commitment toward elaborating a cognitive the-

ory, he did sketch a cognitive-affective parallelism (Piaget 1981). But an understanding of affect <u>and</u> cognition requires (and might present us ultimately with a central justification for) a superordinate structure, the self. Equilibrative structures and processes of the self create a dynamic of affect and cognition and change the focus from how they relate to what common develomental process underlies them. This underlying mediational structure <u>is the self</u>.

4. <u>The self is not an entity but a process--mediating between an organism and its environment</u>. The process of interaction between structures in the organism and structures in the environment is guided by the processes of assimilation and accommodation with a primary "self drive" of equilibration or self-consistency. Equilibration is the basis for self-environment integrations and attempts at synthesis of contradictory demands, influences, and social conditions. Mastery is the psychological expression of the orientation toward equilibration, since it presupposes the processing of environmental demands and opportunities in a way the person can respond to productively as well as ways to solicit from the environment adequate "self nurturance". The problem any self theory runs into is the idealistic interpretation of social reality, in which the "structures in the environment" remain abstract and ultimately borrow their form through biological metaphors. The self is a product of social contradiction and its attempts to find meaning and balance within a context of social disequilibration. Thus, a theory of the self will not place the whole burden of social inequality <u>into</u> the person but rather has to reconstruct this process of self-environment equilibration <u>socially</u>. This includes an understanding of the social structure and the place of the person in the social opportunity system. In this context, self psychology becomes a theory of interpersonal and intersystemic interaction in which the meaning constitution <u>between</u> people and institutions becomes the expression and the base of the self-mediation between "organism and environment".

5. <u>The self is "object-relational"</u>. From the first moment after birth, and perhaps even before, the self develops out of social interactions. The British object-relations school has taught us of a primary object-relations "drive". The extension of Piaget into social cognition has presented us with new ways of studying the self as object-relational. Kohlberg, Selman, Gilligan, Kegan, Noam, and others have ex-

plored ways to understand differentiations and integrations
of the self in relation to important "others". These descriptions entail new ways of understanding processes of internalization (and externalizations) of important relationships to the degree that they represent an integration of
social-cognitive theory and social-emotional theory. The integration of Piagetian structuralism and Eriksonian functionalism will expand our knowledge of different forms of
internalization, like incorporation, imitation, identification. Specifically, these processes can be placed in a developmental hierarchy addressing different form and meaning
during different stages of self-development.

6. The self processes—and distorts. The self implies
processes of distortion and projection. The defenses have
been studied with great detail and clinical relevance in the
psychoanalytic tradition. Ever since Anna Freud's The Ego
and the Mechanisms of Defense, an elaborate system has been
available. Recently, formulations have also led to research
methodologies and made structural-developmental interpretations more desirable. Vaillant's work has demonstrated a
sequence of processes of adaptation and distortion, although
both (A. Freud's and Vaillant's) theories still lack a satisfying elaboration of the underlying logic of the hierarchy
and the relationship to structural-developmental transformations. It is our position that distortions are partly cognitive in nature and follow the developmental line of the
self (even though structural-developmental self theory lacks
a concept of defenses and systematic distortions at this
time). Psychoanalytic defense theory will become more important as academic developmentalists turn with greater intensity to areas of clinical application and the study of
psychopathology. Of particular interest will be the interpersonal processes of distortions like projective identification and delegation. These mechanisms have mainly been understood as ways to delegate onto important others primitive
and "early" forms of thinking and feeling unacceptable to
the conscious ego. Structural-developmental self psychology
points more dramatically toward developmenal distinctions of
such a concept: what is the developmental level of the self
and what form of projection is used with which construction
of the other? This approach is one that places more emphasis
on present self-other balances (Noam & Kegan 1982). At the
same time defenses are the example par excellence of forms
of earlier mental functioning expressing itself at later de-

velopmental periods  and can guide our explorations  of this
topic.

7.  The self is historical--the  developmental  genesis
is biography. The history of internalizations of important
others, relationships with others and  experiences  of  self
are  "forever  with us" (Erikson 1968). Although the transformational model states that earlier structures  are  integrated  into  later  ones, clinical knowledge and theoretical
considerations force us to take a closer look at how certain
experiences and demands trigger forms of  earlier  functioning,  be  it through the recurrence of symptoms, old ways of
thinking, or forms of ego  fragmentation.   These  "self-re-
gressions"  are  always  within  the  context of the highest
forms of organization, which can serve as the observing  capacity  over the earlier functioning. The self-dynamic will
always be the attempt to restructure old experiences in  the
face  of new ones. The process will lead to recapitulations
of old ways of seeing the world while  being  superseded  by
the  new  one. Each one of these brings a potential reinte-
gration, which is always in relationship to the highest form
of mental functioning. Psychoanalytic psychotherapy (with a
focus on "transference") and a developmentally based  theory
of clinical intervention can be interpreted from such a view
that traces development historically.

8.   The self has strength and esteem. Structural-developmental  self  theory and Loevinger's structural-functional
model present a transforming self which renews its  relation
to  the  world  through  reconstructions of "meaning structures". These processes are not directly related to concepts
of ego strength and self-esteem, although of  course,  certain
constructions of the self and  the  world  make  the  person
especially  vulnerable  to  certain stresses and rejections,
while certain victories and not others reach the core of the
self.  "Stage 3" in Loevinger and Kegan's  models,  for  instance,  brings  the  interpersonal  interpretation of mind,
self, and society. Any rejection of the person by  a  loved
one, any accusation of not being trustworthy or lovable, any
expulsion from a group that served as a referent, leads to a
loss  of self. The self is its context--the group conformer,
the lover and trusted one. At the next stage, the self would
experience these events as painful, like any loss and rejection, but it would not threaten the foundations of the  self
structure.   The  self  at "Stage 4", less vulnerable to the
opinions of others, would be able to internally mediate  the

assault.    Loevinger has repeatedly cautioned not to use her
ego model as a conceptualization of ego strength and self-
esteem.     An integrated self model will have to take account
of the esteem and strength of the self   and   reconstruct   it
structurally and developmentally.     Erikson's insights can
give us invaluable guidance.    His ego theory builds on a
model of crises, their resolutions, and the resulting
self-esteem and strength.

Each virtue in Erikson's schema is a new form of esteem
and strength (hope, will, purpose, competence, etc.) and
each crisis also results in ego weaknesses (mistrust, shame,
guilt,   inferiority,   etc.). What   is missing from Erikson's
account, however, is an underlying developmental logic   that
gives a rationale for a developmental hierarchy of different
types  of  esteem and purpose. As sketched earlier, the logical sequence could be based on an   interpretation of   each
strength  being part of an underlying "attention-will" cluster.   At first the ego stages capture general   qualities   of
will,   while later the values become more "moral will".   Fidelity, love, and caring, the adult virtues   represent   successive   capacities for ethicality, commitments freely made,
but ethically binding.     Further   theoretical   investigation
will   shed   light on these questions.   For now our statement
must suffice that a self model without a theory of   strength
and esteem is an incomplete model.

9.   The self develops throughout the life cycle. One of
the   great changes in psychology and the mental health field
where the concept of self holds a respected   place,   is   the
recent   interest in adult development.   Ever since Erikson's
redefinition of the psychoanalytic ego from a   retrospective
childhood ego to a prospective adolescent and adult ego, the
ages of man also became the ages of the processing and experiencing ego.  But Erikson's insights, acknowledged and respected as they were, have   only   now,   thirty   years   after
their   first   publication,   reached the center stage of psychology.  Life-span-developmental psychology has   become   an
important   subspecialty; psychology has devoted its methodological know-how to an exploration of all ages, and theories
in general have begun to account for cognitive   and   personality transformations beyond adolescence.   In fact, what the
1950s and 1960s were for the study of adolescence, the 1970s
and   1980s   are   for an advance of interest in and knowledge
about adult development.   Unfortunately, many   theories   and
studies   still describe the "egoless adult"--the person with

many traits, interests, behaviors, occupational aspirations, mid-life crises, and empty nests. Where, in all these descriptions, is the organizing principle, the referent, to which these experiences, choices, commitments, and feelings connect? Where is the self-structure that coordinates the divergent demands of intimacy and identity over the life cycle? It is in this area that an integrated model of self development might provide the most important insights both for "normal" development and clinical applications. When clinical psychology and psychiatry have incorporated a knowledge of expectable transformations in adulthood, reconstructions of childhood vulnerabilities will not become superfluous, but integrated into a model of the mind that encompasses the earlier experiences into a process orientation of personality development in adulthood. This renewed look at development beyond adolescence, occurring simultaneously in psychoanalysis and academic developmental psychology, also puts into question a traditional way of understanding development as a progression from symbiosis to individuation and it suggests, rather, a life-long attempt to resolve paradoxical needs for affiliation and individuation.

    10. The self--a last glance at structural and functional distinctions. While Erikson organized his stages of the ego around ages of development, a structural self theory conceptually dissociates stages and ages; stages refer to processes and levels of understanding of self in relationship to others, while age refers to a culturally organized timetable of tasks. Can there be an adequate self theory without both a structural and a functional description of development and social time?

    The theoretical dissociation of stage and phase of development opens important possibilities for understanding the self. Adolescence is not one form of identity crisis. Erikson's crisis presupposes a high level of cognitive development and occurs in the transition from defining self through peer group (Kegan & Loevinger, Stage 3) to defining the self more internally, based on more self authored standards (Stage 4). The adolescent world can include much earlier constructions (e.g., usually concrete or early formal operations), which center around impulse-guided behavior and direct gratifications of needs, as well as strong group identifications. In other words, many adolescents do not reach the higher levels of self-development. On the other hand these different "stage worlds" are united by certain

(age-period) functional requirements which adolescence poses
in our culture: turning from family to peer group, organizing a life-plan, developing intimate ties, and so forth.  On
the other hand, Stage 3 will have a very definite shape during  the life-phase of adolescence than in adulthood and old
age, although the underlying "self-logic" is the same. Function and structure cannot be dissociated in real life.   The
"stageness" will  always  be  exposed  in  relationship  to
thoughts and feelings about life tasks.  Thus function  and
structure  create  a  dynamic  that  we  see as an important
approach to the study of the self across the life span.

## ACKNOWLEDGEMENTS

The authors wish to thank John Broughton,  Lynn  Clark,  Ann
Fleck  Hendersen,  Robert Kegan, Benjamin Lee, Merrill Mead,
David Miranda, and Maryanne Wolf for invaluable comments.

## ENDNOTES

1.  In this chapter we use ego and self interchangeably, despite earlier  psychoanalytic  distinctions,  since  we  are
mainly  concerned with those aspects of the person that deal
with overarching systems of meaning that give  coherence  to
development.

2.  A variety of projects are presently under way by A. Colby, A. Henderson, A. Hewer, R. Kegan, G. Noam, S. Parks, and
L.  Rogers  studying  the  relationship  between  Piagetian
psychology and a clinical theory of the self.

## REFERENCES

Abrahamson, M.  1978.  Functionalism.  New Jersey: Prentice
          Hall, Inc.
Althusser, L.  1965.  Pour Marx.  Paris: Maspero.
Angell, W.  1930. "The Functional Approach". In C. Murchison
          (ed.), Psychology of 1930.  Worcester: Clark Uni-
          versity Press.

Anthony, E. J.  1976.  "Freud, Piaget and Human Knowledge:
    Some Comparisons and Contrasts".  Annual of Psy-
    choanalysis 4 .
Ausubel, D. et al.  1980.  Theories and Problems of Child
    Development.  New York: Grune and Stratton.
Baldwin, J. M.  1902.  Social and Ethical Interpretations in
    Mental Development.  New York: Macmillan. Rep.
    ed. 1906, New York: Macmillan.
Basch, M.  1980.  "Psychoanalytic Interpretation and Cogni-
    tive Transformation".  Unpublished manuscript.
    Chicago: Center for Psychosocial Studies.
Bellak, L., M. Hurvich, & H. Gediman.  1973.  Ego Functions
    in Schizophrenics, Neurotics, and Normals.  New
    York: John Wiley.
Bibring, G., T. Dwyer, D. Huntington, & A. Valenstein. 1961.
    "A Study of the Psychological Processes in Preg-
    nancy and of the Earliest Mother-child Relation-
    ship: II. Methodological Considerations".  The
    Psychoanalytic Study of the Child, XVI:25-72.
Blos, P.  1962.  On Adolescence.  New York: Free Press.
Broughton, J. & M. Zahaykevich.  1977.  "Review of J. Loev-
    inger's Ego Development".  Telos, XXXII:246-53.
Candee, D.  1974.  "Ego Development Aspects of New Left
    Ideology".  Journal of Personality and Social
    Psychology, XXX:620-30.
Colby, A., J. Gibbs, L. Kohlberg, B. Speicher-Dubin, & D.
    Candee.  1979.  Standard Form Scoring Manual.
    Cambridge, MA: Harvard University, Center for
    Moral Education. Repr. ed., in press, New York:
    Cambridge University Press.
Decarie, T.  1978.  "Affect Development and Cognition in a
    Piagetian Context".  In M. Lewis & L. Rosenblum
    (eds.), The Development of Affect.  New York:
    Plenum Press.
Dewey, J.  1911.  Moral Principles in Education.  New York
    Philosophical Library.
Edelstein, W. & G. Noam, (in press).  "Regulatory Structures
    of the Self and Post Formal Operations in Adult-
    hood".  Human Development.
Elder, G.  1975.  "Adolescence in the Life Cycle: an Intro-
    duction".  In S. Dragastin & G. Elder, (eds.),
    Adolescence in the Life Cycle.  New York: Wiley
    and Sons.
Elkind, D.  1978.  Children and Adolescence.  New York:
    Oxford University Press.
Elkind, D.  1878.  "Cognitive Development and Psychopathol-

ogy: Observations on Egocentrism and Ego Defense". In The Child and Society. New York: Oxford University Press.

Erikson, E. 1950. Childhood and Society. New York: W.W. Norton & Co., Inc.

Erikson, E. 1959. "Identity and the Life Cycle: Selected Papers". In Psychological Issues, Vol I 1:133-4. New York: University Press.

Erikson, E. 1968. Identity, Youth and Crisis. New York: W.W. Norton & Co., Inc.

Erikson, E. (in press). Elements of a Psychoanalytic Theory of Psychosocial Development.

Erickson, V.L. 1977a. "Deliberate Psychological Education for Women: a Curriculum Follow-up Study". The Counseling Psychologist, 6(4):25-9.

Erickson, V.L. 1977b. "Beyond Cinderella: Ego Maturity and Attitudes Toward the Rights and Roles of Women". The Counseling Psychologist, 7(1):83-8.

Erickson, V.L. 1977c. "The Domains of Ego and Moral Development". Moral Education Forum, 2(4):1-4.

Erickson, V.L. 1980. "The Case Study Method in the Evaluation of Developmental Programs". In L. Kuhmerker, M. Mentkowski & L. Erickson (eds.), Evaluating Moral Development. Schenectady, New York: Character Research Press.

Fairbairn, W.R.D. 1952. Psychoanalytic Studies of the Personality. London: Routledge and Kegan Paul.

Freud, A. 1936. The Ego and the Mechanisms of Defense. New York: International Universities Press.

Fromm, E. 1976. "To Have or to Be". World Perspective, L.

Gardner, H. 1972. The Quest for Mind: Piaget, Levi-Strauss, and the Structuralist Movement. New York: Random House.

Gilligan, C. 1982. In a Different Voice. Cambridge, MA: Harvard University Press.

Gilligan, C. & M. Belenkey. 1980. "A Naturalistic Study of Abortion Decisions". In R. Selman and R. Yando (Eds.), Clinical Developmental Psychology. San Francisco: Jossey-Bass.

Grant, J.D. & M.O. Grant. 1959. "A Group Dynamics Approach to the Treatment of Nonconformists in the Navy". Annals of the American Academy of Political and Social Science, 322:126-35.

Greenspan, S.I. 1979. "Intelligence and Adaptation: An Integration of Psychoanalytic and Piagetian Developmental Psychology". Psychological Issues Mono-

graphs. New York: International Universities
        Press.
Greenspan, S.T. & R.S. Lourie. 1981. "Developmental Struc-
        turalist Approach to the Classification of Adap-
        tive and Pathologic Personality Organizations:
        Infancy and Early Childhood". In American Jour-
        nal of Psychiatry, 138:6, p. 725-35.
Guntrip, H. 1971. Psychoanalytic Theory, Therapy, and the
        Self. New York: Basic Books.
Haan, N. 1977. Coping and Defending. New York: Academic
        Press.
Haan, N., J. Stroud, & C. Holstein. 1973. "Moral and Ego
        Stages in Relationship to Ego Processes: a Study
        of 'Hippies'". In Journal of Personality,
        XLI:596-612.
Habermas, J. 1975. "Moral Development and Ego Identity".
        Telos, XXIV:41-55.
Hauser, S.T. 1976. "Loevinger's Model and Measure of Ego
        Development: a Critical Review". Psychological
        Bulletin, 83(5):928-55.
Hauser, S.T. 1978. "Ego Development and Interpersonal
        Style in Adolescence". Journal of Youth and
        Adolescence, VII, 4: 333-351.
Hauser, S., W. Beardslee, A. Jacobson, G. Noam, & S. Powers.
        1979. "Longitudinal Studies of Adolescent Ego
        Defenses and Ego Development". Paper presented
        at the annual meeting of the American Psychoana-
        lytic Association, New York.
Hauser, S., A. Jacobson, G. Noam, & S. Powers. (in press).
        "Ego Development and Self-image Complexity in
        Early Adolescence: Longitudinal Studies of Psy-
        chiatric and Diabetic Patients". Archives of Gen-
        eral Psychiatry.
Hulsizer, D., M. Murphy, G. Noam, & C. Taylor. 1981. "On
        Generativity and Identity: From a Conversation
        With Joan and Erik Erikson". Harvard Educational
        Review, LI(2):249-69.
Isaacs, K.S. 1956. "Relatability, a Proposed Construct and
        an Approach to its Validation". Unpublished doc-
        toral dissertation, Chicago: University of Chi-
        cago.
Jacobson, E. 1964. The Self and the Object World. New
        York: International Universities Press.
Kegan, R. 1979. "The Evolving Self: A Process Conception
        for Ego Psychology". Counseling Psychologist,
        VIII, 2: 5-34.

Kegan, R. 1982. The Evolving Self. Cambridge, MA: Harvard University Press.

Kegan, R., G. Noam, & L. Roger. (in press). "The Psychologic of Emotions: a Neo-Piagetian View". In D. Cicchitti & P. Hesse (eds.), Functional Development. In New Directions in Child Development. San Francisco: Jossey-Bass.

Kernberg, O. 1976. Object Relations Theory and Clinical Psychoanalysis. New York: Jason Aronson.

Kohlberg, L. 1969. "Stage and Sequence: the Cognitive Developmental Approach to Socialization". In D. Goslin (ed.), Handbook of Socialization, Theory and Research. New York: Rand McNally.

Kohlberg, L. 1971. "From is to Ought: How to Commit the Naturalistic Fallacy and Get Away With It in the Study of Moral Development". In T. Mischel (ed.), Cognitive Development and Epistemology. New York: Academic Press.

Kohler, K. 1947. Gestalt Psychology. New York: Mentor.

Kohut, H. 1971. The Analysis of the Self. New York: International Universities Press.

Lambert, H.V. 1972. "A Comparison of Jane Loevinger's Theory of Ego Development and Lawrence Kohlberg's Theory of Moral Development". Unpublished doctoral dissertation. Chicago: University of Chicago.

Laughlin, H.P. 1979. The Ego and its Defenses. New York: Jason Aronson.

Levi-Strauss, C. 1962. The Savage Mind. Chicago: The University of Chicago Press.

Levi-Strauss, C. 1963. Structural Anthropology. New York: Basic Books.

Levi-Strauss, C. 1969. The Elementary Structures of Kinship. Boston: Beacon Press.

Lewin, K. 1951. Field Theory in Social Science. New York: Harper.

Loevinger, J. 1969. "Theories of Ego Development". In L. Breger (ed.), Clinical-cognitive Psychology Models and Integrations. New Jersey: Prentice-Hall.

Loevinger, J. 1976. Ego Development. San Francisco: Jossy-Bass.

Loevinger, J. 1978. Scientific Ways in the Study of Ego Development. Heinz Werner Lecture Series, XII. Worcester, MA: Clark University Press.

Loevinger, J. 1978a. "Recent Research on Ego Development". Presented at the Harvard Graduate School of Education. Cambridge, MA.

Loevinger, J. 1979. "Construct Validity of the Sentence Completion Test of Ego Development". Applied Psychological Measurement, 3(3):281-311.

Loevinger, J., Wessler, R. & Redmore, C. 1970. Measuring Ego Development II: Scoring Manual for Women and Girls. San Francisco: Jossey-Bass.

Malerstein, A.J. & M.J. Ahern. 1979. "Piaget's Stages of Cognitive Development and Adult Character Structure". American Journal of Psychotherapy, 23(1): 197-218.

Malinowski, B. 1927. Sex and Repression in Savage Society. New York: Harcourt Brace.

Mead, G.H. 1934. Mind, Self and Society. Chicago: University of Chicago Press.

Merton, R.K. 1947. Social Theory and Social Structure. New York: Free Press.

Noam, G. 1979. "Borderline Psychopathology, Psychoanalytic Object Relations Theory and the Structural Developmental Paradigm". Unpublished manuscript: Harvard Medical School.

Noam, G. 1982a. "Separation Individuation and Affiliation-Integration Across the Life Cycle. A Clinical-Developmental Study of Vulnerabilities". Unpublished manuscript: Harvard University and McLean Hospital.

Noam, G. 1982b. "Dynamics in the Development of the Self: Stage, Phase and Style". Unpublished manuscript: Harvard Medical School.

Noam, G. (in preparation). Social Cognition, Psychodynamics, and Defenses--A Model.

Noam, G. & Kegan, R. 1982. "Social Cognition and Psychodynamics: Towards a Clinical-development Psychology". In W. Edelstein & M. Keller (Eds.), Perspektivitat und Interpretation, Beitrage zur Entwicklung des Sozialen Verstehens. Frankfurt: Suhrkamp Publishing House.

Noam, G., R. Higgins, & G. Goethals. 1982. "Psychoanalysis as a Developmental Psychology". In R. Wolman (ed.), Handbook of Developmental Psychology. New York: Prentice-Hall.

Parsons, T. 1964. Social Structure and Personality. New York: Free Press.

Peck, R.F. & R.J. Havighurst. 1960. The Psychology of Character Development. New York: Wiley Press.

Piaget, J. 1926. The Child's Conception of Physical Causality. 1960 ed., New York: Littlefield Adams.

Piaget, J. 1965, The Moral Judgment of the Child. New York:
        Free Press. (originally published in 1932).

Piaget, J. 1968. Structuralism. 1970 ed., New York: Basic
        Books.

Piaget, J. 1973. "Affective Unconscious and Cognitive
        Unconscious". The Child and Reality. New York:
        Grossman Publishers.

Piaget, J. 1981. Intelligence and Affectivity. Annual
        Reviews Monographs: Palo Alto.

Radcliffe-Brown, A.R. 1935. "On the Concept of Function in
        Social Science". American Anthropologist, XXXVII.

Rapaport, D. 1967. "An Historical Survey of Psychoanalytic
        Ego Psychology". In M. Gill (ed.), The Collected
        Papers of David Rapaport, New York: Basic Books.

Santostefano, S. 1978. A Biodevelopmental Approach to
        Clinical Child Psychology: Cognitive Controls and
        Cognitive Control Therapy. New York:
        Basic Books.

Selman, R.L. 1980. The Growth of Interpersonal Understand-
        ing: Developmental and Clinical Analyses. New
        York: Academic Press.

Selman, R.L. & R. Yando (eds.). 1980. Clinical-Develop-
        mental Psychology. New Directions for Child Psy-
        chology G. San Francisco: Jossey-Bass.

Snarey, J. "The Social and Moral Development of Kibbutz
        Founders and Sabras: A Cross-Sectional and Lon-
        gitudinal Study". Doctoral thesis, Harvard Uni-
        versity, Cambridge, MA, March, 1982.

Snarey, J. & J. Blasi. 1980. "Ego Development Among Adult
        Kibbutzniks: A Cross-Cultural Application of
        Loevinger's theory. Genetic Psychology Mono-
        graphs, 102, pp. 117-57.

Sullivan, H.S. 1953. The Interpersonal Theory of Psy-
        chiatry. New York: W.W. Norton & Co., Inc.

Sutherland, J.D. 1963. "Object Relations Theory and the
        Conceptual Model of Psychoanalysis". British
        Journal of Medical Psychology, XXXVI:109-24.

Vaillant, G. 1977. Adaptation to Life. Boston: Little
        Brown.

Wertheimer, M. 1912. "Experimentelle Studien uber das
        Sehen von Bewegung". Zeitschrift fur Psychol-
        ogie, LXI:161-265.

Winnicott, D.W. 1958. Collected Papers: Through Pediatrics
        to Psychoanalysis. London: Tavistock Publica-
        tions.

Wolff, P.H.   1960.   "The Developmental Psychologies of Jean
        Piaget and Psychoanalysis". <u>Psychological Issues</u>,
        II, 1, Monograph 5.   New York: International Uni-
        versities Press.
Yankelovich, D. & W. Barrett.   1971.   <u>Ego and Instinct: the
        Psychoanalytic View of Human Nature</u>.   New York:
        Vintage Books.

# SENSORIMOTOR EGOCENTRISM, SOCIAL INTERACTION,

# AND THE DEVELOPMENT OF SELF AND GESTURE

Werner van de Voort

Max-Planck Institute

## SUMMARY

J. Piaget has maintained that the development of cognitive structures is accompanied by a parallel development of
structurally isomorphic social schemes and concepts. Whereas Piaget refers to the dominance of physical experiences
for the explanation of the development of this reflective
intelligence, G.H. Mead points to the experiences of the
child in social interactions.

By means of a reinterpretation of Piaget's data on the
sensorimotor phase, I have tried to find some evidence for
G.H. Mead's dialogical model. In the first part I argue
that Piaget's monological model of the elimination of sensorimotor egocentrism is not plausible, and that--contrary to
his central hypothesis on the interdependent development of
the schemes of object and self--Piaget finds a permanence of
the self's activity (the child's hands) long before the permanence of the objects. Based solely upon Piaget's own data,
an alternative dialogical model is proposed to explain this
décalage: after eye-hand coordination is final, the child is
able to follow the activity of his own hands and those of
the caregiver; but his egocentrism hinders him from distinguishing between these different hands. In the social interaction with his caregiver, the child is confronted with the
activity of the other's hands, which are not a part of his
motor experiences and which he cannot control. I argue that
these assimilatory experiences are frustrating for the egocentric child and therefore lead him to a comparison of
hands, and that this process ends up in a distinction between his own hands and other's hands.

In the second part I sketch out the child's elementary social interaction experiences in the sensorimotor phase, and discuss--again, on the basis of Piaget's data--the transition to communication with gestures. Gestures differ from (sensorimotor) actions in that 1) the actor directs them at an external and autonomous interaction partner, and 2) that she/he is conscious of the meaning these gestures have for particular interaction partners. I propose the hypothesis that the ego/alter differentiation implies a novel control of the child's own actions which enables him to become conscious of the meaning his actions have for caregivers, and, accordingly, to use gestures as a means of communication.

### SENSORIMOTOR EGOCENTRISM, SOCIAL INTERACTION, AND THE DEVELOPMENT OF SELF AND GESTURE

Jean Piaget has occasionally pointed out the connections between his theory of cognitive development and particular themes that are the concern of sociology. This has not, however, prompted sociologists to pay close attention to Piaget's theory. Sociological socialization research was interested neither in universal developmental sequences nor in the development of logico-mathematical thought. This paper is intended as a contribution to the sociological interpretation of Piaget. Its main concern is the impact of social interaction on the child's cognitive development in the sensorimotor phase.

### J. Piaget: Biology and Social Interaction

The hypotheses formulated by G.H. Mead (1934), and his conceptions of intelligence and mind, bear resemblance to the argument proposed by Jean Piaget in his early investigations (Piaget 1927, 1928, 1959, 1977; see also Swanson 1974; Oevermann 1974, 1976; Habermas 1972). In his early work, Piaget chose to use (as he himself has put it; 1928, p. 201)[1] the "language of sociology", giving emphasis to the significance of social cooperation in the development of cognitive structures. In the summary and conclusions to Judgment and Reasoning in the Child (1928), on the other hand, Piaget indicated that he intended supplementing this one-dimensional model of explanation by a "biological explanation".

I would argue that the shift to the biological  pattern
of  explanation entails the devaluation of the three sets of
factors developmental theory has held to be the determinants
of mental growth: maturation, and experience of the physical
and of the social environments (see Piaget 1970a).  Invoking
the great variability of social environments, Piaget put into question the explanatory value of the child's social  experience  relative to the universal invariant sequential order of cognitive structures. For Piaget, at  this  stage  of
his work, social experience is important only within certain
limits--that  is, only to the extent that it can explain the
acceleration or retardation of  the  developmental  process.
However this argument is contradicted by another of Piaget's
central  hypotheses, which asserts the isomorphism of interpersonal and intraindividual  operations.   In  these  terms
"there  is  a fundamental identity between the interpersonal
operations and intraindividual operations" (1968a,  p. 129),
and  both "obey the same laws" (1970a, p.729; see also 1967,
1971).

The isomorphism hypothesis implies inferentially  that,
regardless  of  the  observable variability of social structures, there is a common core of  universal  social  structures.  This means that the thesis advanced to devaluate the
relevance of  social  experience  is  not warranted.  It is
therefore not surprising that even after Piaget had effected
a shift in the pattern of explanation to the biological  model,  he  posed the question of whether the cognitive structures described by him are cause or effect of social cooperation (1959, 1977).  The question itself indicates that Piaget has  withdrawn  the  devaluation  of  social  experience
resulting from the shift to the biological model of explanation.

This  is made evident even more clearly by the way Piaget answers the cause-or-effect question (1959,  1977),  for
he  repeatedly goes back to the sociological model of explanation he produced in his early work.  The inference is that
Piaget has not given up the sociological pattern of explanation but has supplemented it by the  biological model.  This
follows from his position that cognitive structures  neither
result  exclusively  from  internal  operative developmental
processes nor are they solely the product of social cooperation.  The intraindividual operative activity of the biological organism and social cooperation are,  rather,  interdependent  aspects  of  one and the same developmental process

(1959, 1977). The conclusions of <u>Biology and Knowledge</u> (1971, secs. 22.5 and 23.5) unmistakably follow a similar pattern of reasoning.

In connection with the thesis of the "bursting of instinct" and the resultant abandoning of the support of the hereditary apparatus in the domain of cognitive development, Piaget observes that "the intelligence gives up the transindividual cycles of the instinct only to adopt interindividual or social interactions" (1971, p. 368). When we examine Piaget's empirical investigations, however, we find that the data focus almost exclusively on the child's instrumental interaction with inanimate objects--that is, on the physical experience--and that very little attention is given to social interaction between caregivers and the child. Thus no hypothesis is produced concerning the significance of universal structures of social interaction for cognitive development. This state of facts is the starting point for my Piaget interpretation. I shall set out the few data on social interaction, and on this basis attempt to sketch a "sociological" model.

The Elimination of Egocentrism<br>in the Sensorimotor Phase

In accordance with Freud, Piaget has pointed out that newborn children do not differentiate between their own bodies and the external world; they have no consciousness of self or of objects. Piaget designated this state "infantile" egocentrism.[2] In his three studies of the sensorimotor period in the child's development--<u>The Origins of Intelligence in Children</u> (1965; hereafter referred to as OI), <u>The Construction of Reality in the Child</u> (1968b; hereafter CR), and <u>Play, Dreams and Imitation in Childhood</u> (1972; hereafter PDI)--Piaget has given a very thorough description and explanation of the elimination of sensorimotor egocentrism. His central thesis is that "intellectual activity, departing from...a lack of differentiation between subject and object, progresses simultaneously in the conquest of things and reflection on itself, these two processes of inverse direction being correlative" (OI, p. 19). In these empirical investigations, Piaget has given priority to data on the child's instrumental or manipulative interaction with inanimate objects; this fact suggests that Piaget devalues the significance of social interaction experiences for cognitive development (cf. Newson & Newson 1975).

Therefore it seems justified to assume that in Piaget's
considered view not only consciousness of objects, but also
the consciousness of self, results from instrumental interaction with inanimate objects. True, Piaget has never explicitly asserted this; and in the chapter on The Development of Causality (in CR), which is relatively rich in data
on social interaction, he sometimes even emphasized the importance of social interaction for the development of self.
It may be argued, nonetheless, that in the investigations
referred to, the importance of social interaction for the
development of reflective intelligence in the sensorimotor
phase has not been sufficiently explored--in view of the
fact that, on the plane of theory construction, the data and
their interpretation have not been taken account of.

On the following pages, I shall first discuss in what
respects I consider Piaget's "monological" explanatory model
of the elimination of egocentrism to be implausible [3]; and
then I shall propose an alternative "dialogical" model in
which the significance of social interaction experiences is
underlined.

PIAGET'S MONOLOGICAL MODEL

Prehension, and later the search for displaced objects,
is central to Piaget's study of the sensorimotor phase. An
observer who has had no training in developmental psychology, watching a child of two or three months stretch out his
hand toward an object within his reach, might well interpret
this act as "an attempt to grasp an object". According to
Piaget, however, the infant does not as yet have a sensorimotor scheme of objects. Rather, in the first months of his
life, the infant experiences objects only as a complex of
tactile, kinesthetic, buccal, and other sensations which
have no permanence for him at this stage. These "objects"
exist only with the child's actions and not yet independently from them.

The concept of "egocentrism", used by Piaget, points to
the child's centration on his own experiences of the action
at a point where a scheme or a consciousness of self does
not yet exist--i.e., the child does not know himself to be

the subject of or his hands to be the source of such sensorimotor experiences, and as a result cannot conceive objects
as  being external  to and independent of the actions of his
own hands.  In addition, Piaget points out that the  child's
visual  and tactile experiences with objects are not yet coordinated with each other.  This means that the visual experience of an object need not correspond to the tactile experience of an object because--as Piaget's observations  indi-
cate--the  activities  of hands and eyes are not yet coordinated with each other or have  not  yet  been  "reciprocally
assimilated" to each other.

The structure of Piaget's explanatory model of the elimination of egocentrism, and thus of the dissociation of the
self[4] from the physical environment, is the following: The
reciprocal assimilation  of these  assimilatory schemes creates  an  increasing number of relationships between actions
and objects, and  this  explains  objectification (see  OI,
p. 415).

This  model,  applied  to prehension, implies that eye-
hand coordination must result in the simultaneous or  parallel objectification of objects in the child's physical environment and of the self. Ocular-manual coordination, according to Piaget, begins at the second stage of the development
of  intelligence  and  is  complete  at the beginning of the
third stage.  During this period the child learns  to  grasp
those  things it sees[5] and, conversely, to look at what it
has grasped.  This coordination between  the  activities  of
the  eye and the hand induces (as Piaget has pointed out) an
important experience: things seen and  grasped  are  endowed
with  "a certain externality"; they are "evidently perceived
by the child as 'external'" (OI, p. 60), and this  is,  precisely,  a  beginning  of  the elimination of egocentrism--a
world split into experiences of the self  and  external  objects  and  thus  a world in which the self is detached from
objects (regarding this model, see OI, pp.143, 172ff., 211).
To be sure, Piaget himself does not assert that  this  elimination  of  egocentrism  is accomplished as early as in the
third stage; he speaks only of an  "incipient  objectivity"
(OI, p. 75).

In  the  chapter on the "Development of Object Concept"
(see CR, ch. 1), Piaget has developed this model of argumentation further.  He starts from the hypothesis that children
form the scheme of the permanent object in  analogy  to  the

criteria scientists have used to define objects (see CR:87).
In these terms, those "phenomena" constitute real objects
which 1) can give rise to anticipation, 2) lend themselves
to distinct experiments whose results are in accordance with
such anticipation, and 3) are connected in an intelligible
way with a causal and spatio-temporal system. At the beginning, the contact between reflex schemes and objects implies
no awareness of the permanence of any object--this even when
the child is able to recognize certain objects.

The problem of object permanence begins to be raised
when a sensation connected with the reflex schemes ends or
does not occur. The child seeks to reproduce sensations or
events experienced in connection with his actions, and to
this end starts to search for and seeks to rediscover the
"lost object"; this gives rise to the anticipation of a
"phenomenon" or an "event". The second criterion for the
construction of objects comes into play in connection with
the reciprocal assimilation of assimilation schemes (e.g.,
of eye and hand). The child coordinates his visual with his
tactile search; this, according to Piaget, in the third
stage "certainly reinforces the consolidation and externali-
aztion of the object (the dissociation between the object
and the action)" (CR, p. 90) and thus contributes to the
elimination of the egocentric illusion, for "the action
ceases to be the source of the external world and becomes
merely a factor among other factors, one that is central, no
doubt, but of the same order as the various elements which
make up his total envirnoment" (CR, p. 92).

I do not find Piaget's argument concerning the beginning of the elimination of egocentrism convincing, because
it does not explain why ocular-manual coordination enables
the child to split his egocentric experiences into consciously perceived own hands and other external objects. The
point is that Piaget analyzes the child's patterns of behavior in her search for lost or hidden objects, but does not
analyze the other aspect which is just as fundamental for
the elimination of egocentrism: the question of the differentiation between own body (say, hands) and objects.[6]

Furthermore, Piaget's own empirical facts run counter
to his argumentation. He states that a first "externalization" of object succeeds completion of eye-hand coordination, and it is at the fourth stage that he speaks of the
beginning of "object permanence"--that is, the stage in the

course of which the child is aware of the existence of an object even when the latter cannot be experienced visually or by touch because it has been hidden behind or screened by another object. In terms of Piaget's central hypothesis concerning "simultaneous" or "correlative" development of schemes of self and objects, formation of awareness of the permanence of his own body or parts of his own body (e.g., his hands) would also start only at stage four. But according to Piaget's own data, this is not the case. In contradiction with the above hypothesis, Piaget maintains that already at the third stage there is permanence of the child's prehensile action--i.e., permanence of a part of the sensorimotor self:

> It is not the object which constitutes the permanent element (for example the coverlet) but the act itself (swinging the coverlet), hence the whole of the situation; the child merely returns to his action.(CR, p. 27; cf. pp. 18, 19, 20, 34, 35, 41)

This is something Piaget did not consider in his explanation of the elimination of egocentrism, and these data are inconsistent with his monological explanatory model. They also support the claim that social interaction experiences are the dominant factor for the elimination of egocentrism.

In connection with the scheme of causality, however, Piaget gives a somewhat more precise description of an action conflict which should induce a beginning of the decentration process because, as Piaget notes, the subject begins to discover that "any object at all can be a source of activity (and not only his own body)" (OI, p. 212). This line of argument is interesting because it differs from the first model: Piaget here refers to the child's understanding that there exists something in common between him and the objects in his environment. He seems to be of the opinion that only on the basis of this understanding can the child differentiate between his own body and objects and thus eliminate sensorimotor egocentrism. The content of this understanding of "something in common" is, according to Piaget, the "source of activity": the child's own body is a source of activity, but so is any object (see CR, p. 270). What Piaget fails to explain, however, is how the child--who, in terms of Piaget's definition of egocentrism, cannot yet know himself to be a source of activity--actually does conceive himself a source of activity.

Piaget (referring to the works of P. Guillaume and J.M. Baldwin) has not, of course, overlooked the similarity between the child's actions and the actions of the caregiver. Hence he cannot but acknowledge that the comparison and contrast with "other bodies" or "sources of activities" is an element in the development of the self relating to the elimination of egocentrism (see CR, p. 233). Nevertheless, the significance of this social element of the child's environment for the elimination of egocentrism is subsequently devalued, and Piaget's starting point continues to be the dominance of the instrumental action experiences with inanimate objects. For this reason, I would like to "test" the plausibility of his monological thesis on the basis of an example taken from Piaget's own data.

Piaget has described many observations of the child playing with the rattle. If one is looking for similarities between the activity of the rattle and the activity of the child's hand—that is, activities and movements perceptible to the child after eye-hand coordination is complete—one can find such action sequences as: quick movements of the rattle/quick movements of the child's hand, the noise of the rattle/the child's vocal acts. I believe that there are no major controversies as to the fact that the above examples do not imply any close similarity between the action of the child and the movements of the rattle. Certainly the child will not identify his own voice or the movements of his hands with the sounds and the movements[7] of the rattle. The difficulty the child has in identifying a similarity between his own actions and the activities of this inanimate object in his physical environment is, in the final analysis, the difficulty of developing his sensorimotor self solely in interaction with inanimate objects.

## EYE-HAND COORDINATION AND EGOCENTRISM

The coordination between vision and hand-movements is not only central to the development of instrumental actions (e.g., the child's ability to reach for, grasp, and handle objects); it is also crucial for the course of the dissociation of the sensorimotor self from the social environment. The completion of ocular-manual coordination enables the child to control his own hand-movements, but not only that; he can also follow the movements of the hands of his partners in social interaction (ie., the caregiving persons).

What actually happens is that as he develops a visual scheme
for hands and hand-movements, the child is able to see something he has in common with another person in his environ-
ment--i.e., hands and hand-movements.

However, in this case, too, the child's egocentrism
must be taken into account. The egocentric child will know
inanimate objects as independent-and-external-objects as
little as he will regard the hand of another person as an
autonomous, external, and permanent center of action located
in the person of someone else. Yet there is one important
difference between the other's hand and an inanimate object:
the likeness that another person's hand bears to the child's
own hand. And the child--according to Piaget's interpreta-
tion--does consider the hand of another person to be "a
similar object" to his own hand (see Piaget's remarks concerning stage 3 and 4 of ocular-manual coordination; OI,
pp.99-116).

In observation 76 (OI, p. 106) Piaget exemplifies this:
at 0;4(4) (see [8] for explanation of numbers), Piaget shows
Lucienne his motionless hand. She regards it attentively and
begins to smile; she finally opens her mouth and puts--as
usual in this stage--her own fingers into it. Piaget reports
that she repeated the same reaction many times. He interprets his daughter's[9] reaction in the sense that she is
"assimilating" his hand to hers, and that this visual image
of his hand makes the child start the sucking-like movements
which are characteristic of the third stage of ocular-manual
coordination. A little later, Piaget formulates this more
precisely when he says "the visual image of my hand is
assimilated to the simultaneously visual, motor, and buccal
scheme of her own hand" (OI, p. 109).

In another set of observations, Piaget develops this
thesis further. He notes that as early as in the second
month, his son Laurent would sporadically join his hands as
soon as they passed into his visual field (the child was
lying in a reclining position), or clasp his hands. Piaget
calls this the "scheme of handclasping". In observation 74
(OI, p. 104f.) Piaget reports that beginning with 0;3(3),
Laurent began to grasp his father's hand as soon as it appeared near his face and entered his visual field. Piaget's
explanation of Laurent's behavior is that Laurent began to
grasp his father's hand because "my hand was visually
assimilated to one of his hands and so set in motion the

scheme of hand clasping" (OI, p. 104). He bases this interpretation on the observation that in the same situation,
when presented with some objects instead of his father's
hand, Laurent did not attempt to grasp them; at this third
stage of eye-hand coordination, the child grasps the objects
he is already sucking or is touching with his hands but not
the objects he is merely looking at. Thus, the grasping of
the father's hand which has been only visually experienced
is a phenomenon which, occurring as it does at this stage of
ocular-manual coordination, needs to be explained.

Piaget holds that these observations can be explained
by saying that the child assimilates the visual image of his
father's hands to the visual image of his own hands, without
confusing them with or distinguishing them from his own (see
also PDI, p. 16). This unconscious (i.e., not yet self-conscious) assimilation, understood within the context of the
Piagetian assimilation theory, raises no problems. Given
the social interaction between the infant and his caregiver,
these data show the developmental psychologist that this unconscious assimilation makes the child visually select something he has in common with another person--although the
child's egocentrism of course prevents him from recognizing
such prehensile activity as a common aspect of two independent and different persons.

Piaget does admit this may seem a somewhat bold interpretation in view of the disproportion between the visual
appearance of an adult's hand and that of the child, and the
difference in their positions. But, for Piaget, the child's
imitation of the hand-movements of another person is an additional indicator of the "fusion" of the two hands (the father's and the child's) or of assimilation by the child of
the father's hands to his (cf., footnote 14 in OI, p. 108,
and PDI, p. 15, obs. 7).

If the data given by Piaget and his interpretation are
valid, then at the age of about four months the child's assimilation activities will have resulted in his visually selecting the prehensile activity as something in common between himself and particular others--although egocentrism
prevents him from differentiating between his own prehensile
activity and that of another person. My hypothesis is that
this state of affairs is the background for the development
of the child's understanding of the difference between his
own hands and another person's hands. Through such distinc-

tion, the child himself begins to discover things he has  in
common  with  others—i.e., that his own prehensile activity
is essentially similar to that of another person.


## WHO IS ACTING?

I should now like to describe a few  observations  from
Piaget's  investigations in which the social interaction between the child and his caregiver might be expected to  provide impulses for the child to make comparisons of the activities of hands within his visual field.

In  observation  72  (OI, p. 103f.), Piaget describes a
situation in which he is holding the child's hand (0;5(12)),
which remains outside the visual field; the  child  attempts
to  free  and  face his hand, but does not look in the right
direction.  At a given moment, she happens to perceive  Piaget's  hand, which is holding her right hand.  She looks attentively at both hands without trying to free her  hand  at
this  exact  moment.  Then she  continues the struggle, while
looking all around her head; she does not look in the direction of her right hand. Piaget's interpretation of this type
of behavior is that it is "egocentric".  The "tactile-motor"
experience or the consciousness of effort is not yet  local-
ized in the visual image of the child's hand.  The hand that
is  seen and watched by the child is still neither the "tactile-motor" hand (see OI, p. 108) nor the hand belonging  to
her (see also OI, p. 110, obs. 77).

In  observation  78  (OI, p. 110f.), Piaget reports that
Lucienne at 0;4(12) attentively regards  her  mother's  hand
while  taking  the  breast.  She then begins to move her own
hand, but continues to look at her mother's hand  until  she
perceives  her own hand.  Her glance begins to oscillate between her mother's hand and her own, then finally she grasps
her mother's hand.

A second observation made on the same day reveals similar patterns of behavior.  In the same  situation,  Lucienne
again  looks at her mother's hands.  She then lets go of the
breast and inspects the mother's hand, while moving her lips
and tongue.  She directs her own hand near her mother's, but
then puts her own hand in her mouth and sucks it.  A  moment
later  she takes her hand out of her mouth, while looking at
her mother's hand.  Piaget notes  that,  as  on  an  earlier

occasion, the child is assimilating her mother's hand to her own. But, Piaget comments, "this time the confusion does not last; after having removed her hand from between her lips, she moves it about at random, haphazardly touches her mother's hand, and immediately grasps it. Then, while watching the spectacle most attentively, she lets go the hand she was holding, looks alternately at her own hand and the other one, again puts her hand in her mouth, then removes it while contemplating the whole time her mother's hand and finally grasping it and not letting it go for a long moment" (OI, p. 111).

Contrary to the preceding observation, in this case the child might have begun distinguishing her mother's hand from her own hand--insofar as she is confronted by the problem of how an inert hand, which she has managed to keep within her field of vision, nonetheless can induce interesting and pleasant sensations in her mouth. This seems improbable, as, according to Piaget, there is as yet no coordination between the visual images of the hand and these other sensory experiences. The fundamental point in the observation just quoted is that it indicates a beginning of comparisons between own hands and others' hands. That the child has begun making this comparison[10] between hand movements will become more evident from the analysis of the social interaction between the child and particular others which follows later. In the chapter on "The Development of Causality", (CR, p. 249ff.) Piaget examines "causality by imitation". In connection with the imitation characteristic of the third stage, the child, by imitating the actions of another person, tries to make the latter repeat various movements he has made.

In observation 137 (CR, p. 250), Piaget relates that the child, Jacqueline (0;7(27)), sits in front of a big quilt and her mother is striking the quilt with her hand. Jacqueline imitates this, and for a moment both of them are striking the quilt together. This unison seems to please Jacqueline; she reacts to it with peals of laughter. Then Jacqueline stops striking the quilt, to look at the movements of her mother's hand. The latter, after, a few more seconds, also stops striking the quilt. But this does not make Jacqueline stop: while looking at her mother's hand, and without looking at her own, she again resumes tapping the quilt --at first gently, then harder and harder. Piaget's interpretation is that Jacqueline is trying to force her mother to recommence striking the quilt. The indicator for this interpretation is the following sequence of imitations: The mother starts striking the quilt again, and Jacqueline stops; then her mother ceases tapping, and Jacqueline resumes the activity--her eyes again fixed on her mother's hands.

In the same observation, Piaget cites other examples of such sequences-in-imitation; and observations 138 and 139 (CR, p. 250f.) report similar interactions with the other two children observed. To support the interpretation according to which the child imitates other persons to make them continue the activity they started, Piaget points to the expression of desire and expectation with which the child looks at other persons--as well the way in which the child regulates his activity according to that of the adults (the child starts his activity normally, then goes faster and faster until the adult starts again; then the child stops).

In connection with these observations, Piaget discusses the question of whether in such interactions the child has already overcome his egocentrism, and whether he conceives the self as detached from the action of other persons. He answers this question in the negative, and points out that there is nothing to indicate that the child views the other person as an autonomous and external center of action. Rather, the observations cited serve to show that the egocentric illusion is maintained by which the child continues to believe his own activity to be the causal factor producing the activity of another person (without being able to distinguish the two centers of action).

However, in light of these observations of a special form of social interaction (i.e., imitation), Piaget feels prompted to make a number of statements--hardly consistent with his thesis on the primacy of instrumental experience--concerning the importance of persons for the child's cognitive progress: he states that in the child's eyes, persons are the most active centers of action, stimulating the child much more than inanimate objects. And he adds that he has the impression that the child is reserved for a moment when a person appears, ready to follow in the direction indicated, thus attributing to a person a certain spontaneity. But in relation to the third stage discussed here, Piaget is not ready to regard this as an indicator of the child's dissociation of the self from the external universe. On the other hand, Piaget makes the assumption that "contact with persons plays an essential role in the processes of objectification and externalization", so that the person and not the inanimate objects "constitutes the primary objects and the most external of the objects in motion through space" (CR, p. 252).[11]   Thereafter, Piaget notes, it is probable that "another person represents the first of these centers and contributes more than anything else to dissociating causality from the movements of the child himself and objectifying it in the external world" (CR, p. 252).

According to Piaget, the prominence acquired by the person relative to inanimate objects results from causality by imitation, which "leads the child toward this externalization" (CR, p. 254)--since causality by imitation enables the child to compare his own actions with those of another person.[12]  However, during the third stage "another person barely begins to be analyzed through imitation" (CR, p. 233); and because elimination of egocentrism is an evolution progressing only gradually, this comparison and contrast with other persons is still very far away from constituting a consciousness of the self (see CR, p. 252f.).

## A DIALOGICAL MODEL OF THE ELIMINATION
## OF EGOCENTRISM

At this point I would like to put--a little more sharply than Piaget himself has done--the problem the child is confronted with in this special form of social interaction. One of Piaget's criteria for Piaget's completion of ocular-manual coordination is the coordination between the visual image of the hand and the kinesthetic or tactile-motor experiences of the hand. Thus when, for example, the child's hand is retained outside his visual field, as a result of such coordination--and differently from previous occasions --the child immediately looks in the direction of the hand which is being held by another person. From this, Piaget concludes that the seen hand is also the tactile-motor hand (see OI, p. 116 obs. 86; p. 117f., obs. 90). This behavior pattern is observable with Jacqueline and Lucienne at about two months and with Laurent at about seventeen days, prior to the beginnings of "causality by imitation". If this Piagetian data and the aforementioned interpretations are valid, it is possible (at least hypothetically) to sketch out the problem which confronts the child in its social interactions with others.

Let us consider observation 137 (CR, p.250), Jacqueline is seated in front of a big quilt. Her mother strikes the quilt with her hand and Jacqueline immediately imitates her, so that both of them strike the quilt together. The egocentric child, however, is still not able to distinguish the activities of her own hands from those of her mother. This differentiation seems to be induced through social interactions--for example, the mother's hand stops striking the quilt and the child, while looking at her mother's hand, stops striking the quilt; the child, while looking at her mother's hand, again begins striking the quilt with her own hands. Given Piaget's assumption that for some time already the child has coordinated the visual image of her hand with the kinesthetic experience of her hand, the problem that presents itself to the child can be put in the form of the question: who is actually striking the quilt? Or put in another way: how do hands that are motionless (in the present case, the hands of the mother) still produce interesting kinesthetic and auditory events?

The question is justified, because in a situation where the child stops moving his hand and looks at the activity of the hands of another person, he is experiencing an incomplete (deficient) form of a multisensory complex. While the moving of his own hands implies tactile, motor, and visual experiences, the child can rely on only the visual dimension when he inspects the activity of another person's hands. Moreover, the child experiences this deficiency not only in situations produced by "causality by imitation", but also in normal social interactions in which action and response are not identical--although both can be controlled by the child through application of assimilatory schemes. The situation of the child looking at the activity of the hands of another person will, for an egocentric child, inevitably produce feelings of frustration because it is not his own hands which he perceives. It may be frustrations of this type which make the child consider the hands striking the quilt as "external".

Another inevitably frustrating experience is that the child fails to control the movements of the other person's hands. Piaget points out that as a result of eye-hand coordination, the child "becomes conscious of the purposefulness of his movements and of the reality of his power over his hands" (CR, p. 233). But in the case of another's hands (which, because of his egocentrism, the child cannot as yet distinguish from his own hands), the frustrating experience will consist of the discovery that there exist movements of hands which--however great his own efforts--he cannot direct. Possibly, conflicts of this kind make the child apprehend that there exist autonomous centers of action.

Succeeding the completion of eye-hand coordination--and this is the thesis of my paper--the egocentric child will inevitably be involved in the identity problem described above, which he can resolve only by extricating himself from his egocentric illusion. On this view, the elimination of egocentrism starts as the child's involvement in social interaction, bit by bit, makes him perceive that the hand movements which he thus far perceived as identical are not identical. And this in turn is the crystallization point for a beginning differentiation of autonomous sensorimotor centers of action--i.e., for a beginning of the dissociation of the self from another person. Through the specific negation of the hand movements of others as non-identical with his own, the child can separate the so-far undifferentiated movements of hands in his visual field into "own movements" and movements of "others" hands--and as a result, perceive the self as distinct from another person. Only after realizing these differences can the child consciously perceive a resemblance between his own hands and the hands of another person.

## THE DEVELOPMENT OF THE SELF

Piaget actually does state that with the beginning of the fourth stage, children distinguish their own hands from those of another person. For example, he reports (CR, p. 259, obs. 141) that his daughter pushed away the hand of her father when he was trying to grasp a duck at the same time as she was. This observation illustrates that the infant has begun to develop a scheme in which self is a source of sensorimotor actions. In subsequent observations Piaget reports that the child begins, for example, to use the hand of the adult as an intermediary to reproduce a result the child is not able to perform himself.

These observations show that the child comes to realize that the external and independent hand can perform comparatively more complex actions than he himself can as yet produce.

In the social interaction with his caregivers, in which the child is constantly confronted by relatively more complex sensorimotor actions performed by others, this scheme of the self is the basis for the successive elaboration of a complete sensorimotor self, whose different stages one finds sketched out by Piaget, especially in the chapters on causality and imitation. Discussing the fourth stage of imitation, for instance, Piaget states that this insight into the greater competence of the particular other motivates the child to imitate those relatively new or different models which are not too remote from the already existing sensorimotor actions of the child (see PDI, p. 50f.). In the same stage the development of a complete sensorimotor self is enhanced by the imitation of even those actions, (e.g., facial movements like tongue protrusion) the child cannot visually control except on the body of another person.[13]

## DIALOGICAL VERSUS MONOLOGICAL MODEL

Piaget's statements about the significance of social interaction for the destruction of egocentrism--and, in this sense, about the importance of social interaction for the development of sensorimotor intelligence--are not consistent with his predominantly monological model of cognitive development and with his orientation to instrumental action. Therefore, Piaget tries to devalue the relevance of social interaction by the argument that it is the formation of intelligence which governs the development of imitation (see CR, p. 319).

I must admit that this argument does not convince me, because the significance I attribute to imitation within the proposed dialogical model does not contradict Piaget's assumption that the child can only imitate such acts as those for which he has already developed assimilatory schemes.

Further, I would question whether, for the dialogical model I have outlined, imitation has this overriding importance. I would suggest that imitation is overrepresented in Piaget's data. To be sure, the data accumulated by Piaget constitute the basis I myself started from in building the dialogical model; but I doubt that it is exclusively the imitation situation which confronts the child with an identity problem. I believe that the normal social interaction process also forces the child to differentiate between self and others.

However, Piaget implicitly--more than explicitly--devalues the importance of social interaction in eliminating egocentrism: he has neither systematically taken account of his own social interaction data nor does he furnish data to support his hypothesis about the "décalage" between the constitution of object-permanence and person-permanence. To back up the alternative thesis that the dissociation of self from other persons in the process of social interaction starts earlier that the constitution of object-permanence, I would like to point to the following facts:

1. As soon as ocular-manual coordination is completed, the child is confronted by an identity problem. The comparable problem of the dissociation of the self from inanimate objects, however, can present itself only later. This is because, as Piaget has indicated, at the third stage the child's action and its results constitute a "single scheme" (see PDI, p. 26; CR, p. 42; and also what is said regarding stages 1, 2, and 3 in chapter I "The Development of Causality" in CR). Additionally, at this level, according to Piaget, the child focuses more on his own actions than on objects (see OI, p. 197, 209, 241, 258). Only during the fourth stage (during which the child's interests shift to the results of actions, to objects, and to novelties) can the child--in his actions on various inert bodies--be confronted with the problems which lead him to differentiate his hands from external, inanimate objects.

2. Contrary to his central hypothesis, Piaget has found that a beginning of permanence of prehensile activities precedes the permanence attributed to objects (see CR, pp. 18, 19, 20, 27, 34, 35, 41). Piaget has given no plausible explanation of this statement, and I doubt that it is possible to explain these facts within the frame of a monological model of cognitive development. In terms of the alternative dialogical model proposed in this presentation, this explanation is unproblematic. On the contrary, starting from the dialogical model, one would even feel compelled to propose the hypothesis that after eye-hand coordination

is final, the child will be confronted by an identity problem in his social interactions which involve prehension.
Solving this problem requires that the child differentiates
hand-movements thus far undifferentiated within his visual
field. For the child, hand-movements of another person relative to his own visually controlled hand-movements are deficient because they lack certain sensorimotor experiences and
consciousness of power over the actions of his own hands.
But to explain the development of the child's understanding
of the "permanence of action", one must refer to yet another
experience of the child. Whereas the child can reproduce
the full complex of experiences which correspond to the actions of his own hands via reproductive assimilation, the
hand movements of another person are to him only passing
visual episodes. It is reproductive assimilation which enables the child to "return to" his own actions in space and
time--and thus to distinguish them as "permanent", in contrast with the actions of another person.[14]

However, to avoid any misunderstanding, I would like to
make clear that I do not call into question the significance
Piaget attributes to manipulative and instrumental action
for the formation of the physical schemes described by him.
My hypothesis is limited to the proposition that, within the
frame of this theory and on the basis of the data provided
by Piaget, it is possible to reach a diametrically opposed
assessment of the relative significance social and physical
experiences have for cognitive development. Piaget seems to
regard the elaboration of social schemes as a derivate of
the development of cognitive competences engendered by the
experience acquired in instrumental action. I should like
to show that starting from his investigations, it is possible to advance an hypothesis which has as great plausibility
as, but is antithetical to, his own--that is, to argue that
social interaction experiences are the precondition for the
elaboration of physical schemes in the process of manipulative and instrumental action with inanimate objects.

For instance, in the fourth stage the child seeks to
set aside obstacles to his goals--i.e., to perform the secondary schemes he has learned at the third stage by reciprocal assimilation of secondary schemes to means-ends schemes.
According to Piaget, this subsumption of means-to-an-ends
scheme implies "intentional coordinations of the schemata"
(OI, p. 211) which (contrary to those of preceding stages)
have the character not of "fusion" (OI, p. 232) but of "mo-

bile" (OI, p. 238) coordination, so that the various schemes can not only be relatively easily intercoordinated but also dissociated from each other.

Interrelation of these schemes into more or less complex means-ends schemes, however, presupposes that the child can distinguish between them; and this implies conscious sensorimotor negation (cf. OI, p. 235). I would argue that such conscious negation, described by Piaget but never explained by him, is the outcome of the ego/alter distinction in social interaction--because it is here that the child faces the problem of recognizing that a pair of hands, within his visual field are not his own hands and in this sense comes to use sensorimotor negation.

## DEVELOPMENT OF THE SOCIAL INTERACTION SCHEME

### Social Interaction and Development of the Causality Scheme

There seems to be no social interaction between caregiver and the young child at the beginning of the sensorimotor phase; in any case, this is the impression given by Piaget's studies of sensorimotor development. In a number of other investigations (1949, 1968a, 1977), Piaget refers to the child's social environment, but without investigating the specific structure of social interaction between the young child and caregiver. The data presented in the studies of the sensorimotor phase (as well as the interpretation placed upon them) suggest that only when eye-and-hand coordination is final, and with the first "intentional" actions in stage three, does anything that may be termed child-caregiver social interaction come about.

Following Piaget's line of reasoning, the interpretation suggests itself that eye-hand coordination provides the child with progressive awareness of his power over his hands and the purposefulness of his hands' movements (cf. CR, p. 232). In conjunction with the visual and tactile experiences gained by the child in manipulating inanimate objects, this induces causality experiences which are subsequently generalized by the child and applied to interaction with social objects. This interpretation accords also with the pattern of reasoning in Play, Dreams and Imitation (1972, p. 207 f.), where Piaget explicitly postulates the ontogeny of

the permanent object as the precondition for the development
of the "personal scheme": the object scheme applied to the
interpersonal sphere and endowed with the affective element
gives rise to the "personal scheme".

This thesis has met with scepticism in socialization
research. Watson (1972), on the basis of a series of his
own experiments, agrees with Piaget that manipulative interaction with inanimate objects in his physical environment
provides the child with causality or "contingency" experience; but, contrary to Piaget, Watson has pointed out that
caregivers play a more important role in the development of
this scheme, since people--unlike inanimate objects--provide
the baby with such causality or contingency experiences intentionally. This view has gained currency in socialization
research (cf. Watson 1972; Goldberg 1977). It proceeds from
the assumption that the parents respond to the baby's vocal
and visual signals, and in this way create social interactions by which, at around two or three months the baby
learns that his "behavior will be effective".[15]

The social interactions Piaget refers to in his data on
the development of the causality scheme look like this: At
about seven months, Laurent looks at his father's hand and
laughs loudly when the latter snaps his middle finger
against the base of his thumb. When his father stops doing
this, he "arches himself, waves his hands, shakes his head
from side to side"; then, once he has managed to grasp his
father's hand, he strikes it and shakes it. Piaget calls
this behavior "magico-phenomenalistic procedures", and reports that he understood Laurent's actions with his hand to
mean that Laurent wanted him to continue the activity (CR,
p. 244, obs. 133).

A similar interpretation is given in the data on "causality by imitation". Piaget narrates situations in which the
child starts imitating the caregiver's hand activities when
the latter has stopped them. He is struck by the expression
of expectation with which the child looks at the caregiver,
and the acceleration of the child's imitative movements until the caregiver resumes his own activities. Caregivers
interpret (as does Piaget) such imitations as attempts to
make them continue their activities.

This interpretation of the child's actions by the caregiver is, however, problematic in the light of Piagetian theory: the child at this stage is still egocentric, and is incapable of distinguishing own actions from those of another person. Accordingly, the child cannot as yet convey any request to another person, nor can he be aware of the communicative meaning of his own action for the communication partner. In brief, we may say that it is not only caregivers untrained in developmental psychology whose interpretations of their children's behavior do not always adhere to the facts obtained by developmental psychology; Piaget too credits the young child with abilities that he does not as yet possess.

Wolff's (1969) investigations on crying and other vocalizations in early infancy, in which he has spectrographically differentiated four different types of cries, point to a similar tendency among parents and caregivers. The different patterns of crying in the young infant are distinguished with relative precision by the caregivers. In the same way as the parents who recognize different types of cries and take them to indicate different needs, Piaget not only seeks to identify the various crying patterns but in his interpretation assigns particular emotions, needs, or intentions to each type (see OI, p. 78, obs. 40). And he proceeds in a similar manner when analyzing, for instance, the first use of the sucking reflex: he does not merely describe the movements of the infant's head but interprets these as a "searching"; similarly, he explains the alacrity with which the infant rejects the eiderdown quilt (the coverlet which he began to suck) and his reaction to his father's finger by the remark "he tries to nurse and not merely to suck" (OI, p. 36).

A more penetrating interpretation of the young child's elementary impulses and actions as "intentional" or "communicative" actions[16] than that provided by Piaget on the basis of his investigations is offered by several more recent contributions. Brazelton, Koslowski, and Main (1974, p. 68) point out that most mothers are not able or not willing to regard the sensorimotor behavior or vocalizations of the neonate as "meaningless" or "unintentional". "They perform as if highly significant interaction has taken place when there has been no action at all". Sylvester-Bradley and Trevarthen (1978) interpret the child's expressive behavior as "communicative" or even as "speaking" (Trevarthen 1977b).

That caregivers tend to interpret the infant's (social-ly) relatively unstructured behavior as if it were socially structured is not a particularly novel insight. Escalona (1953), for instance, made similar observations relative to gratifying the child's elementary psychological and physio-logical needs. The increasing interest in analyses of pre-verbal social interactions, and the problem posed by such analysis (that is, to understand how there can actually be any social interaction between caregiver and the neonate who is not yet able to communicate verbally), have resulted in more attention being given in the literature to overinter-pretations of this type (see among others, Wolff 1969; Ryan 1974; Newson & Shotter 1974; Newson & Newson 1975; Newson 1977, 1978; Snow 1977; Dore 1978; Lock 1978b; Clark 1978; Shotter 1978).

The caregiver's contributions to communication, which compensate the as yet undeveloped level of verbal communica-tion, is something that Piaget's theory of cognitive devel-opment leaves out of account. G.H. Mead, however, has made these compensations the point of departure for his "dialogi-cal" model of the ontogeny of reflective intelligence; he starts from the general hypothesis that there exists an ef-fective process of social interaction between the newborn and his mother/caregiver (see Mead 1934). In his analysis of the ontogeny of reflective intelligence in humans, Mead starts from hypotheses about the elementary preconditions of communication. Piaget, on the other hand, seeks to specify the foundation of the same developmental process by an anal-ysis of the elementary internal organization and regulation principles of the individual organism.

Thus, for instance, Piaget restricts his empirical research to analysis of the development of visually guided reaching and search for (animate and inanimate) objects, and he disregards the interindividual or communicative dimension of these activities. In contrast to Piaget's conception of sensorimotor activities (and even of imitation, according to Mead), the neonate's prehensile activity takes on an impor-tance that goes beyond the individual organism; according to Mead, this activity induces other exemplars of the species to respond to the child's action with socially meaningful actions. In selecting socially meaningful reactions to the actions or vocal gestures of the egocentric child, the care-givers provide compensating communicative contributions which Mead calls "sympathetic response" (see 1934, p.

364ff.). The adults regard the socially relatively unstructured sensorimotor activities of the infant as if they were
communicative actions or gestures, and this fiction enables
them to respond to the infant with a socially meaningful action.

Thus the baby's "hunger cry" will be not considered an
unpleasant noise, and the response to it will not be a shut-
ting of the door so as not to hear it. It will be regarded
as a gesture the child has made to communicate to the caregiver that he is hungry, and it is this fiction which will
make the caregiver prepare food for the baby and feed him.
This communication process ensures gratification of the infant's elementary needs by the caregiver.[17] In addition,
it provides the infant with the experience of relatively
highly structured communicative action-sequences (cf. Lock
1978a, Clark 1978) in which the meaning of his actions is
already a social reality (insofar as the caregivers react to
them in a socially meaningful way) long before these meanings become a psychological reality for the child.


                        ACTION AND GESTURE

With his limited cognitive endowment, the egocentric
child cannot yet even adequately assimilate the relatively
complex social interaction structures described above. According to Piaget's analysis, these first social responses
of the caregiver are experienced by the child as an extension of the tactile, postural, kinesthetic, etc. impressions
which are a part of the child's own assimilatory activities.
These various bodily sensations fuse with the perception of
the response or reaction of the interaction partner into an
indissoluble union in which the child's own contribution to
the social interaction has not as yet been distinguished
from that of the interaction partner. Piaget has designated
this early interaction scheme[18] "egocentric", and characterizes it with reference to the child's physical experience
--as a union of the "feeling of efficacy" and "phenomenal-
ism" (CR, p. 228).

In the third stage, little by little, the two elements
of this union begin to be detached from one another because,
with the completion of eye-and-hand control, the young child
can control the movements of his own hands and thus gain a
clearer perception of his contact with "phenomena". It is

at this stage that Piaget speaks of the child's "progressive
awareness of his power over the particular objects that are
his hands" and of a "sort of reflection on the purposefulness of those movements" (CR, p. 232). Because in the first
stages the child has made multiple social interaction experiences which he wants to reproduce (reproductive assimilation), and because he is now endowed with visual control of
his hands, he can better than heretofore perceive the effects on the external environment; Piaget, referring to the
third stage, speaks of the "purposefulness" of the child's
movements. Such intentional movements, however, are not as
yet gestures. The caregiver may interpret them as gestures
(that is, as actions intended to have an effect on the actions of another person), but for the child himself such intentional movements have no such communicative meaning.

   In the third stage, the child still cannot distinguish
between the two sensorimotor action sources in the caregiver
-child dyad, and activities remain socially relatively unstructured. Piaget terms these activities "magico-phenomenalistic procedures" such "as if the gesture[19] itself were
charged with all the necessary efficacy" and could produce
the "phenomena" (CR, p. 235). Two examples from his observations provide very good illustrations of this: Piaget reports (CR, p. 246, obs. 135) that he is standing facing
Laurent, and shakes a rattle hanging in front of him. He
then withdraws his hand, but leaves it in the air close
enough to Laurent's own hand and to the rattle so that Laurent can push his father's hand toward the rattle. Laurent
(0; 7 (22)) finds this an interesting effect and, employs
"magico-phenomenalistic procedures" to reproduce the movement and noise of the rattle--that is, he arches himself,
shakes his head, his hands, etc. while looking at the
rattle, without paying any attention to his father's hand.
Only when the desired effect is not reproduced does Laurent
finally start examining his father's hand. Piaget points
out that Laurent does not use magico-phenomenalistic procedures to act on his father's hand. Only when he again looks
at the rattle does he repeat this ineffective behavior pattern. In another situation, however (CR, p. 244, obs. 133),
for instance when his father snaps his middle finger against
the base of his thumb (i.e., without the presence of an object which produces the interesting sights), the child tries
through magico-phenomenalistic procedures (i.e., by striking
and shaking his father's hand) to also act on the hand--and

this in such a way as if it were not a source of action independent of his own hand.

Only in the fourth stage do these patterns of behavior
start changing; and this, beginning with the fifth stage,
leads to the emergence of gestures. In the fourth stage,
the child no longer tries to get another person to do something by his own global body movements. He touches or
grasps only the hand or arm (of the social interaction partner) which has produced the interesting phenomenon, and he
obviously inhibits his own activities as if he perceived the
caregiver's hands or arms as relatively autonomous sources
of action: Piaget reports (CR, p. 259f., obs. 142) that Jacqueline (0;8(19)) watches her father with interest while he
is alternately spreading his index finger and thumb apart
and bringing them together again. When he stops this activity, she lightly pushes either the finger or the thumb. Piaget interprets this as a call to make him continue. In
another interaction scene (CR, p. 260, obs. 143), Piaget
tickles his other daughter's belly and then places his hand
on the edge of the bassinet. Lucienne (0;11(7)) seems to
have enjoyed being tickled and she waves her own hand, then
touches her father's hand. When he intentionally does not
react immediately, she tries to push his hand; finally she
grasps it and brings it to her belly.

In the data cited by Piaget, two changes become evident: the child begins to regard the hand of caregiver as an
"independent source of action" (CR, p. 260), and he begins
to change the patterns of behavior by which he seeks to produce actions on the part of the caregiver. Nonetheless, I
still do not think it makes sense to call the changed behavior patterns of the child at this moment "gestures"--this
because the child still finds it necessary to touch the hand
or the arm of the interaction partner, and this, according
to Piaget, indicates that he still perceives the other's activity as depending on his own activity.

Only in the fifth stage of sensorimotor development, as
described by Piaget, do interaction sequences occur in which
the child uses gestures. Piaget reports, among other examples (all those in this paragraph are from CR, p. 275, obs.
152), that he is blowing his daughter's hair and she wants
the game to continue. In contrast to the fourth stage, however, to achieve this she no longer (as formerly) tries to
push her father's arms or lips; she now shows her interest

in continuing the game by placing herself in position, head
tilted, as she was when the game began. In another social
interaction situation Jacqueline (1;3(30)) tries to open a
box but does not manage it. She then holds the box out to
her mother, who pretends not to notice. Then she transfers
the box to her left hand, with her free hand grasps her
mother's hand and puts the box in it; her mother interprets
both gestures as an appeal to her to open the box. In addition Piaget reports on other interaction scenes in which,
unable to reach an object, Jacqueline calls, cries, points
to the object with her finger, and the adults view such actions as obviously intended to get them to bring the child
the respective object or help her carry out some project.
The gestures used by the child in these examples are calls,
cries, pointing to something, giving someone else an object
or putting it in their hand, and placing the body in a certain position.

I myself shall designate as "gestures" the intentional
sensorimotor actions during which the actor 1) is aware that
they are directed at an external and autonomous sensorimotor
center of actions and 2) perceives the sensorimotor meaning
his action has for the recipient. In the fifth stage of
sensorimotor development, the child is conscious of these
two aspects. In none of the situations described above is
the child attempting to use either global body movements
(stage three) or touching the arm, hand, or mouth, etc. of
the caregiver (stage four) to induce the anticipated reaction as if it were in any way dependent on the child's own
activities; instead she behaves as if she regards the caregiver as a completely autonomous and external source of action with whom no physical contact is necessary to induce
him to undertake some activity (see CR, ch. III, sec. 4).

The indicator for the second characteristic aspect of
gestures, anticipation of the sensorimotor reaction of the
interaction partner, is the selection of relatively differentiated actions. In the first illustration, the child--as
a reminder--places herself in the original body position
with which the game started spontaneously, and uses this as
a gesture. In the other social interaction situations cited,
even when her efforts fail she does not take recourse to undifferentiated "magico-phenomenalistic" procedures; rather,
she makes use of vocal, expressive, and manual sensorimotor
activities to make the caregiver do what she wants him to
do.

Such indicators are certainly not strong enough evidence for the hypothesis that the child has selected these gestures on the basis of knowing the sensorimotor reaction of the caregiver.[20] We shall perhaps be on safer ground if we confine ourselves to the gesture of "giving", implied in the second interaction example (Jacqueline wants her mother to open a box which she places in her hand). According to Clark's (1978) data, the infant is not initially capable of structuring the movements of her hands so as to be able to "give" an object to the mother. As Clark describes the situation, the child sees the object in her hand and close to it her mother's hand held palm upwards; but as she does not place the object into the mother's hand, the mother must herself take hold of the object. Clark assumes that the child learns to drop things or propel things on purpose, and that the mother—anticipating this—catches such objects in her open palm; and this would seem well suited to the development of a communication structure where the child places an object into the mother's upturned hand. At any rate, according to Clark, it is only toward the age of one year that the infant can structure his own hand movements in such a way that it is possible to speak of "giving" in the communicative sense of the term. For it is only then that the child holds the object and waits for the mother to cup her palm before he places the object into it.

We shall find that Piaget's second example allows us to draw similar conclusions: the child fails in her attempt to open a box and wants to "give" the box to her mother, who, however, pretends not to notice this. If the hypothesis that, with her gesture of "giving", the child has associated the anticipation that her mother would open her hand and offer her upturned palm was valid, then in this example the child would have to perceive that the anticipated reaction and the actual reaction differ—and react accordingingly herself. Opening her mother's hand and putting the box into it is the reaction which I would take as an indicator supporting the assumption that with this gesture of "giving", the child actually has anticipated the mother's reaction. And without this assumption, the child's reaction, would be understandable.

Whether the child also expected the mother to open the box cannot be decided by reference to the empirical indicators of this observation. We might discover what the child's expectation was if we could observe her reaction to an intentionally. "false" reaction on her mother's part. The same qualification applies, of course, also to the third observation—that is, to the situation when the child uses calls, cries, pointing to something to induce specific reactions on the part of the caregiver. Pointing[21], in particular, seems to be a gesture concomitant to the manifest expectation that the caregiver will perform a certain action. However, these intuitions can be backed up by some of Piaget's theoretical considerations.

Pointing is a gesture presumably originating in prehension. The child no longer attempts to grasp the object, but uses the original grasping movement in a different way. It is a movement directed at an object but aiming at an external and independent center of action; pointing at the object is intended to make this external source of action perform in a certain way. Underlying this expectation is certainly the ongoing process of social interaction during which the caregiver, having seen the child's futile grasping efforts, hands him the objects he cannot reach. Knowing the caregiver's reaction enables the child to select from his behavioral repertoire the prehensile action transformed into pointing. To achieve this, the egocentric child might turn to undifferentiated magico-phenomenalistic procedures; the no-longer egocentric child turns to the caregiver for assistance. In so doing, the child not only regards the caregiver as an external and independent center of actions who can satisfy his wish; he has also gained a realistic assessment of his own capabilities and has become able to reflect on his limited capabilities.

The point of reference of such reflection is the caregiver. It is through the actions of this particular other that the child can learn to regard himself as a comparatively less competent sensorimotor source of action (see above, Chapter I). Such reflection is but a special case of the "reversibility" that, according to Piaget, emerges in the fifth stage and which enables the child to gain the insight that he is a source of action located in space among other sources of action (or, as Piaget has termed it, "sources of causality"), so that he "feels his dependence on the external world as well as his power over it" (CR, p.290). According to Piaget, (CR, ch. III, sec. 4), reversibility not only enables the child to order visible displacements of inanimate and animate objects in space and time but also to order various actions of a "causal", in relation to a "social", interaction sequence in space and time—and in this way, starting from the specific reaction of an independent center, to identify the appropriate own action needed to elicit this reaction.

In short, my hypothesis is that Piaget's analysis of sensorimotor intelligence in the fifth stage supports the assumption that the child's actions are gestures. It should be noted, however, that these gestures—at least according to Piaget—do not as yet imply any mental images and are not, therefore, symbolic gestures. The child still has no mental representation of the sequence of action and reaction; he is still tied down by immediate perception, and anticipation proceeds in terms of immediately preceding experience.[22]

## THE TRANSFORMATION OF ACTION INTO GESTURE

It was possible to reconstruct the emergence of gestures as outlined above with reference to the chapter on the genesis of the causality scheme (CR, ch.III) because in this chapter Piaget also gives data concerning the social interaction between caregiver and the young child. I have already pointed out that the impression one gets from reading this analysis is that Piaget regards the development of the social interaction scheme, in relation to gestures, to be a derivative of sensorimotor intelligence which emerges in manipulative or instrumental interaction of a solitary child with inanimate objects. In contrast to this monological conception, I shall attempt to sketch out a dialogical model of the path from action to gesture.

The distinctions to be made between the child's gestures and intentional sensorimotor actions are: first, that the gesture is a conscious act oriented toward a social interaction partner; and second, that it involves a newly emerged reflective control of his own actions. This new reflective control enables the child, starting from certain anticipated reactions of the caregiver, to select from his repertoire of sensorimotor behavior those actions which will produce the anticipated reactions. The ontogeny of these two characteristics of the gesture thus have to be explained.

Following Piaget's argument, once eye-hand coordination has become complete in the third stage, the child is able to visually follow the movements of the caregiver's hands as well as those of his own hands. But what the child cannot do is to differentiate these two pairs of hands within his perceptual field and to attribute them to different centers of action. What the child experiences in social interaction is solely a global relation of action and reaction; he does not as yet recognize that the source of the reaction to his own action is an external and independent center of action.

Now I have argued that in this situation the child is confronted by an identity problem--an identity problem he can resolve only by coming to recognize his own actions as similar to, but at the same time distinct from, the actions of the caregiver. Such differentiation of sensorimotor actions in the process of social interaction is made possible, for instance, by the experience that the movements of the hands of another person are "deficient" in comparison to the activities of his own hands; while the activity of his own hands provides experience of the unity of tactile, motoric, and visual sensations, the experience of the movements of the other's hands is confined to the visual domain. The child will inevitably undergo a second experience: he is visually perceiving hand-movements which--in contrast to his own hand-movements--he cannot (however much he tries) direct or control, because these are the caregiver's hands.

These experiences compel the child to differentiate the hands within his visual field into "own" hands and "not-own" hands and, starting from this, to distinguish between these two centers of action in this primary communication process. This leads to the development of the sensorimotor scheme of self and of other person, which is the elementary precondition for the ontogeny of gestures. True, gestures are also sensorimotor actions; but beyond that, they possess a communicative dimension: they imply the intention to act on an external and independent center of actions to make it perform in a certain way. The ontogenetically preceding sensorimotor actions lack this special intent; they were produced to achieve a certain event or "phenomenon".

What still needs to be explained is how reflective control of the child's own actions, as shown in Piaget's investigations, actually comes into being--especially as it is also constitutive for the development of the gesture. To lay ground for this argument, I shall reformulate the problem just discussed (the question of the differentiation between two sensorimotor centers of action) in terminology more appropriate for analysis of social interaction; to this end, I shall first make some conceptual distinctions.

I draw upon G.H. Mead's concepts of "taking the perspective of the other" in relation to "role taking", as well as Piaget's concept of the "sensorimotor roles" in relation to "taking the sensorimotor perspective of the other".[23] In the specific sense that "sensorimotor role" or "sensorimotor perspective" is used here, it refers to the sensorimotor aspects of actions in social interaction--demarcating them, for instance, from the mental aspects in social action. While the main question relative to the developmentally more advanced "role taking" processes in the transition to the concrete operational phase is "What does the other think?" (see Shantz 1975), the analogous question for the sensorimotor phase is "What is the other doing?". Therefore, the contents of sensorimotor "role taking" are sensorimotor and actions such as vocal acts, and activities of the hands, arms, and legs, and other parts of the body.

The elementary visual control of the hand-movements of the caregiver by the egocentric child in the third stage can, in terms of this definition of the "sensorimotor role", be described as follows: The child can perceive the sensorimotor role (or aspects of that role) of the caregiver. The imitations in the following observation are accordingly based on this perception of the sensorimotor role of the particular other. At the age of 5 months and 5 days, Jacqueline attentively watches her father several times bring his hands close to each other and then separate them again. According to Piaget's report, she then imitates these movements three times (PDI, p. 15f., obs. 8). But the child is still egocentric and, according to Piaget, she hence cannot yet differentiate these sensorimotor roles (i.e., detach her own sensorimotor role from that of the caregiver). There is an identity of similarity, of which the child is not as yet conscious, between the child's sensorimotor role and that of the caregiver (cf., Chapter I); the actions of the two interaction partners appear so similar to her that she cannot distinguish them from each other. In this observation, the child thus oscillates between the two sensorimotor roles.

I referred earlier to this oscillation between roles as the identity problem the child faces in this situation. I shall try to describe the way the child resolves this identity problem (see CR, Chapter I) as a process in which the child comes to differentiate the sensorimotor roles referred to. Such oscillation cannot last indefinitely: the hands and the hand-movements of the caregiver and the child are different, and as time goes on, the child who is moving to and fro between the two roles will come to perceive these differences. This is described by Observation 137 from Piaget's data already referred to in Chapter I (CR, p. 250):

Observation 137. At 0;7(27) Jacqueline is seated in front of a big quilt. Her mother strikes the quilt with her hand and Jacqueline immediately imitates her amid peals of laughter. For a brief moment both of them strike together and this unison seems to delight Jacqueline. Jacqueline stops striking the quilt to look at her mother's hand. The hand continues to strike for a few more seconds and then stops. Then Jacqueline, while staring at her mother's hand (and without looking at her own once during the observation), begins to strike the quilt, at first gently and then harder and harder, exactly as if she were trying to force her mother to recommence. Her mother yields, Jacqueline stops (which shows that the imitation was entirely inherent in causal procedure), then when her mother's hand is again motionless, Jacqueline resumes tapping while looking at it.

In this interaction scene, the egocentric child visually controls both her mother's sensorimotor actions and her own; and she thus perceives the sensorimotor role of both interaction partners—without, however, yet distinguishing them from one another. In the sequence described, the child first perceives the sensorimotor role of her mother and subsequently can also look at her own hands striking the quilt. Regarding her mother, Jacqueline has only a visual image of moving hands; a moment later, she is again looking at hand movements (her own), but now further sensory and motor impressions are added to the visual experience. The egocentric child experiences similar visual images of moving but dissimilar sensorimotor global impressions. In the sequence at the beginning of the observation cited, the child perceives the sensorimotor role of her mother and, by virtue of this, can begin to understand her own actions as richer sensorimotor experiences. These are richer because they involve not merely visual but also other sensorimotor sensations. However, the young child achieves this understanding only through a comparison between the actions of the two sensorimotor centers of action. This comparison, in turn, is based on the child's perception of the sensorimotor role of the caregiver; this enables the child to objectify in a totally new way her actions, which thus far were not yet her "own" actions.

By this visual experience of the sensorimotor role of the mother, the child will discover that only these richer actions can be acted on directly or can be directly controlled--while this is not the case for the actions which can be experienced only visually. Similarly, the child's awareness of the "permanence" of her actions, discussed in Chapter I, is made possible only after she has perceived the sensorimotor role of the interaction partner. By perceiving this role, the child can understand that the "richer" and "controllable" actions are reproducible and therefore can be rediscovered in time and space; they are "permanent" while the corresponding actions of the caregiver are but passing visual episodes which--however great her assimilatory efforts--the child cannot reproduce and is thus unable to find again in time and space.[24] It is this awareness, reached by the child, of the differences between sensorimotor actions which gives rise to the distinction between ego and alter and at the same time brings about the development of conscious sensorimotor role-taking.

The process of role-taking initially is confined to activities of the hands, but in the fourth and fifth stages it is extended to other parts of the body (see Piaget's discussion of imitation, PDI). These new insights obviously change the child's understanding of the social interactions brought about by the compensating communicative contributions of the caregivers. In direct contrast to the egocentric child, this new awareness enables the child to differentiate the two sensorimotor roles in this early dialogue; and this means that the child now can attribute the sensorimotor actions occuring in the early interaction sequences to two different persons. With this insight, the child is on the brink of the reversible ordering of action and reaction, in relation to reflective control of actions, such as is constitutive for the emergence of the gesture.

Because in a given ongoing social interaction with the caregiver the child experiences the sensorimotor role of both interaction partners, he can--whenever he wishes--make the interaction partner repeat a given reaction and he can, by taking the other's sensorimotor role, "rediscover" (i.e., reproduce) his own action in the given interaction sequence. Reproduction of actions achieved in this way is a "gesture" which, depending on the child's previous interaction experiences, will be associated with more-or-less realistic expectations that now the other person will also reproduce the anticipated sensorimotor role.

To explicate how the child achieves awareness of the meaning of his gestures, Piaget has introduced his monological theory of sensorimotor intelligence. But within the frame of the "dialogical" model outlined here, following G.H. Mead (1934) [25], this is a research strategy we would not need. It would suffice to incorporate into the dialogical model Piaget's general biological presuppositions (1971) --that is, the innate social abilities of the neonate to participate in social interactions (see, among others, Trevarthen 1977b; Trevarthen, Hubley, & Sheeran, 1975; Stern 1977; Schaffer 1977; and, programmatically, Richards 1978). Given the assumption of the species-specific organic endowment of the neonate, one would then conceptualize the ontogeny of the understanding of the meaning of gestures as resulting from the process of differentiation between self and other in social interactions--that is, of social interactions in which the meaning of the child's actions was already a "social" reality (Mead 1934) or "instrumental" reality (Vygotsky 1966) for the caregivers long before it becomes a psychological reality for the child.


ACKNOWLEDGMENTS

I would like to thank John Broughton for having assisted me in editing the final version of this paper, Susanne Libich (Max-Planck Institute) for her translation, and Gottfried Seebass for his critical comments.

## ENDNOTES

1.  References without author's name always refer to Piaget.

2.  Bower 1974, Trevarthen 1974, and Trevarthen & Hubley 1978 call into question the existence of sensorimotor egocentrism.

3.  Stern (1977) has expressed doubts that the elimination of egocentrism has been adequately explained. See also Hamlyn 1978.

4.  Lewis & Brooks (1975) distinguish between the development of an "existential self" and a "categorical self".

5.  Bower, 1974; Bower, Broughton, & Moore 1970a, 1970b; Bruner & Koslowski 1972; and Bruner 1973 report on eye-hand coordination in much younger children than those observed in Piaget's study. They do not however call into question Piaget's data, as they make a distinction between early eye-hand coordination and Piaget's "visually guided reaching". Field 1977; Ruff & Halton 1978 report failures in replicating some findings of Bower et al.

6.  Replication studies on object permanence are in this respect even more one-sided, as they have simply ignored the question of the constitution of the sensorimotor self. I also do not intend to take up questions raised by the research literature in which the development of self is examined primarily in terms of "self-recognition" measured by observing the child's reactions when looking into a mirror (see Amsterdam 1972, Bertenthal & Fischer 1978, Dixon 1957, Zazzo 1975), because this approach also does not elucidate the process of the development of self in social interaction.

7.  True, Piaget also describes imitations of the movements of inanimate objects (i.e., the swinging of a rattle), but his data concerning imitation undeniably indicate that it is primarily in his contacts with the caregiver that the child discovers something he has in common with another person, rather than in connection with inanimate objects brought into movement as part of the game.

8.  The age of the child: years;months(days).

9.  Piaget's observations on sensorimotor development were made with his own children.

10. For an analysis of the relation between the similarity of stimuli and their comparison by young children, see Ruff 1975.

11. The investigation by Jackson, Campos, & Fischer (1978) expresses doubts about Piaget's hypothesis and the research literature in support of it.

12. The importance of imitation for ego/alter differentiation is mentioned by Uzgiris (1973).

13. In contrast to Piaget, Meltzoff & Moore (1977) report that even neonates are capable of this stage IV imitation. Moore & Meltzoff (1978), in discussing their results, delimit the imitations observed by Piaget in stage IV (from imitations by neonates) by affirming that children in stage IV are conscious of the similarity of the two actions.

14. It remains to be pointed out that this identity problem is experienced not only in connection with prehension but also with other coordinated assimilation schemes. For example, the egocentric child will face the problem of differentiating his own vocal activities from those of the caregiver. In the case of the "thalidomide" children (Decarie 1969) or in blind children, it may be assumed that vocal activities may play a relatively more important role in the elimination of egocentrism than it would in a population of non-disabled children.

15. For a critical view of this thesis, and strong arguments for the existence of innate communicative abilities, see Trevarthen 1977b, Trevarthen et al. 1975. (Also see Schaffer 1974 1977; Stern 1977.) Trevarthen is doubtful about this scheme having been acquired through learning processes in social interaction. The qualification should, however, be made that these persuasive maturationist hypotheses were derived from data on the child's expressive behavior, and do not relate

to all forms of social interaction--in particular,    not
to those discussed in what follows.

16.    Bates, Camaioni, & Volterra (1975) have surveyed the
       preverbal expressions of the young child   in   terms   of
       speech-act   theory.    Regarding a critique thereof, see
       Dore 1978.

17.    Regarding the interpretation   of   difficulties   experi-
       enced by caregivers see Escalona 1953, Ryan 1974, Gold-
       berg 1977.

18.    As the present study views   the Piagetian theses on the
       structural isomorphy of cognitive and social schemes as
       unproblematic,   I   would   hold   that--in analogy to the
       scheme of causality--it is plausible   to   speak   of   an
       "interaction scheme".   For criticism of the isomorphism
       thesis,   see   Trevarthen   1974, and Trevarthen & Hubley
       1978.   In contrast to Piaget (CR) and Spitz (1976)--and
       in accord with   findings   of   Brazelton   et   al. (1974,
       1975)--they   report   that against the background of the
       expressive communication with   caregivers   investigated
       by them,   infants can distinguish objects from persons
       even prior to the end of the first three weeks of life.

19.    In contrast   to Piaget,   I use the term "gesture"   more
       restrictively.   In the given case, I would speak of an
       "action" and not of a "gesture".   Regarding this   limi-
       tation of the term, see also Clark 1978.

20.    Clark   (1978)   speaks of similar difficulties in trying
       to find empirical indicators for   this   transition;   in
       the   interaction   exemplified   by   Lock (1978b), these
       indicators are not made clear.

21.    C.M. Murphy (1978), who   has studied the development of
       "pointing" when the child is looking at a picture-book,
       doubts that the   indicator   "visual   contact   with   the
       caregiver"   used   by Bates, Camaioni, & Volterra (1975)
       is meaningful for the definition of "gesture".

22.    Regarding the difficulties encountered in   demarcating
       the two stages of gestures, see Piaget, CR:293ff.

23.    The   use   of the concept   of "role" or "perspective" in
       this   context   may   be surprising, but the term is used

also by G.H. Mead in relation to such elementary sensorimotor actions. Use of this term makes sense to me
because the observations analyzed here involve ontogenetically early stages of social interaction.

24.  This comparison of sensorimotor action experiences may,
     because the child is oscillating between the two senso-
     rimotor roles, also proceed inversely in that the
     child's sensorimotor perspective constitutes the start-
     ing point. In that event the child will, starting from
     his own sensorimotor action experiences, regard the ac-
     tions of the caregiver as "deficient", non-control-
     lable, and "non-permanent".

25.  In this connection, attention should be paid also to
     the hypotheses formulated by L.S. Vygotsky (1966), as
     well as several essays in the reader edited by Lock,
     Action, Gesture and Symbol (1978a).

## REFERENCES

Amsterdam, B.  1972.  "Mirror Self-Image Reactions Before
          Age Two".  Developmental Psychobiology, B, 4:
          297-30.
Anderson, G.C.  1978.  "Der Ursprung der Intelligenz und die
          Sensomotorische Entwicklung des Kindes".  In Pia-
          get und die Folgen (ed. by G. Steiner, Die Psy-
          chologie des 20 Jahrhunderts, Bd. VII), pp. 94-
          120.
Bates, E., L. Camaioni, & V. Volterra.  1975.  "The Acquisi-
          tion of Performatives Prior to Speech".  Merrill-
          Palmer Quarterly, XXI:205-26.
Beilin, H.  1971.  "The Development of Physical Concepts".
          In Cognitive Development and Epistemology (ed. by
          T. Mischel, New York: Academic Press), pp. 86-
          119.
Bertenthal, B.I. & W. Fischer.  1978.  "Development of
          Self-Recognition in the Infant".  Developmental
          Psychology, XIV, 1:44-50.
Bower, T.G.R.  1974.  Development in Infancy.  San Francis-
          co: Freeman.
Bower, T.G.R., J.M. Broughton, & M.K. Moore.  1970a.  "The
          Coordination of Visual and Tactual Input in

Infants". Perception and Psychophysics, VIII, 1:51-3.

Bower, T.G.R., J.M. Broughton, & M.K. Moore. 1970b. "Demonstration of Intention in the Reaching Behavior of Neonate Humans". Nature, CCXXVIII: 679-81.

Brazelton, T.B., B. Koslowski, & M. Main. 1974. "The Origins of Reciprocity: The Early Mother-Infant Interaction". In The Effect of the Infant on Its Caregiver (ed. by M. Lewis & L.A. Rosenblum, New York: Wiley), pp. 49-76.

Brazelton, T.B., E. Troniok, L. Adamson, H. Als, & S. Weise. 1975. "Early Mother-Infant Reciprocity". In Parent-Infant Interaction (Ciba Foundation Symposium 33, Amsterdam), pp. 137-54.

Bruner, J.S. 1973. "Organization of Early Skilled Action". Child Development, LXIV:1-11.

Bruner, J.S. & B. Koslowski. 1972. "Visually Preadapted Constituents of Manipulatory Action". Perception, 1:3-14.

Clark, R.A. 1978. "The Transition from Action to Gesture". In Lock 1978a, pp. 231-57.

Decarie, T.G. 1969. "A Study of the Mental and Emotional Development of the Thalidomide Child". In Determinants of Infant Behaviour, IV (ed. by B.F.Foss, London: Methuen), pp. 167-87.

Dixon, J.C. 1957. "Development of Self Recognition". Journal of Genetic Psychology, XCI:251-56.

Dodwell, P.C., D. Muir, & D. DiFranceo. 1976. "Responses of Infants to Visually Presented Object". Science, CXCIV:209-11.

Dore, J. 1978. "Conditions for the Acquisition of Speech Acts". In The Social Context of Language (ed. by J. Markova, New York: Wiley), pp. 87-111.

Esacalona, S. 1953. "Emotional Development in the First Year of Life". In Problems of Infancy and Childhood: Transactions of the Sixth Conference, March 17-18, New York, N.Y.. (ed. by M.J.E. Senn, New York: Josiah Macy), pp. 11-92.

Field, J. 1977. "Coordination of Vision and Prehension in Young Infants". Child Development, XLVIII:97-103.

Goldberg, S. 1977. "Social Competence in Infancy: A Model of Parent-Infant Interaction". Merrill-Palmer Quarterly, XXIII, 3:163-77.

Habermas, J. (1972). "Notizen zum Begriff der Rollenkompetenz". In Kultur und Kritik: Verstreute Auf-

sätze (ed. by J. Habermas, Frankfurt: Suhrkampf),
pp. 195–231.

Hamlyn, D.W.  1978.  Experience and the Growth of Understanding.  London: Routledge and Kegan Paul.

Jackson, E., J.J. Campos, & K.W. Fischer.  1978.  "The Ques-
tion of Decalage Between Object Permanence and
Person Permanence".  Developmental Psychology
XIV, 1:1–10.

Lempers, J.D., E.R. Flavell, & J.H. Flavell.  "The Development in Very Young Children of Tactic Knowledge
Concerning Visual Perception".  Genetic Psychology Monographs, XCV:3–53.

Lewis, M. & J. Brooks.  1975.  "Infant's Social Perception:
A Constructivistic View".  In Infant Perception:
From Sensation to Cognition: Perception of Space,
Speech and Sound, Vol. II (ed. by L.B. Cohen &
P. Salapatek, New York: Academic Press),
pp. 101–48.

Lock, A. (ed.).  1978a.  Action, Gesture and Symbol: The
Emergence of Language.  New York:  Academic
Press.

Lock, A. (ed.).  1978b.  "The Emergence of Language".  In
Lock 1978a, pp. 3–18.

Mead, G.H.  1934.  Mind, Self and Society.  Chicago: University of Chicago Press.

Meltzoff, A.N. & M.K. Moore.  1977.  "Imitation of Facial
and Manual Gestures by Human Neonates".  Science,
198, 7 Oct.:75–8.

Moore, M.K. & A.N. Meltzoff.  1978.  "Object permanence,
imitation, and language development in Infancy:
Toward a Neo-Piagetian Perspective on Communicative and Cognitive Development".  In Communicative and Cognitive Abilities: Early Behavioral
Assessment (ed. by F.D. Minifie & L.L. Lloyd,
Baltimore: University Park Press), pp. 151–84.

Murphy, C.M.  1978.  "Pointing in the Context of a Shared
Activity".  Child Development, XLIX:371–80.

Newson, J.  1977.  "An Intersubjective Approach to the
Systematic Description of Mother–Infant Interaction".  In Studies in Mother–Infant Interaction:
Proceedings of the Loch Lomond Symposium Ross
Priory, University of Strathclyde, September 1975
(ed. by H.R. Schaffer, New York: Academic Press),
pp. 47–61.

Newson, J.  1978.  "Dialogue and Development".  In Lock
1978a pp.31–42.

Newson, J. & J. Shotter. 1974. "How Babies Communicate". New Society, August 8:354-57.

Newson, J. & Newson. 1975. "Intersubjectivity and the Transmission of Culture: On the Social Origins of Symbolic Functioning". Bulletin of the British Psychological Society, XXVIII:437-46.

Oevermann, U. 1974. "Zur Programmatik einer Theorie der Bildungsprozesse". MS., Max-Planck-Institut für Bildungsforschung, Berlin.

Oevermann, V. 1976. "Programmatische Überlegungen zu einer Theorie der Bildungsprozesse und zur Strategie der Sozialisationsforschung". In Sozialisation und Lebenslauf: Empirie und Methodik Sozialwissenschaftlicher Persönlichkeitsforschung (ed. by K. Hurrelman, Reinbek bei Hamburg: Rororo Studium), pp. 36-67.

Parsons, T. & R.R. Bales. 1968. Family Socialization and Interaction Process. London: Routledge and Kegan Paul.

Parton, D.A. 1976. "Learning to Imitate in Infancy". Child Development, XLVII:14-31.

Piaget, J. 1927. La Causalité physique chez l'enfant. Paris: Alcan.

Piaget, J. 1928. Judgment and Reasoning in the Child. New York: Harcourt and Brace.

Piaget, J. 1947. La Psychologie de l'Intelligence. Paris: Librairie Armand Collin.

Piaget, J. 1959. The Language and Thought of the Child. London: Routledge and Kegan Paul.

Piaget, J. 1965. The Origins of Intelligence in Children [cited in present paper as OI]. New York: International Universities Press.

Piaget, J. 1967. Etudes Sociologiques. Genèva: Librairie Droz.

Piaget, J. 1968a. Six Psychological Studies. With an introduction, notes, and glossary by David Elkind. New York: Vintage Books.

Piaget, J. 1968b. The Construction of Reality in the Child [cited in present paper as CR]. London: Routledge and Kegan Paul.

Piaget, J. 1970a. "Piaget's Theory". In Carmichael's Manual of Child Psychology, Vol. 1 (ed. by P.H. Mussen, New York: Wiley), 703-32.

Piaget, J. 1970b. "Nécessité et signification des recherches comparatives en psychologie géné-

tiques".  In Psychologie et Epistémologie.
    Paris: Gonthier.
Piaget, J.   1971.  Biology and Knowledge (an essay on the
    relations between organic regulations and cogni-
    tive processes).  Chicago: University of Chicago
    Press.
Piaget, J.   1972.  Play, Dreams and Imitation in Childhood
    [cited in present paper as PDI].  London:
    Routledge and Kegan.
Piaget, J.   1977.  The Moral Judgment of the Child.  Har-
    mondsworth:Penguin.
Richards, M.P.M.  1978.  "The Biological and the Social".
    In Lock 1978a, pp. 21-30.
Ruff, H.A.  1975.  "The Function of Shifting Fixations in
    the Visual Perception of Infants".  Child Devel-
    opment, XLVI:857-65.
Ruff, H.A. & Halton, A.  1978.  "Is There Directed Reacting
    in the Human Neonate?" Developmental Psychology,
    XIV, 4:425-26.
Ryan, J.  1974.  "Early Language Development:  Towards a
    Communicational Analysis".  In The Integration of
    a Child into a Social World (ed. by M.P.M. Rich-
    ards, New York: Cambridge University Press),
    pp. 185-213.
Schaffer, R.  "Behavioral Synchrony in Infancy".  New
    Scientist, 4 April:16-18.
Schaffer, R.  1977.  Mothering.  London: Fontana.
Serafica, F.C. & I.C. Uzgiris.  1971.  "Infant-Mother Rela-
    tionship and Object Concept".  Paper presented at
    the annual meeting of the American Psychological
    Association, Washington, D.C.
Shantz, C.U.  1975.  "The Development of Social Cognition".
    In Review of Child Development Research (ed. by
    E.M. Hetherington, Chicago: University of Chicago
    Press).
Shotter, J.  "The Development of Personal Powers".  In The
    Integration of a Child into a Social World
    (ed. by M.P.M. Richards, New York: Cambridge
    University Press), pp. 215-44.
Shotter, J.  1978.  "The Cultural Context of Communication
    Studies: Theoretical and Methodological Issues".
    In Lock 1978a, pp. 43-78.
Snow, C.E.  1977.  "The Development of Conversation Between
    Mother and Babies".  Journal of Child Language,
    IV, 1:1-22.

Spitz, R.A.  1976.  Vom Dialog: Studien über den Ursprung
        der Menschlichen Kommunikation und ihrer Rolle in
        der Persönlikeitsbildung.  Stuttgart: Klett.
Stern, D.  1977.  The First Relationship: Mother and Infant.
        Cambridge: Harvard University Press.
Swanson, G.E.  1974.  "Family Structure and the Reflective
        Intelligence of Children".  Sociometry, XXXVII,
        4:459-90.
Sylvester-Bradley, B. & C. Trevarthen.  1978.  "Baby Talk as
        an Adaption to the Infant's Communication".  In
        The Development of Communication.  (ed. by
        N. Paterson & C. Snow, New York: Wiley),
        pp. 75-92.
Trevarthen, C.  1974.  "Conversations With a Two-Month-Old".
        New Scientist, LXII, 896:230-35.
Trevarthen, C.  1977a.  "Descriptive Analyses of Infant
        Communicative Behaviour".  In Studies in Mother-
        Infant Interaction: Proceedings of the Loch
        Lomond Symposium Ross Priory, University of
        Strathclyde September 1975 (ed. by H.R. Schaffer,
        New York: Academic Press), pp. 227-70.
Trevarthen, C.  1977b.  "Instincts for Human Understanding
        and for Cultural Cooperation: Their Development
        in Infancy".  M.S., Werner-Reimer-Stiftung Sym-
        posium on Human Ethology, Bad Homburg, 25-29
        Oct. 1977.
Trevarthen, C. & P. Hubley.  1978.  "Secondary Intersubjec-
        tivity: Confidence, Confiding and Acts of Meaning
        in the First Year".  In Lock 1978a, pp. 183-229.
Trevarthen, C., P. Hubley, & L. Sheeran.  1975.  "Les
        Activités Innés du Nourrisson".  La Recherche,
        VI, 56:448-58.
Uzgiris, I.C.  1973.  "Patterns of Vocal and Gestural Imi-
        tation in Infants".  In The Competent Infant (ed.
        by L.J. Stone, T. Smith, and L.B  Murphy, New
        York: Basic Books), pp. 599-604.
Vygotsky, L.S.  1966.  "Development of the Higher Mental
        Functions".  In Psychological Research in the
        USSR, Moscow: Progress Publishers.
Watson, J.S.  1972.  "Smiling, Cooing and "The Game"".
        Merrill-Palmer-Quarterly, XVIII, 4:323-39.
Wolff, P.H.  1969.  "The Natural History of Crying and Other
        Vocalizations in Early Infancy".  In Determinants
        of Infant Behaviour, IV (London:  Methuen), pp.
        81-109.

Zazzo, R. 1975. "Des Jumeaux devant le mirroir: Questions de méthode". Journal de Psychologie, IV:389-413.

# THE SELF AND COGNITION: THE ROLES OF THE SELF IN THE ACQUISITION OF KNOWLEDGE, AND THE ROLE OF COGNITION IN THE DEVELOPMENT OF THE SELF

Augusto Blasi

University of Massachusetts at Boston

In the past several decades there has been a growing interest in applying the general principles and findings of cognitive development to the understanding of personality and personality development. Kohlberg may be the best known representative of this movement, the one who most systematically attempted to articulate a new theory of personality based on cognitive foundations (Kohlberg 1969); however, the other members form a long list, and their contribution is by now very substantial.

While it is now obvious that the merits of the cognitive developmental approach to personality are considerable and its achievements lasting, some of its weaknesses are also becoming apparent. A central one, in my opinion, concerns its inability to account for the centrality of the self and of consciousness (Blasi 1975, 1976; Blasi & Hoeffel 1974). This is true even though the concept of the self, or related concepts, have been assigned important functions within a cognitive view of personality. Kohlberg's overall work and, for instance, Elkind's (1978, Elkind & Bowen 1979) recent research on self consciousness indirectly acknowledge the necessity of including the self in a cognitive theory of personality inspired by Piaget. It seems, however, that these attempts are a result more of a sensitive reading of the empirical object than of a consistent application of the most basic structural principles of cognitive developmental theory to the subject matter. It is as if the constructs of self and consciousness are convenient appendages, which neither clarify the theory of cognitive developement nor are clarified by it.

It will be argued in the present paper that the limitations of the cognitive developmental approach to personality

189

do not derive from its being cognitive, but rather from having inherited a mutilated conception of cognition. The outlines of a different view of cognition will be presented,
and the essential roles that the self plays in it will be
discussed. As an indirect result of this analysis, a con-
ception--still inarticulate and vague--of the self should
also emerge.

The ideas presented here are not new, though they have
been neglected and are far from fashionable.[1] The analysis, in many respects, will be more philosophical than psychological, but there are undoubtedly clear implications for
psychological research; unfortunately, these cannot be elaborated in this context.

## THE CONCEPTUAL VIEW OF COGNITION:
## THE SELF AS SELF-CONCEPT

To some extent, most (if not all) psychological theories of the self are cognitive--at least in the sense that
they view the self in terms of self-concepts, self-percepts,
self-images, and similar cognitive constructs.[2] However,
not all theories acknowledging the role of self-concepts in
human functioning are cognitive theories. Only some views
of cognition consider the most fundamental process by which
the individual organism comes to terms with the surrounding
world, gives meaning to it, and ultimately adjusts to it.
Cognitive activity and cognitive tools, then, by necessity
affect all other processes and all other human characteris-
tics--including the self.

Two outstanding examples of this type of view are
Kelly's theory of personal constructs and Piaget's theory of
cognitive development. The first is more informal, more com-
monsensical; the second is broader in scope, putting a much
heavier emphasis on those developmental processes and forms
that are universal across the human species. However, these
two theories have remarkable similarities in their general
understanding of cognition and of its function.

According to Kelly (1955), one of the basic needs of
human beings is to predict environmental events. As a consequence of this need, events are ordered according to con-
structs--that is, dimensions of similarities and differences. The specific constructs, their number, their inter-

nal organization, their degree of generality and flexibility, vary from person to person. Invariant, however, is
their function of providing order and meaning to all experience, and thereby guiding behavior. One such construct,
common to all normal human beings after a certain age, is
the self/not-self dimension. Though its importance is
clearly recognized, both as the center of a conceptual subsystem and as the source of the feeling of control over
one's action, Kelly emphasizes that the self/not-self construct is only one of many constructs, a part of the overall
system--determined, as in all other cases, by the general
need for predicting and controlling events.

Piaget's account starts, even more basically, from the
nature of the interaction that all living organisms have
with their environment; through this interaction the environment is defined by the organism's behavioral and conceptual structures, and these structures are then adjusted to
and modified by the environment. The available schemes,
categories, or concepts are also organized so as to insure a
degree of equilibrium and stability. On the basis of complexity, type of organization, and, particularly, degree and
form of equilibrium, it is possible to recognize qualitatively different structural levels in cognitive activity;
the formal relations between these levels form the basis for
the invariant order in the developmental sequence. The concept of self is one result of the general developmental differentiation: it is dependent in its nature on the general
cognitive structure available to the individual, and it is
subject to change as a result of later cognitive development.

One difference between Piaget and Kelly, probably a
result of their different interests in personality, is that
the concept of individual psychological self is not nearly
given as much attention by Piaget, who, instead, writes of
the "epistemic subject"--some kind of ideal, transcendental,
species-wide, cognitive structure. The implications of the
Piagetian principles for a cognitive theory of personality,
and specifically for a theory of the self, were later
explored by other cognitive-developmentalists--most prom-
inently by Kohlberg (1969). Other important sources of
inspiration in this project of extension were J.M. Baldwin
and G.H. Mead. According to this broader view, the self
would be constructed directly through social interaction and
in dialectical exchange with other people; social interac-

tion could itself be understood according to the general
Piagetian principles of assimilation, accomodation, and
organization. Moreover, logical-conceptual structures would
maintain a crucial role in defining which social interactions will be meaningful, and which self-concepts will be
possible.

The major characteristics of the explicit or implicit
cognitive theory that underlies the above approaches to the
self can be summarized as follows:

1.  The fundamental cognitive unit and the basic cognitive
tool in interacting with the world is the concept, the behavioral category, the construct, or the discriminatory di-
mension--terms that are understood here as equivalent. This
is also true in Piagetian theory, where schemes and concepts
are themselves determined by the superordinate cognitive
structure.

2.  Consciousness and the symbolic nature of cognition are
recognized, but they are not given by either Kelly or Piaget
the cognition-constitutive function that other theories
assign to them. The effect of constructs and cognitive
structures can be observed perfectly well in behavior
without introspection; in principle, then, human cognitive
processes can be replicated by appropriately programmed
computers.

3.  Judgments, ultimately, can be reduced to complex or
composite concepts. There is no real difference between the
concept "black man" and the statement "This man is black."
nor between the concept "woman self" and the judgment "I am
a woman.". The label conceptualism is meant to indicate the
almost exclusive emphasis on concepts that characterizes
this cognitive view. The fact that nothing cognitively
specific is recognized in judgments should be particularly
stressed, as the alternative cognitive view to be presented
takes judgments and not concepts as the primary cognitive
facts.

4.  One final characteristic is that concepts and categories
are acquired or changed in immediate contact with the physical and social world--without any need for conscious search,
debate, or choice--as a result of the most basic need to
adapt to one's environment. Again, consciousness may exist,
but it does not exercise any constitutive function in stimu-

lating and directing one's development. Both cognition and development simply happen, once the appropriate conditions are present. For development also, as well as for cognition, the essential processes can in principle be replicated by computers. In other words, conceptualism is frequently associated with a functionalistic view of cognition.

These seem to be the general cognitive principles that underlie present cognitive-developmental approaches to the self. Of course, the nature of the self as a general concept must be affected by them. From a conceptualistic perspective, obviously, only the self as object will be available to theory. Moreover, there seems to always be a gap between the generality of cognitive structures and the specificity of self-concepts. This is particularly so because there is a de-emphasis of the judgment--namely, of that cognitive process in which concept and experience, the general and the specific, merge together. The most important limitation perhaps is that when the self is not given any role in the cognitive process itself, it is difficult to make sense of and to account for the characteristics of direction and control that our experience of self seems to provide. This point will be particularly elaborated in the next section.

### COGNITION AS JUDGMENT: THE ROLE OF THE SELF IN THE COGNITIVE PROCESS

It is possible to think of cognition and of the cognitive process as what one does and what one finds out when asking questions about the world and about oneself--rather than the application to the environment of an organized set of discriminative dimensions, concepts, or constructs. From this point of view, the judgment becomes the basic irreducible unit and acquires the same building-block function that is assigned to the discriminative category within the conceptualistic approach. The differences are important, both for a theory of cognition and for a theory of the self.

Cognition is stimulated by experience (i.e., by internal and external sensations and perceptions), but it formally begins when one asks two general questions: "What is this?" and "Is this really so?". The first "this" refers to the object of one's experience and the question about it is a search for understanding, aimed at inserting the experi-

ence within one's conceptual and theoretical network: this
sensory complex of colors, odors, and shapes is a rose--a
special kind of flower, which, like similar flowers, grows,
blooms, etc. Concepts and categories, then, whether in isolation or as parts of larger structures,still play an essential function; their function, however, is only one of several in the overall cognitive process and is subordinated to
the judgment, the natural outcome of cognition.

The "this" of the second question refers directly to
the conceptually understood experience, the answer to the
first question. The second question is reflective and critical and aims at a judgment grounded on evidence--one that
is understood to be very different from impression, opinion,
preference, or prejudice. To use the above example, the
judgment "This is a rose" leads to further questions such as
"Is it really a rose?", "Could I be wrong?", "Could it be a
different, though similar, kind of flower?", "How do I know
what it is?", etc.

It is central in this view that these three aspects
(i.e., experience, understanding, and judgment) form a dynamic unity, each step requiring to be completed by the others and urging towards the completion of the process: experience asks for understanding; understood, meaningful experience asks for correctness and truth; and truth in judgment
is about one's understanding of experience.

It is important to stress both the relation of whole to
part between judgments and concepts, and the irreducibility
of the former to the latter. Essentially, the function of
cognition is truth--namely, to inquire about the type of
existence that characterizes the object, and to affirm the
reality of the object as independent of the cognitive process that recognizes it. Concepts, as such, deal with definitions, meanings, and their reciprocal relations; they have
nothing to say about existence and reality. Concepts, categories, and theories have exactly the same characteristics
whether they are derived from the surrounding world or
dreamed--whether they are revealed by God, transmitted
through oral tradition, or fantasized during a paranoid epi-
sode or an LSD trip.

It is easy to miss the cognitive limitation of conceptual structures because concepts are usually created, used,
and transformed in the course of and within the context of a

broader cognitive activity that has reality as both its
starting and its concluding points. But it would be a mistake to attribute to the part what is only a characteristic
of the whole. The important consequences of such a mistake
are to not recognize the existential, reality-oriented
thrust of human cognitive activity and the role that the
individual self plays in it.

It may be easier to appreciate the fundamental limitation of concepts by asking which aspects of the cognitive
process can and which cannot be replicated by computers.
This is not only a useful exercise--we will resort to it
again in the course of this essay--but an appropriate one,
considering the widespread acceptance that the computer
model has gained among cognitive psychologists (including
those of a Piagetian persuasion), and considering, in addition, that this model seems to epitomize the conceptualistic
approach to cognition. Computers, at least in principle,
appear capable of replicating rule-regulated responses, "decisions" according to specified strategies, classificatory
discriminations--in sum, much of what is implied in conceptual and operational functioning when considered in isolation.

Two aspects of human cognitive functioning, however,
seem to be "program-resistant", to use Gunderson's (1971)
terminology. One is consciousness, or the awareness of cognitively operating and knowing.[3] The other is the reality
function of human cognition--namely, the affirmation that
the object exists and has a reality independent of the cognitive process. Neither the computer's ability to consistently differentiate what belongs to the computer and what
belongs to the noncomputer, nor the ability to change its
programs (its conceptual structures) in order to adjust to
the changing nature of its input (if these abilities were
ever to be present in computers), imply affirmation of reality or a truth function. In order to know what truth is,
one needs to know what untruth is; in order to know reality,
one needs to know unreality. These achievements seem to go
beyond the capacities of computers, as well as beyond the
capacities of conceptual structures.

The implications of this view of human cognition, particularly concerning the role that the individual self plays
in it, can be elaborated by looking at the meaning of asking

questions about an object and of answering them through judgments.

Asking a question about an object or about the world is different from adapting to the world, or even from behaviorally discriminating and anticipating events in the world. While some may think of the process of asking and answering questions as being adaptive, many adaptive processes (e.g., digesting one's food and all the discriminatory processes involved in it, memorizing nonsense syllables, developing a fine hand-eye coordination) may have nothing to do with questions. The difference is not simply one of activity—as adaptation is a very active process, bringing the organism's own structures to bear on the interaction with the environment. The crucial difference is that a question is formally about an object, thus requiring a conscious differentiation between the subject asking the question and the object about which the question is being asked; it also requires that the distance between subject and object be represented symbolically in the conscious act of puzzling and inquiring.

Consciousness of self in action, which is the fundamental constitutive of the self, is therefore at the origin of cognitive activity.[4] Here is one of the major themes of the present approach: cognition is about objects, but objects are not objects unless the conscious subject plays a fundamental role in the cognitive process.

The cognitive questions, as stated above, can only be fully satisfied by judgments. These can take several forms, but their essential structure is the affirmation that something is true of an object (this is a rose., Johnny kicked me., etc.). When one analyzes the structure of the judgment, two further implications appear concerning the role of the self in cognition. The first is that the person answering the questions "What is this?" and "Is it really so?" is required to learn the distinction between opinion and fact, dream and reality; between correct and incorrect statements; between what one wants, likes, or wishes and what is.

It is not being suggested that these differentiations are present from the very beginning, nor that from the very beginning the child understands what are valid and satisfactory grounds for judgments. Rather, these differentiations and the concern for grounding one's judgments seem to be demanded by the dynamics of the cognitive process and are virtually, germinally, present in the fundamental cognitive differentiation of self from not-self. If object is different from self and if knowledge is about the object as different, then one needs to learn to distinguish what belongs to the object from what belongs to the self, one needs to learn about wishes and fears, about points of view, about biases and prejudices. The critical attitude, and the concern with standards, criteria, and cognitive norms, have their origin in these discriminations of self and reality, of subject and object.

The reciprocal influence of self and cognition, and the dialectical nature of the whole process, should be emphasized: on one side, the dynamics of the cognitive activity presses on the self a more differentiated consciousness and a more differentiated self; on the other, the kind of self that asks the cognitive question determines the kind of cognitive answer that he will give himself.

Some of these ideas are, in some way, present in the Piagetian notions of egocentrism and decentration. These concepts, however, are understood as referring to the balance between assimilation and accomodation and to the degree of reversibility, for instance, among all the possible perspectives in the "3-mountains task" or among the possible transformations of a conservation experiment. Self and consciousness are not required; all that is required for decentration is a rule regulating the transformation from one to another perspective.

The second important implication for the self derives from the fact that a judgment is essentially affirmation or denial. The discriminatory response to the object cardboard apple (the response may consist of refraining from eating it, or of putting on a board, next to each other, a plastic symbol for cardboard and a plastic symbol for apple) is, psychologically, quite different from stating "This apple is of cardboard." or "This is a cardboard apple.". This seems to be true even if the practical consequences of the two kinds of responses were exactly the same, or if one can find in both cases the same conceptual categories of "cardboard", "apple", and "cardboard apple". The crucial difference is that a judgment is precisely a judgment and not a response-- namely, a statement, a stand, of the subject vis-a-vis the object, where the emphasis is on "stand" and on the possessive "of".[5] The subject is responsible for his judgment;

he is a witness for truth and a guarantor that the object is
what he claims to be.

In order to more fully realize the extent of the role
of the self in cognition, it may be helpful to take the term
judgment quite literally. One can imagine a science-fiction
situation in which legal decisions are made by a computer-
ized machine. The machine would be duly programmed with all
the legal principles and the relevant rules of the nation;
in each concrete case, then, the details of actions and
events would be fed into it. The outcome, finally, would
consist of the machine dictating (or, better yet, announcing
through a sound apparatus): "John Doe guilty", "Carl Smith
not-guilty". The addition of the copula "is" would not
change the automatic, unconscious, nature of the verdict.
This verdict, in sum, would be an output and not a judgment.
One may even decide that such an output is more efficient,
more economical--and, in the end, fairer--than a judgment.
None of these characteristics, however, would transform an
output into a judgment, if the essential characteristic of a
self consciously pronouncing it as his own is missing.

To summarize what has been said thus far, this account
of cognition stresses a necessary reciprocal influence between the self and knowledge. On the one hand, no knowledge
can originate and progress without a parallel origin and
development of the self. The conscious differentiation
between self and not-self makes it possible to focus on the
object as such; the discovery of the subjective (i.e., of
personal biases and viewpoints) makes it possible to look
for what is independent of biases and viewpoints; the realization that the self is cognitively responsible makes it
possible to become concerned with the truth of one's judgments. On the other hand, the dynamics implicit in the cog-
nitional process affect the development of the self. This
development does not simply proceed along the lines of a
richer, better-organized and integrated self-concept, but
along such qualitatively different steps as: a rudimentary
differentiation between self and not-self; the understanding
of the relations between subjective viewpoint, bias or prejudice, and reality; the understanding of and the concern
with the effect of the self on knowledge, and thus the
development of a self who is responsible vis-a-vis one's
knowledge and the community of knowers.

It may be worth mentioning, very briefly, the necessary

relation between the self and that controversial but (-at least to some of us) essential characteristic of knowledge—namely, objectivity. Objectivity is not given in immediate experience, in the sensual primordial contact of person and thing; rather, it is the result of the continuing and increasing role of the self in the overall cognitive process. Paradoxically, only at the height of self consciousness—only in the conscious unity of the self's activity of experiencing, understanding, and judging—can there be objective knowledge. Two examples from psychology, or from social sciences in general, can be used to illustrate this point. One is the realization that one's research depends on one's assumptions; the second concerns the findings and the theorizing on "experimenter's bias". Aside from the unwarranted relativistic conclusions that are occasionally derived from these examples, there is no doubt that in both of them there has been an increased awareness of our self-as-scientist in action, and that, as a result, the chances for more objective experimental findings were improved.


## THE SUBJECTIVE CONTEXT OF COGNITION

A point made in the earlier section is that one of the tasks of the self in cognition consists of differentiating what belongs to the object from what belongs to himself—dreams, wishes and fears, viewpoints and expectations. Human consciousness (contrary to what Sartre (1957) seems to have imagined) is not disembodied, simple, and transparent—with the immediate capacity of piercing through illusions and appearances to arrive at the substance of reality. Our self is cognitive only to the extent that our cognitive activity falls, as it must fall by definition, under the purview of our unifying consciousness. But other activities and movements are also in consciousness, in experience if not in understanding and judgment.[6] To the extent that a child experiences in consciousness the pains of hunger, the pleasure of being hugged and caressed, the desire for more love and the fear of not getting enough, the security in the expectation of the parents' support or the corresponding insecurity, and unifies all these experiences, in the very process of experiencing as leading and belonging to him—to that extent this child has a multifaceted self, one that perhaps may not yet be clearly differentiated in a set of self-concepts.

Several important points can be made about the fact
that the consciousness of oneself in action is the ground
for a rich and multifaceted self. The most obvious is that
this experience opens a new world--the world of the sub-
ject--to one's cognitive activity. Like experience of the
object, so experience of the subject demands to be understood and made sense of in terms of concepts and categories;
its understanding demands to be evaluated and judged in
terms of reality. It is thus possible to appreciate the
major limitations of traditional psychological approaches to
the self as self-concept. In agreement with conceptualistic
premises, self-concepts are rarely (if ever) seen as arising
out of a responsible inquiry about experience; instead,
these categories--whether acquired through imitation, reinforcement, or social interaction--are cut off, on either
side, from their experiential referent and from the responsibility of the questioning self. It is hard to see how a
conceptualistic approach can avoid making the self slave of
one's cultural categories and, ultimately, a conformist.

Another important point, related to the fact that cognition is only one of the activities of the self, can only
be mentioned. On one hand, cognition cannot but be contextual and subordinated to the whole which is the self in all
of his relations. This idea (which seems to be having a
revival in contemporary philosophy of science) involves the
principle for the limitations of cognitive and scientific
activity, and the criticism of the notion of "pure" knowl-
edge--at least of some of its meanings. On the other hand,
cognition, though limited, occupies a privileged position
because it is concerned, by its very nature, with evaluating
and ordering in terms of objective reality.

A third implication deserves closer attention. The
other side of the coin of the multifaceted self is--to use
other metaphors--the self's thickness, murkiness, and ambiguity; parallel with it, there is a thickness, murkiness,
and ambiguity in the object to be known. These metaphors
are meant to refer to the fact that various self-aspects are
not neatly ordered and classified in nice compartments: the
self as knower (with his curiosity for the object and his
desire for objectivity) is not clearly separated from the
self as needing love and acceptance, from the self as hungry
for security, etc. Rather, the same curious and competent
self is also the fearful, needy, insecure, aggressive self.
A two-year-old boy (to take one of Margaret Mahler's exam-

ples) may begin to accumulate plenty of evidence that he and
mother are quite different and, to a large extent, separate
from each other. On the other side, he feels so small, so
needy, and so helpless (being together was so safe and
warm!), that the clues of his separateness cannot be real—
especially if he manages to stay close to mother and to appear helpless.

These are the concrete, contextualized conditions under
which human knowledge is acquired. What was said earlier
about the nature of judgment and about the responsibility of
the self in the cognitive process now receives a fuller and
much sharper meaning. In other words, objective knowledge
requires to be differentiated from defensive processes. But
this differentiation can be accomplished only by a responsible self in control of cognition.

A typical solution in psychoanalysis consists of carv-
ing out a "conflict-free" area, and in limiting the objectivity and reality value of knowledge to such an area. The
obvious problem with this solution is that it may not be so
simple to determine which areas are conflict-free; more importantly, this solution seems to imply that we are excused
from bias and irresponsibility whenever our conflicts and
emotions interfere with our cognitive tasks.

Within cognitive developmental psychology the problem
has either not been acknowledged, or has been explained
away. The reasons for this state of affairs lie not only in
the implicit rationalism of the prevalent approaches, but
also in the explicit conceptualistic view of cognition—one
in which the concrete psychological self is not assigned any
essential role. An example supporting this conclusion is
Piaget's discussion of the unconscious and of defense mechanisms; another example is Langer's account of the inconsistency between moral judgment and moral behavior.

Relatively early in his career, when he was interested
in comparing Freud's views to his own, Piaget (1962) gave an
account of unconscious defensive processes. Already, his
definition of the unconscious symbol as characterized by
"the ignorance of its meaning" (p. 191) suggests a static
approach to the unconscious. The dynamic and intentional
aspects—two central features of the psychoanalytic unconscious—have been completely left out. The ignorance of
meaning is then explained in terms of the standard concepts

of assimilation and accommodation. A symbol is unconscious
if it is used purely for assimiliation: not being available
for accommodation to the external reality, it becomes impossible to distinguish what, in it, corresponds to the object;
the symbol is thus given a "mistaken", non-objective meaning.


   To finally explain why the unconscious symbol is not in
contact with reality, Piaget does resort to the notion of
repression which he interprets, however, as the automatic
result of incompatible structures. Under the organization
principle, it is in the nature of structures or schemes,
that they tend to assimilate one another; when this is not
possible, they tend to exclude one another. The victorious
structure will regulate the adaptive contact with the world;
the weaker structure, not in contact with the world (repressed), will become purely assimilatory--leading thus to
"mistaken" symbols. Many details are far from clear (one
could ask, e.g., what determines strength or weakness in a
structure.). It seems clear, however, that there is no need
of self, of a super-structural (if we may say so) agency
regulating the interplay among structures; there is no intention, the whole process being automatic. Under these
conditions, a defense can only be a mistake.[7]

   Langer (note 1) looks at the problems of moral inconsistency as one specific instance of a more general issue
concerning the equilibrated coordination of observation,
judgment, and action. Both observation and action are
sources of information about the world; this information is
fed back to the cognitive structure, which can then accommodate to reality and adapt to the environment. A lack of
coordination is simply an instance of disequilibrium, which
is the dynamic source for a renewed cycle of assimilation
and accommodation and for a change in one's concepts, categories, expectations, and similar structural tools. This
may be an elegant and economic analysis of the relations
between action and thought, at a very general level--of why,
in other words, when we think of opening the door, we actually open the door. Langer's approach, however, seems to be
most unsatisfactory when it is applied to questions such as
"Why should I help others?" or (his own question) "Why don't
I help when I see clearly that I should help?". The case of

prolonged inconsistency--such is frequently the inconsistency between moral judgment and moral action--cannot be easily understood within a disequilibrium model, in which automatic adaptation is expected and in which the self is not assigned a central role in the regulation of cognitive transactions.

There is, finally, a general solution to the problem of dealing with individual sources of bias and distortion, which resorts to social interaction and communication and which gives social exchange a cognitive corrective function. There are several varieties of this mode of thinking, depending on the general cognitive theory that one subscribes to, on the mechanisms of social interaction and on the kind of dynamics that one hypothesizes; the most general features, however, are the same, and one can easily recognize in them a functionalistic--and, ultimately, conceptualistic--approach to cognition.

There is, of course, an element of truth in resorting to social correctives for individual distortions--truth that is contained in the old, common sense, wisdom that two people can see better than one. Others can contribute tremendously to one's cognitive tasks by bringing new information, by reinterpreting information already available and thus challenging one's dogmatism, by pointing to areas of selective inattention or to patterns of blindness. All this is obvious, noncontroversial, and not specific to any one view.

But the cognitive process does not stop here. The self still needs to consider, without distorting it, the information brought up through social interaction; he still needs to inquire about its basis, he needs criteria for resolving differences that appear in a dialectical exchange, and, finally, he needs to arrive at a judgment of which he alone is responsible. Not to recognize this central role of the self would mean not to realize that concepts and categories (whatever their origin) are cognitive only in a limited sense, that cognition goes beyond a functionalistic adaptation to the environment, that there are also cultural and social sources of bias and distortion, and that communication may frequently be motivated by power rather than by the pure interest in knowledge.

## ATTENTION, THE SELF'S COGNITIVE LEVER

In the foregoing account of cognition, a very central role was reserved for the self. In particular, there was an emphasis on the importance of a self that would be responsible for the direction of cognitive activity and that would be accountable for the truth of his judgments. But, if the use of the notion of responsibility is more than metaphorical, then a responsible self should enjoy a certain degree of freedom in accomplishing his cognitive tasks. If there should be in the self a certain degree of fidelity to the object (not an improper way of characterizing the human cognitive enterprise), three different kinds of freedoms seem needed: freedom from those cognitive structures that are embedded in one's culture; freedom from one's own individual cognitive structures, from the general philosophical outlook that was acquired through the accumulation and the organization of one's experience; finally, freedom from the conditioning and the interferences of one's individual emotional and affective makeup.

Before giving up the cognitive task as utterly hopeless, it may be wise to add that only a limited kind of freedom seems to be necessary. What seems required is the possibility to realize, at least occasionally and gradually, what influences tend to determine one's judgment, and in some degree to subtract one's cognition from their hold.

As in many cases of freedom, the self does not exercise his by directly acting on the sources of constraint--cultural, cognitive, or emotional that they might be--but by maintaining for himself, in relation to the objects of experience that are to be understood and judged, a certain amount of extra space and a certain degree of flexibility. In this attempt, the self uses attention as perhaps his most important tool.[8]

There is a sense in which both the control that the individual exercises on one's attention and the importance of attention for the construction of knowledge are also part of common experience. One of the clearest examples, of course, concerns the manipulation of perception in the case of reversible figure-ground patterns. But the process is most general: in a specific interpersonal relationship, for instance, one may focus on a class of events X, keeping his attention away from events Y and Z; as a result, a meaning-

ful construction of the other individual and of the relationship is put together; this construction, in turn,
changes the subject's actual experience of the relationship.
Of course, one can, in principle, focus one's attentions on
events Y and Z, together with or without focusing on events
X. The resulting constructions and experiences would then be
quite different.

To realize how powerful attention can be in affecting
one's cognitive structures, one could try to focus on each
object and each movement as separate entities and then, for
each object or movement, concentrate on each minute detail
of shape, color, or texture. The effects are dramatic: the
world begins to lose all meaning then, as the music is lost
when the listener isolates every note. Attention not only
can change one's cognitive structures, but can, at the limit, literally destroy them.

Frequently the selective power of attention is used
precisely to keep out-of-focus, in darkness, what does not
fit with one's wishes or one's prejudices, what arouses
anxiety, or what contradicts the picture of the world that
one finds comfortable and comforting. Defensive maneuvers
essentially consist of an irresponsible use of attention for
the purpose of aborting the cognitive process.

But to recognize that attention can be abused, and that
its abuse can be intentional in some genuine meaning of the
word also implies that attention, with all its power, has
its cognitive uses. Going back to the original issue concerning freedom of the self in exercising his cognitive
tasks, the function of attention is to bring back, in a
reflective movement, one's understanding to the object of
experience. One's understanding can be and is affected by
cultural categories, expectations, and world-views; by social pressures, fashions, and stereotypes; by one's own past
experience and the kind of logic one finds useful; at times,
by one's fears and desires. Attention explores the object
of one's understanding for elements that have been missed,
and brings into focus this or that detail to test the stability of understanding; most importantly, it brings into
the limelight what is in the darkness, what seems to be contradictory or unaccountable.

However, when one looks to scientific psychology for
clarification and enlightenment of these common experiences,

one is bound to be disappointed. Though the past two decades marked (after a long period of neglect) a renewal of interest on attention, this work has been concerned with rather simple processes quite removed from the epistemologist's problem of whether the knower can notice and use events that conflict with his cherished expectations. More importantly, our contemporary scholarship has little or nothing to say about one aspect of our everyday experience, namely, the control that we seem to have on attentional processes.

Wright and Vliestra (1975) recently proposed a distinction that may seem to be relevant to the issue of control. They differentiate between exploration and search behaviors: the first is a simpler process, nonsystematic and play-like, instigated by curiosity and boredom, consummatory in nature; the latter is systematic, planned, implying--in its continuity--intention and purpose, is instrumental and work-like, and is based on the relevance and informativeness of the cues. More central to our topic, exploration is passive and is controlled by the properties of the stimulus field, whereas search is active and is controlled by the logical characteristics of the task and, ultimately, by the conceptual structure that the individual brings to the task.

That these two kinds of behavior exist and can be differentiated seems obvious; that they have different developmental characteristics seems plausible, considering the evidence gathered by Wright and Vliestra. However, neither of them seems capable of explaining or clarifying those experiences that suggest a certain degree of control of attention by the self: shifts in the construction of the perceptual field, the intentional selectivity of defense mechanisms as well as the intentional control of defensive processes, the destruction and restoration of meaning, as were described earlier. What appears to be present in these cases is not only the control of attention by certain structural characteristics of the organism--cognitive or affective that they may be--but also the control of the structural constraints controlling attention. For this reason Wright and Vliestra's active attention cannot establish any degree of cognitive freedom.[9]

It is important to recognize the intrinsic relations between cognitive structures and active or selective attention, as understood by contemporary psychology, selective attention simply being the basic interactive property of a structure. It is not surprising, then, to learn that Gestalt psychology, in contrast with older schools, had little use for attention.[10] In a parallel way, the cognitive limitations of this kind of active attention are essentially the same as the limitations that earlier in this paper were attributed to the conceptualistic view of cognition: both seem ultimately to derive from ignoring the role of the self in the acquisition of knowledge. In fact, we should go one step further and conclude that structural attempts to understand the self (e.g., Frondizi 1971) are destined to fail, as they make the self completely dependent on cognitive processes and do not recognize any control of the self on the functioning of cognitive structures.

When one looks at cognition, attention, the self, and their reciprocal relations in the way that has been suggested in this paper, one gets very close to the concept of will. This would be a tempting direction to take, one that cannot be pursued in this context. We may simply remind ourselves that a similar direction seems to have been taken by Wundt, who placed the concepts of attention and voluntary action at the center of his psychological system of "voluntarism".[11] This reminder, of course, is not an invitation to accept Wundt's views (which, especially around the problem of the self, do not seem to have achieved a high degree of coherence); it is rather an invitation to maintain our curiosity and concern open to those processes and experiences that do not disappear, even when we cannot account for them in terms of prevalent views or when they conflict with philosophical and methodological taboos.


CONCLUSION


It is not surprising to observe how widely conceptualistic premises are shared by otherwise quite different approaches to the self (see this volume). Obviously, conceptualism is not limited to Kelly and Piaget or to the views inspired by Piaget; it includes those views derived from French Structuralism (Levi-Strauss in particular), those inspired by Mead's symbolic interactionism or relying on linguistic theories, those brands of psychoanalysis that are influenced by either Lacan's structuralism or by Mead.

It was not the purpose of this paper to belittle the contribution of conceptualistic approaches to the self or of the research body on self-concept, its development and its cultural variations. My main purpose, rather, was to point out their limits and their potential risks. It is the central characteristic of conceptualistic approaches to focus on the self as object of perception and categorization without looking at the self that perceives and categorizes. The self, then, becomes a static, fixed, cognitive conclusion, rather than the active process of cognizing self and others, of wondering, of asking questions, gathering information and evidence, doubting one's conclusions, or changing self percepts and self definitions. In so doing, what seems most central to the self—the standing back and apart even from his knowledge, the control of it—is hopelessly lost.

Perhaps even more importantly, what appear to be overlooked are the processes leading to the formation of the self-concept, to its change and its development, as well as the relations between these processes and the nature itself of the self. It would be unfair to accuse the prevalent conceptualistic views of being unconcerned with the change and development of self-concepts. Cognitive developmentalism of the Piagetian kind looks at the individual general cognitive structures as the major factor determining the self-concept and underlying its changes; Mead's symbolic interactionism was meant to be a theory of self-concept formation; linguistic structuralisms view the conceptual structures embedded in a culture's language as the major matrix in the formation of the self; anthropological approaches apply one or the other of these theories to the explanation of cultural differences in the self. The changing of self-concepts as a result of cultural changes is one of the main questions that has been, and is being, explored.

And yet there is not recognition, in either theory or research, that all these factors—internal cognitive structure, language and cultural structures, family interactions and socio-economic conditions—are ultimately processed by an individual who does not simply have a self but is a self. The main thrust of this paper was to point out that the formation of the self (concept) is not the result of an automatic interaction of internal structures and of sociocultural structures, whether external or internalized. Instead, the self itself is actively cognizing before it forms a self-concept—i.e., he asks questions about himself,

about the truth of his own or of others' concepts of himself; tries to sort out from his own delusions and distortions as well as from cultural stereotypes what is actually
there, what he actually is.

This paper attempted to argue that the self is primarily and originally in knowing and acting and that the self-
concept is a derivation of it in the same way that any concept is an attempt to grasp, encapsulate, make comprehensible, a complex and live reality. One corollary is that differences in the self (individual and developmental) can be
profitably studied by analyzing the relations that the individual has with his own conceptual structure, with the cultural, social, and linguistic structures, in the process of
deriving and developing a self-concept. The central questions concern not simply what individual self-concepts are
like, nor what cultural or cognitive factors account for
them, but also how the self relates to his own self-concept.
The nature of this relation (the self's ability to stand
back, to question his concern with the truth of his perception) would tell something very central to the self which
the self-concept alone could not possibly capture.


ENDNOTES

1. Several of the ideas presented in this essay have been
   directly and heavily influenced by the philosophical
   work of Bernard Lonergan (1957, 1967, 1968). My debt
   to him cannot be overestimated.

2. Of William James's (1890) well-known division between
   self-object (Me) and self-as-agent (I), only the first
   has achieved within American psychology its own niche,
   a reasonable interest, a body of research, and some
   degree of conceptual clarity. The second (the I) has
   been entirely neglected. The major reason is that the
   self-as-subject can be reached only indirectly, through
   consciousness in action (Blasi 1978). And this approach
   is not available to a conceptualistic view of cognition.

3. A distinction is assumed here between operational "reflection", available to any cybernetic machine, and
   that awareness of acting that is implicit in experience

proper and establishes the possibility of conscious awareness (See Blasi 1978.)

4. Piaget does recognize the differentiation of self from not-self and acknowledges the developmental importance of this step. However, consistent with his conceptualistic position, this differentiation seems constituted less by consciousness than by differentiation among sensorimotor schemas; moreover, the differentiated self is not constitutive of cognition—since cognition as equilibrated adaptation existed before, and since the conscious self does not seem to play any role either in the description of or in the explanation of later cognitive development.

5. This brings up another point of comparison with Piagetian theory and methodology. The question is sometimes raised (e.g., in conservation research) as to whether one should trust the simple discriminatory judgment ("This quantity is the same or is larger."), or whether reasons justifying one's judgment are desirable or needed. Whichever solution is accepted, it appears that reasons are simply given a methodological role, as indication that the discrimination is due to genuine conservation, and not a constitutive cognitive function. If one were able to set up the experiment in such a way that the decision about quantities could not be due to accidental factors (as is sometimes done in research with monkeys; Woodruff, Premack, & Kennel, 1978), reasons would have no importance at all. Ultimately this is because language, in Piagetian theory, has no other function but to represent concepts. And concepts, as such, do not need a self.

6. For a brief account of the way the multiplicity of consciousness is constitutive of the multifaceted unity of the self, see Blasi 1978.

7. A similar approach to defense mechanisms is also present within Kellian theory, at least as represented by Mancuso (1970). Mancuso reminds us that, from the perspective of personal constructs theory, "the theorist can eliminate the need to distinguish 'reality' from

'unreality'" (p. 363). Under certain circumstances, it may be "useful" to speak of defense mechanisms. The difference, however, between a defensive and a non-defensive approach to the world does not lie along the lines of reality, distortion, or even of successful adaptation. Both modes of interaction are equally constructive and may be equally successful.

8.  Kelly (1955) also writes about freedom in cognition. This freedom, to him, is based on the fact that events never completely determine one's construction, as well as on the fact that the inevitable ambiguity of the universe allows one to imagine, in every case, more than one construction ("constructive alternativism"). The role played by imagination in guaranteeing cognitive freedom may indeed be very important, though it hardly elicits any interest in contemporary psychological theories of cognition. Kelly himself, however, does not avoid ambiguity. On one hand, by insisting that constructions are made in order to predict the real world, he seems to acknowledge an objective anchorage for cognition; on the other, by not determining the sources or the range of freedom, he may be stressing subjectivity at the expense of responsibility to the object.

9.  The distinction between passive and active attention is an old one and was suggested, for instance, by Jaspers (1913) and by Vygotsky (1930/1978). However, the meaning and the degree of convergence of these concepts are far from clear. The notion of active is particularly ambiguous, as it frequently seems to be a negative notion and to refer to the absence of environmental control.

10.  Rizzoli (1979) reports that in 1926 Rubin presented a paper in Jena entitled, "On the Nonexistence of Attention".

11.  The work of Wundt is undergoing a radical reappraisal. Two recent papers are relevant to the issues raised here; Blumenthal 1979 and Robinson 1979.

## REFERENCE NOTES

1.  Langer, J. Equilibration of Moral Conduct. University
    of Berkeley, unpublished, undated manuscript.

## REFERENCES

Blasi, A.   1975.   "Role-taking and the Development of Social
            Cognition". Paper presented at the Annual Con-
            vention of the American Psychological Associa-
            tion, Chicago, Ill., September.
Blasi, A.   1976.   "The Concept of Development in Personality
            Theory". In Ego Development: Conceptions and
            Theories (ed. by J. Loevinger, San Francisco:
            Jossey-Bass).
Blasi, A.   1978.   "The Varieties of Consciousness and the
            'Bicameral' Consciousness". Paper presented at
            the Annual Convention of the American Psycho-
            logical Association, Toronto, Canada, August.
Blasi, A. & E. Hoeffel.   1974.   "Adolescence and Formal
            Operations". Human Development, XVII:344-63.
Blumenthal, A.L.   1979.   "Wilhelm Wundt--Founding Father".
            Paper presented at the Annual Convention of the
            American Association for the Advancement of
            Science, Houston, Texas, January.
Elkind, D.   1978.   The Child's Reality: Three Developmental
            Themes. Hillsdale, N.J.: Erlbaum.
Elkind, D. & R. Bowen.   1979.   "Imaginary Audience Behavior
            in Children and Adolescents". Developmental
            Psychology, XV:38-44.
Frondizi, R.   1971.   The Nature of the Self: A Functional
            Interpretation. Carbondale, Ill.: Southern
            Illinois University Press.
Gunderson, K.   1971.   Mentality and Machines. Garden City,
            N.Y.: Anchor Books.
James, W.   1890.   The Principles of Psychology. New York:
            Holt & Co.
Jaspers, K.   1913.   Allgemeine Psychopathologie. Berlin: De
            Gruyter.
Kelly, G.A.   1955.   The Psychology of Personal Constructs, 2
            Volumes. New York: Norton.
Kohlberg, L.   1969.   "Stage and Sequence: The Cognitive-
            developmental Approach to Socialization". In

Handbook of Socialization Theory and Research
(ed. by D. Goslin), New York: Rand McNally.

Lonergan, B.  1957.  Insight: A Study of Human Understanding. New York: Philosophical Library.

Lonergan, B.  1967.  Collection.  New York: Herder.

Lonergan, B.  1968.  The Subject.  Milwaukee, Wisconsin:
Marquette University Press.

Mancuso, J.C.  (ed.) 1970.  Readings for a Cognitive Theory
of Personality.  New York: Holt, Rinehart, &
Winston.

Piaget, J.  1962.  Play, Dreams, and Imitation in Childhood.
New York: Norton.

Rizzoli, A.A.  1979.  "L'attenzione e i Suoi Disturbi in
Psichiatria".  In Giornale Italiano di
Psicologia, VI:45-64.

Robinson, D.N.  1979.  "Minds, Brains, and Selves: The
Leipzig Approach".  Paper presented at the
Annual Convention of the American Psychological
Association, New York, September.

Sartre, J.P.  1957.  The Transcendence of the Ego: An
Existentialist Theory of Consciousness.  New
York: Noonday.

Vygotsky, L.S.  (1930).  Mind in Society, 1978 ed. (ed. by
Cole, M., V. John-Steiner, S. Scribner, & E.
Souberman), Cambridge:Harvard University Press.

Woodruff, G., D. Premack, & K. Kennel.  1978.  "Conservation
of Liquid and Solid Quantity by the Chimpanzee",
Science, 202:991-94.

Wright, J.C., & A.C. Vliestra.  1975.  "The Development of
Selective Attention: From Perceptual Exploration
to Logical Search".  In Advances in Child Development and Behavior Vol. 10 (ed. by H.W. Reese),
New York: Academic Press.

# THE COGNITIVE-DEVELOPMENTAL THEORY

# OF ADOLESCENT SELF AND IDENTITY

John M. Broughton

Teachers College
Columbia University

In recent years, a cognitive-developmental approach to psychology has grown out of the work of Dewey, Baldwin, Piaget, Mead, and Kohlberg. Despite difficulties of both a theoretical and an empirical nature, this approach has matured to a point where it rivals psychoanalytic theory in scope. Although it is a general psychology of development, recent post-Piagetian work has permitted the emergence of a specific theory of adolescent phenomena.

However, the casual reader stumbling upon cognitive-developmental writings would have to be forgiven for not recognizing that they represent a single viable perspective. Rather than a clearly defined and well-articulated school of thought, one finds an accumulation of diverse views and positions. Part-theories are constantly appearing, undergoing revision, and then entering into a collective consciousness that is rarely examined in detail. There is a ragged diaspora of post-Piagetians with little in the way of a cohesive identity. Individual researchers pick up widely varying sub-domains of the cognitive realm, and rarely worry about putting their special concern back into relation with a developmental whole. To make matters worse, they tend to be vague about the ages or periods of the life-span under study; they slide from childhood into adolescence, and elide adolescence with adulthood. There are the makings of a theory, but serious attempts at theoretical synthesis are still lacking. Furthermore, there is virtually no historical reconstruction of these various vicissitudes of the cognitive-developmental approach, and few systematic literature reviews. One is left with the impression of a family in disorder, a home in disarray.

Nevertheless, with a little effort, one can shape up the domestic scene sufficiently to discover an interesting account that speaks to the issues of self and identity in a unique and provocative manner. It is a narrative of the youthful years, replete with unusual ideas and embellished with suggestive findings. In unraveling this story, however, it becomes clear that adolescence as such has become harder, not easier, to define.

The cognitive-developmental approach has roots that are not simple to trace. Certainly, it owes much to Kantian philosophy, especially as that body of thought was refracted through Hegel and Darwin. However, at the European end, there were many important intermediaries who share only the weakest of family resemblances: Janet, Claparède, Brunschvig, Reymond, Durkheim, and Bergson, for example (see Gobar 1968). At the American end, a single, major trend is easier to discern. Hegelian and Darwinian thought appear to have lent a diachronic flavor to psychology via the mediation of the mental philosophies of Dewey, Mead, and Baldwin. The latter fused Hegelian genetic idealism with an evolutionary naturalism, in a pragmatist or instrumentalist synthesis. This integration was much in tune with the growth of economy and technology in the liberal atmosphere of progressive reformism in turn-of-the-century America (Broughton, 1981a).

The three American philosopher-psychologists subscribed to an action-based constructivist epistemology in which knowledge was the emergent consequence of qualitative step-by-step development. While all were significantly influenced by Hegel, in the end none of them wanted to be thought of as Hegelian. All capitalized upon the powerful convergence of scientific neo-positivism, evolutionary theory, and sociology to transform their mental philosophy into a functional social psychology. Their new science of mind was intended and received as a revolt against the absolute metaphysical character of Hegel's scheme. The ideas comprising modern developmental psychology are very different from Hegel's, and from those of the early neo-Hegelian proponents of a genetic perspective. One example of this discrepancy is that, in its most recent form, cognitive-developmental theory has exhibited a particular concern with adolescence. Hegel never evinced any special interest in this period, and neither did Dewey or Baldwin. Mead (1936) gives adolescence a passing mention, but otherwise conforms to the same pattern.

In fact, their genetic schemes were not couched in terms of life-span periods at all.

The modern focus on adolescence per se is less characteristic of the neo-Hegelians than of the positivistic social Darwinists like G. Stanley Hall (1904). For Hall, adolescence was "the infancy of man's higher nature", a time when biological plasticity received an inherited boost. The added momentum lent the potential to carry the adolescent beyond the constraints of present-day civilization to a new and higher level of evolutionary adaptation. However, Hall's biologism entailed a certain "storm and stress". Plasticity equally implied heightened vulnerability and susceptibility to disturbances, whether they were new ones or eruptions of turmoils sedimented at some point in the racial past. Adolescence was "the stage of natural inebriation", and on this account required scrupulous supervision.

## PIAGET'S THEORY

Early and Late Versions. Though still positivistic, biological, and evolutionary in orientation, Piaget departed markedly from Hall in making only occasional reference to adolescence. In Piaget's early work, adolescence served merely to mark the final disappearance of cognitive "egocentrism". By "egocentrism" he meant a total cognitive orientation to the world, visible in children's scientific, metaphysical, and epistemological beliefs about physics, psychology, life, and the cosmos. It was a state of relative self/world and self/other fusion, giving way through gradual transformations to a "decentered" view that acknowledged both the differentiation of subject from object that makes knowledge of reality possible and the differentiation of subject from subject that reveals the relativity of multiple perspectives on reality.

In Piaget's later work, it became more apparent that he was adopting the Kantian program of accounting for the form that rationality takes in mathematics and science. He updated the latter in terms of formal logic and the modern hypothetico-deductive model of inquiry, claiming that these were the natural outcome of development to adolescence. Going beyond Kant's design, he used the logico-scientific structure of "formal operations" to account for the broader epistemological and social orientations of adolescents:

their penchant for theory construction and their concern
with utopian personal and social futures. Furthermore, he
claimed that the formal operational calculus is the functional equivalent of self-consciousness in adolescence.

From Self to Reality in Adolescence. The notion of
egocentrism predominant in the early writings is not lost in
the later work. Rather, Piaget proposes that at each stage
of logico-scientific development, a new higher form of egocentrism emerges and is overcome (see Elkind 1967). Thus,
at adolescence, the first flush of hypothetical thinking
brings with it an unconstrained surge of possibilities which
submerges reality (an idea that Piaget borrows from Char-
lotte Bühler). The ideal overwhelms the actual in an excess
of assimilation. Environmental and social reality are lop-
sidedly and uncritically fitted to the mold of the young
adolescent's self-constructed hypotheses, ideals, and imag-
inations. In a subsequent, second sub-stage, the balance of
epistemological activity is restored by compensatory accommodation. In mid-adolescence, the individual thinker acquires the ability to test out the practicability of ideals
and the truth of hypotheses against brute fact and against
the theories of others. The diverse creations of the newfound subjectivity are constrained selectively in the direction of objectivity.

Piaget refers to the first sub-stage, but not the second, as the appearance of the "adolescent ego". This pejo-
rative use of "ego" fits a pattern detectible in both his
early and his late work: scant mention is ever made of the
development of the self or self-consciousness, and whenever
mention is made, it is negative in tone. This is because
Piaget views development as precisely the movement away from
the self:

> The self is at any rate relatively primitive. It is like
> the center of one's own activity and is characterized by
> its conscious or unconscious egocentricity. (Piaget 1940;
> 1967a reprint, p. 65)

Both Piaget's early and late accounts indicate that the
development of formal operations is completed by early
adolescence (12-15 years). However in a relatively recent
article (Piaget 1972), he corrects himself, estimating that
completion occurs in late adolescence (15-20 years). In
fact, certain statements in the earlier Inhelder and Piaget
book (1958) already suggested this interpretation of formal

operations as stretched out over the duration of adolescence, the following being the clearest example:

> The focal point of the decentering process is the en-
> trance into the occupational world or the beginning of
> serious professional training. The adolescent becomes an
> adult when he undertakes a real job. It is then that he
> is transformed from an idealistic reformer into an
> achiever. In other words, the job leads thinking away
> from the dangers of formalism back into reality.
> (p. 346)

Thus we can take Piaget's two-phase account of adolescence
as a potential way of defining theoretically the transitions
into and out of adolescence (Gilligan & Murphy 1979), noting
that the latter centrally involves a non-cognitive societal
factor that the former does not, i.e., the assumption of a
work role, or what Piaget (1967a, p. 65) elsewhere called
"the auto-submission of the self to some kind of discipline". This latter event marks the beginning of adulthood;
Piaget has no place in his theory for adulthood per se. He
does not give adulthood a cognitive characterization, since
he believes its nature is such as to be entirely beyond scientific understanding (Piaget 1972, pp. 11f.; see also
Broughton 1979a).


The Moral Dimension. In Piaget's earlier work, adolescence was not differentiated into initial and consummatory phases. Neither was the final stage of development
spread over the span from early to late adolescence. Rather, subjectivism and relativism were two aspects of a single
cognitive organization, achieved at about 10-12 years of
age. It is not surprising that his account of moral development, produced during the same period (Piaget 1932), exhibited a similar view. An infantile "motor" orientation
was succeeded in childhood by an "egocentric-cum-realist"
respect for rules, which eventually gave way to a "co-opera-
tive" ethics in preadolescence. In this final phase, the
egocentric tendency to appropriate rules for the satisfaction of the child's own interests was replaced by a hypothetico-deductive approach to ideas and facts concerning
rule-construction and application, backed up by a reciprocal
imitation of peers. On the side of consciousness, the "the-
ocratic" tendency to conceive of rules literally as sacred
and unalterable adult ethics, carrying an inexorable objective responsibility to conform, was replaced by a democratic
view. According to the latter, rules deserve respect, out
of loyalty, yet are subject to alterations within a group of

peers through a process of mutual consent (Youniss, forthcoming). At the same time, in this movement toward democ-
racy, a belief in immanent justice, literal obedience, and
retributive punishment yielded to an egalitarian notion of
justice as distributive and social solidarity as premised
upon mutual forgiveness and understanding.

In this fashion, Piaget conceived of childhood's end at
the transition from heteronomous to autonomous morality. It
is this transition that impressed Piaget. The culmination
of development in concepts of cooperation and justice was
not presented in the form of a theory of adolescence. In
fact, especially on the practical level of rule-usage, mutuality is already appearing in late childhood. All the drama
of change, be it practical or conceptual, is complete by
preadolescence (11-12 years). Thus, at most, the elaboration
of the moral dimension offers only a sketch of certain conditions that might attend the dawning of the adolescent period.

A somewhat puzzling fact about the life-history of
Piagetian theory is that its early moral dimension has been
largely repressed. However, in an interesting return of the
repressed, Piaget and Inhelder's (1969) theoretical overview
revives the earlier account. In this book, the detailed
analysis given in the earlier work is dropped out, and only
the most general and abstract envelope of the earlier concepts is delineated. This eases the task of integrating the
"old" theory within the "new". As Gruber and Vonèche (1977,
p. 154) point out, the treatment of morality is now subordinated to the overall account of operational development.
The child is egocentric because non-operational. This differs somewhat from Piaget's earlier views, in which intelligence and moral cooperation were mutually formative (see
1932, p. 172). Instead, in the modern revision, autonomy and
mutual respect are construed as moral norms parallel to and
reflecting the organization of normative operations in the
logical sphere.

In addition, Piaget and Inhelder stress that part and
parcel of moral autonomy is the emergence of rational, social, and aesthetic ideals premised upon the newfound capacity for hypothetical formal thought. Here we hear echoes of
their earlier (1958) work, in which socio-moral development
radiated out from a logical core, eventuating in commitment
to a work role. However, in their overview, Piaget and

Inhelder speak alternatively of a parallelism in which the development of the moral norms comprises the affective side of a single principle of genetic decentration that unifies the differentiation of subjects with the coordination of their diverse viewpoints.

As is the Piagetian wont (Afrifah & Broughton, ref. note 1), there is thus a good amount of ambiguity concerning the connection between logical and social, cognitive and affective, leaving us unclear as to whether social value and moral reasoning are derivative from logical cognition or parallel to it. The lack of clarity is perhaps not surprising, given the persistence of Piaget's relative lack of concern with adolescence per se.

## THE NEO-PIAGETIAN THEORY OF ADOLESCENCE

<u>Kohlberg & Gilligan on the Adolescent Self</u>. The "completed" genetic theory that we were left with at Piaget's demise essentially offers a two-stage account of development which starts with the sensorimotor and culminates in the operational (Gillieron 1980). Adolescence marks only the terminus of this second stage, rather than its center or core, which lies instead in the concrete operational period. It is the latter, not the formal cognition of the adolescent, that has dominated Piaget's attentions. It is the concrete period in which the reversible character of thought (prepared during the pre-operational period) finally emerges, and the basic Kantian categories of the understanding take their appropriate form. What is the formal stage but an abstract propositional reflection upon those very categories and their relations? Thus it is with good reason that adolescence never captured the imagination of "le Patron" in quite the way that childhood did, leaving him as the definitive child psychologist at the historical watershed between traditional "child psychology" and the broader modern field of lifelong "developmental psychology".

Nevertheless, others claiming to be in the Piagetian tradition have found something in the formal power and scientific control of the more abstract adolescent operations that make this phase of life following childhood seem more vital to their endeavors. It might not be inappropriate to view this as a species of Piagetian revisionism. Perhaps the best example of such revisionism is to be found in the work of Kohlberg and his school. Kohlberg and Gilligan stress that

> what is of special importance for understanding adoles-
> cents, however, is not the logic of formal operations,
> but its epistemology, its conception of truth and real-
> ity....At its extreme, adolescent thought entertains sol-
> ipsism and the cogito, the notion that the only thing
> real is the self.[1] (1971, pp. 1063f.)

They depart from the Piagetian account of adolescence, and complicate it, in several respects. First, in response to empirical evidence that was available then although only later made public (Kuhn, Langer, Kohlberg, & Haan 1978), they see adolescence as not occurring in every individual, despite its universality in potential (Blasi & Hoeffel, 1974; Broughton 1979c). This clearly runs counter to Piaget's (1972) statement that formal thought is "common to all individuals" (p. 11). Second, there is much less emphasis on the relativism ("multiple perspectives") moiety of Piaget's account (the differentiation of subject from subject), and much more attention to the subjectivism moiety (the differentiation of subject from object). Following therefrom, the third departure is their much more positive treatment of adolescence. They emphasize that it involves an increase in objectivity through the discovery of subjectivity as a sector of reality. Rather than bemoan the new possibility of solipsism, they stress the fact that this discovery can be equated with the creation of a "self."[2] It is unclear whether or not the individual who never reaches formal operations, and therefore never experiences a real adolescence, is meant to have no self. Perhaps by "self" they mean self-consciousness.

At any rate, the positive view of the appearance of "self" leads them to stress the relative independence of adolescent cognitive structure from its logical base. In a fourth departure, then, they prefer instead to highlight its more reflective, epistemological nature. This base of reflective subjectivity, they claim, is necessary to the eventual emergence of the mature adult's contemplative and aesthetic relation to nature, art, and experience in general. Here we discern a fifth departure. Adolescence is deprived of its ultimacy, and is cloaked instead in the garb of penultimacy. It is viewed in the context of its being not the final stage but merely a preparation for a later, adult stage which will supersede it.

Consequently, there is less need to draw adolescence to a close. This combination of a rosy dawn and no dusk implies the sixth departure: a minimization of the two-phase assimilation/accommodation model of adolescence as a necessary egocentric error that must be hastily corrected. Thus, seventh, they are more consistently cognitive-developmental and less sociological in approach, de-emphasizing the societal imposition of work roles. They thereby simplify Piaget's more heterogeneous account, while losing some of its richness. They could also be said to mitigate what has been a locus of male bias in the Piagetian account, since it was clearly young men that Piaget had envisaged as the trusty souls willingly autosubmitting to the yoke of society's established division of labor. An eighth and final departure from the Piagetian theory could therefore be seen in Kohlberg and Gilligan's implicit anti-sexism. Although, as we saw above, their first departure from the cognitive-developmental position on adolescence reduced its generality, this last revision compensates for the loss by a virtual doubling of the population to which their account can apply.

   The Relationship Between Logical and Moral Development. As we have noted, in his later theorizing, Piaget's socio-moral psychology was progressively eclipsed, was recalled only on occasions, and even then tended to be viewed as an extension or application of general intellectual structures to the value domain. At best, the role of moral thought in the general course of development was vaguely defined. This state of affairs contributes to the latterday characterization of adolescence in largely epistemological terms, as a period of "thinking about thinking".

   Kohlberg, on the other hand, came to adolescence as an epistemological phase in his need to ground a moral development theory that was of preeminent importance to him. Within the Piagetian vision, it is quite possible to see the young person's moral concepts as existing in a relation of "horizontal décalage" with corresponding structures of general intelligence. In contrast, Kohlberg has stressed that moral judgment is a different domain, its progress having its own principles of development and its own criteria of validity.

   Kohlberg's position is based on the Kantian argument for the distinctiveness of theoretical and practical reason (Broughton & Riegel 1977). However, he does not argue that moral judgment is independent of intellectual structures. In fact, he gives the impression that the latter obtain a considerable amount of their significance from the role they play in funding moral structures. For example, Kohlberg and Gilligan (1971) say that the significance of attaining formal operations is that it is a necessary (though insufficient) condition for entry into the conventional moral thought that typically governs the social orientation of adolescents. This position supplanted a previous claim (based on intuitive parallels that appeared to be confirmed by Colby's 1973 dissertation) that formal logic corresponded to principled moral judgment.[3] This prior claim had become unfeasible when changes in Kohlberg's theory and scoring system (Colby 1978) had resulted in the finding that real principled thought could no longer be found before adulthood (Kohlberg 1973, Murphy & Gilligan 1980), and that there were no instances of stage 6 before 35 years of age (Kohlberg 1978). This shift reflected the significant discovery that what Kohlberg had been scoring, since his 1958 dissertation, as "principled" moral judgment was in fact only an advanced form of conventional reasoning, the orientation of "a member of society" to "conscience" and "moral law."

Kohlberg and Gilligan's revised claim recently became subject to even further revision. Concomitant conceptual advances and empirical findings indicated that it was not formal operations proper that were necessary for conventional thought, but "beginning formal operations", i.e., a transitional phase between concrete and formal thought during which concrete operations are elevated from the level of directly manipulable objects to the level of verbal discourse (Colby & Kohlberg 1976, ref. note 2).[4] Many cognitive-developmentalists blithely assume the veridicality of the "necessary-but-not-sufficient" relation. True, there is some support for Colby and Kohlberg's claim (Walker & Richards 1979, Walker 1980). However, six other studies have produced varied results; and four of them succeeded in confirming the most out-of-date hypothesis, that formal thought is necessary but not sufficient for principled reasoning (Broughton 1981, ref. note 3).

The Beginning and End of Adolescence. One of the consequences of the most recently postulated "necessary-but-not-sufficient" relationship between logical and moral development is that it leads us to expect a certain incidence of formal operational but preconventional subjects. In

other words, the joint proposition that adolescence is relatively universal, and is structured by the advent of conventionality no longer seems tenable. Moral stages no longer
hold any potential for characterizing or explaining adolescence in any straightforward way.

The finding that even some adults may not have attained
formal thought presented the first major obstacle to a simple cognitive-developmental explanation of adolescence. At
least, it pressed for a more complicated account which recognized two cognitive types of adolescence: "formal" and
"concrete" (Blasi & Hoeffel 1974). The additional findings
of moral conventionality lagging behind logical formality
exacerbates the problem, and presses for the recognition of
two moral types of adolescence: "conventional" and "preconventional".

If we look at the American longitudinal studies of moral development (Kohlberg 1978, Holstein 1976, Kuhn 1976),
we find a consistent pattern of age-related stage development. There are no conventional thinkers before age 10 and
no preconventional thinkers after age 17, regardless of sex
or class; half the subjects have attained the conventional
level by 13 years of age (with a slight lag in the working-
class subjects). Thus, unless adolescence is to be confined
to a certain proportion of individuals, it is hard to use
Kohlbergian theory to account for it, much as it is hard to
use Piagetian theory of logical development to account for
it. Since some subjects had developed a conventional morality by age 10 while others did not do so until age 17, the
advent of conventional moral judgment could explain either
the beginning or the end of adolescence!

In addition, as we have noted, it is difficult to
explain the end of adolescence and the beginning of adulthood in terms of the advent of principled thought, given the
more recent empirical findings that reveal how difficult it
is to attain that level of reflective judgment. As argued by
Kohlberg (1973), another possibility, suggested by Perry's
(1968), Kramer's (1968) and Turiel's (1974) studies, was to
see the relativistic questioning of conventional morality as
the end of adolescence. Kohlberg (1973) has called this relativistic transition between levels "stage 4-1/2" (Broughton
1978a, 1978b; Gibbs 1979), while Boyd (1976, 1980) entitled
it "sophomoritis". In this phase, the rejection of conventional moral judgment supposedly begins with the realization

that the given, consensual standards of right and wrong in a
specific society constitute a morality that is only one
arbitrary choice among a variety of possible alternative
systems (Turiel 1974).[5]

However, outside of Harvard undergraduate and compar-
ably privileged populations, relativistic stage 4-1/2 "sophomoritis" does not seem to be the norm, or even to be common. For example, Berkowitz, Gibbs, & Broughton (1980)
found not a single case in a sample of 100 undergraduates in
a state university in Detroit. They found that even stage-4
subjects were infrequent. Although Murphy and Gilligan
(1980) appear to have located a considerable number of Harvard students in transition between conventional and post-
conventional levels, it is unclear that "sophomoritis" is
prevalent even in the Harvard population. Inspection of the
original intervention study by Boyd (1976, pp. 6137f.), upon
which the "sophomoritis" notion is based, discloses that not
a single subject in either experimental (n=11) or control
(n=10) groups was scored as transitional between stages 4
and 5 on the pre-test, and only two subjects in the experimental group were so scored on the post-test. In fact, 18
out of the 21 were at conventional stage 3 on the pre-test.

This conflicting set of claims and findings has been
partially clarified by Colby (1978), who makes the useful
conceptual and empirical distinction between those "sopho-
moritic" subjects who are in transition between conventional
and post-conventional levels, and those who are merely involved in a transition within the conventional level. The
latter, moving from stage 3 to stage 4, she has dubbed
"stage 3-1/2" (cf. Ries 1978). Despite the fervent self-
questioning and turmoil experienced by these subjects, their
skepticism is confined to the relativity of individual opinions, not social conventions. The relativity of opinion that
Colby describes is reminiscent of the "decentered" relativism described in Piaget's early work as emerging in the pre-
adolescent period. Turiel (1980) finds a parallel relativistic view of conventions in the 11- to 13-year-old age group,
and Selman (1980) finds an equivalent form of perspective-
taking in a comparable, though slightly younger, age range.
Since the age of attaining Kohlberg's third (conformist)
stage is so variable (from 9 to 17), it seems feasible that
relativistic questioning succeeding it would appear at
equally variable points across the adolescent period.

An early manuscript of mine (Broughton 1969, ref. note 4) pointed out two features of related interest in the original Kramer (1968) data that had stimulated the notion of "sophomoritis". First, with the scoring system employed at that time, each subject received varying percentage scores spread across a wide range of stages (often as many as four or five). Second, careful inspection in the original data of the pattern of percentages (not reported by Kohlberg & Kramer), and its shift over the college years, revealed that some of the subjects in question showed a substantial proportion of stage-3 thought throughout, in addition to stage-4 and 5 thought. For instance, "case G" had been scored as having 27% of stage-3 thinking at age 16 and 33% at age 19; for "case W", the corresponding figures were 22% and 16%. At that time, I queried some of the scores and suggested that they might be disguised forms of stage-3 reasoning. Although there have been changes in the scoring system, the scoring of stage 3 has not been altered drastically. Therefore, these and similar cases of "sophomoritis" might well have been instances of the more primitive, individual relativism within conventionality that Colby has dubbed "stage 3-1/2".

Finally, Colby's distinction receives some additional corroboration from my finding of two distinguishable levels of relativism (individual and societal) in epistemological thought (Broughton 1979b). Nevertheless, the circumstantial evidence in favor of Colby's speculations does not resolve our problem of aligning the phenomenon of adolescence with cognitive development and its boundaries with specific stage transitions. Given the nature and variability of the age-by-stage data from cross-sectional and longitudinal studies mentioned above, it does not seem likely that even Colby's "3-1/2" will serve to mark consistently the end of adolescence.

## RECENT COGNITIVE-DEVELOPMENTAL THEORY: KOHLBERG'S DUAL SYSTEM

Contrasting Types of Experience. Kohlberg's original dissertation (1958) was a study of adolescents, and therefore could not answer questions about adulthood. Since adolescence involves physiological, hormonal, and hypothalamic changes, he was unable to show that the transformations of moral cognition were not simply reflections of biological maturation. The "sophomoritis" issue raised, for him, the question of whether there were any stages that were to be found only in adulthood. At first (Kohlberg & Kramer 1969), he answered this question in the negative. Only later (Kohlberg 1973), when he was able to elevate the criteria for post-conventionality and distinguish real principled thinking from pseudo-principled (sophisticated conventional) thinking, was he led to give an affirmative answer. At that time he was able to define adulthood cognitively in terms of post-conventional reason, (albeit with the corollary that most adults were trapped in an endless adolescence!).

The finding of cognitive transformations peculiar to adulthood, and reserved for that period alone, at last enabled Kohlberg to firmly establish the cognitive-developmental approach as a psychological theory that was irreducible to biology, a task that had troubled him for some time (Kohlberg 1968). As he had realized since Kramer's dissertation (1968), structural transformations exclusive to adulthood were immensely significant because "such change must be the result of experiential interaction with the environment rather than being linked to biological maturation" (Kohlberg 1973, p. 179). Thus the discovery of truly adult stages promised a disentangling of the roles of maturation and experience in generating stages and stage-change, something that Piaget was never able to achieve (cf. Kohlberg & de Vries 1971).

Over the years, and with the help of some of his students and colleagues (see Boyd 1976; Perry 1968; Colby 1978; Gibbs 1977, 1979; Belenky et al. 1979; Murphy & Gilligan 1980), Kohlberg has been considering the possibility that something is qualitatively different about development beyond conventionality that makes it a distinct and more difficult phase than development up to conventionality. In the 1973 paper, we can see that his quest for "adulthood stages" has matured into the search for processes of transformation that may be peculiar to adult development. The approach is no longer formulated in terms of a contrast between adolescent and adult structural "types"; rather, it is couched in terms of the opposing forms of "experiential interaction with the environment".

Kohlberg advances the intriguing hypothesis that adulthood is distinguished from adolescence not only by novel structural stages but, more fundamentally, by the emergence

of a qualitatively new mechanism of developmental progress,
located in a different ontological and epistemological
sphere.  He proposes that, in becoming adults, individuals
shift the basis of their development from criteria of cognitive consistency to reflective understanding and appropriation of concrete personal experience.  Thus is exposed to
view, lying behind the structural stage approach, the theory
of the role of experience in development:

> The nature of the experiences leading to adulthood devel-
> opment are somewhat different than those involved in
> childhood and adolescent movement to the conventional
> stages of moral reasoning. Development of moral thought
> in childhood is an increasingly adequate comprehension of
> existing social norms and social ideals. Accordingly, it
> develops through the usual experiences of social symbolic
> interaction and role taking. In contrast, construction of
> principles seems to require experiences of personal moral
> choice and responsibility usually supervening upon a
> questioning period of "moratorium". (Kohlberg 1973,
> p. 180)

Adolescent development, like its childhood precedent,
"is largely cognitive and symbolic and does not require
large amounts of personal experience....The experiences
which generate stage movement have a strong general and
symbolic component; they are experiences involving thinking"
(ibid., p. 193).  The experience of emotionality and personal involvement in those social experiences (e.g., discussion) that engender conflict and stage advance is actually
only serving as "a trigger to general thinking. It is a
different matter than the role of emotion in a concrete
moral choice, the consequences of which one lives with and
emotionally experiences long afterward" (p. 194).  Through
discussion and other experiences of cognitive conflict,
developing adolescents come to choose the next stage up as a
general mode for resolving moral problems on non-personal
cognitive grounds, regardless of the particular experiences
associated with either stage solution.  The adolescent is
involved in the relatively impersonal task of learning to
perceive the established social system.  The process of
role-taking that mediates this learning is an abstract,
symbolic mental activity. Enhancing the opportunities for
role-taking brings about developmental change even in the
most depersonalized of contexts (Turiel 1966; Berkowitz,
Gibbs, & Broughton 1980). However, on its own, it cannot
move stage-4 thinkers to stage 5.

Principled thinking, on the other hand, is not a more adequate perception of the current social system. It is a postulation of principles to which both society and self ideally ought to be committed. Development in the post-conventional phase is dependent upon "the experience of sustained responsibility for the welfare of others and the experience of irreversible moral choice" (Kohlberg 1973, p. 196). Thus, here Kohlberg emulates Piaget's later theorizing in attempting to bring development to a close by introducing a non-cognitive or supra-cognitive societal factor. For Piaget it was the assumption of a work role that was decisive; for Kohlberg it is the assumption of individual responsibility.

The college years are situated at the interface between these two phases of development, the "cognitive" and the "personal". During this in-between period, the self's new experiences of freedom and responsibility to make its own choices are sufficient sources of reflection, questioning, and commitment to set the late adolescent/young adult on the path to principled reasoning.

The Problems of Dualism. Kohlberg's (1973, ref. note 5) dualism of experiential modalities is supported at the theoretical level by a dual explanatory system. Childhood and adolescent development are accounted for by standard cognitive-developmental or "structural" theory. Adult development, on the other hand, is made comprehensible through "functional" theories, such as those of Erikson (1968) and Fowler (1978, 1980). Those theories deal with life-history in terms of conscious choices, or uses by an ego of new functions which serve to organize personal experience in a stable and purposive form (See Noam, G., L. Kohlberg & J. Snarey in this volume).

A manifold of problems greets Kohlberg's (1973) "integrative" approach to lifespan development. Some of these have been considered in detail by Gibbs (1979) and Philibert (in press). Let me summarize what I think are the basic difficulties. Kohlberg attempts to resolve the apparent contradiction between the two types of experience by positing an area of "overlap" or "interaction". However, in the absence of either conceptual or empirical analysis of how this takes place, such a tertium quid turns out to be no more than a euphemistic restatement of the problem. It is about as successful as was the act of positing a "mental-physical substance" in resolving the dilemmas of the famous mind-body dualism. Much as interactionist psycho-physics has tended to degenerate into monisms that are nothing more than disguised materialism and idealism (Young 1970), so too, the transitional bridge of "college experience" has tended to dissolve into its more cognitive or personal vectors. At times, Kohlberg seems to claim that functional adulthood is the committed choice of a late adolescent self. Why, then, should we typify adolescence as a structural development? In other statements, especially vis-a-vis Boyd's (1976) pedagogic intervention, he argues that cognitive-structural advance precipitates adult development. Why, then, should we typify adulthood as a functional development? And if structural and functional are somehow combined in development to maturity, how can this be explained in the face of the antagonistic and even contradictory presuppositions of the two theoretical orientations (Broughton 1981b, ref. note 6)?

Anyway, is it not intuitively clear that cognitive and personal experience are never divorced from one another in any phase of development, with the possible exception of pathological cases? No wonder, then, that Kohlberg (1973, p. 198) himself admits his inability to satisfactorily integrate the two moieties of his new lifespan theory. It is the way in which he has segregated the two halves of both experience and explanation that defeats the purpose and leads him to an awkward position of resigned eclecticism. If theory is to develop, and so do justice to the development of thought and personality, it must reconsider the ground on which it stands so that it can then take a step forward.


## THE END OF ADOLESCENCE: IDENTITY FORMATION

The Theoretical Need for "Identity". As we have seen, there are certain consistencies in the cognitive-developmental approach to adolescence, regardless of whether it is the Piagetian or Kohlbergian version that we are considering. For example, what originally appeared to be a convincing theoretical explanation of the psychological inception of adolescence (formal or conventional thought) turned out to be empirically untenable due to unexpectedly wide variations among subjects in the age of attaining the specified stages. At the other end of adolescence, there was a similar empir-

ical problem, with no particular structural transition in
development identifiable that could clearly and uniformly
define or demarcate the end of adolescence.

Regardless of this empirical situation, there is the
conceptual problem that cognitive-developmental theory has
consistently been inadequate to the task of explaining how a
purely cognitive change could represent accession to adulthood. It has had to submit to the indignity of importing a
non-cognitive factor from outside the theory in order to
make its developmental explanation appear complete. Gilligan
and Murphy, Elkind, and others have followed Inhelder and
Piaget's example in invoking employment or vocational training as the crucial factor terminating adolescence. Kohlberg
and, to some extent, Perry have appealed to another sociological phenomenon: the separation, independence, and
responsibility for self of the individual consequent upon
attendance at an institution of higher education. As in the
case of Piagetian theory, the explanatory role of this
supra-cognitive factor has assumed much greater significance
as the empirical research has revealed the lack of correlation between specific stages and specific ages after childhood.

In both cases, Piagetian and Kohlbergian, the central
factor marking and shaping the transition to adulthood is a
normative change in social location, one connected to work,
the reproduction of the division of labor, and thus the
reproduction of societal structure and stratification. The
cognitive-developmental position has no place for social
structure in the formation of thought, except as the pure
object of knowledge. This is what is implied by the cognitive developmentalists' term "social cognition", meaning
knowledge that takes the social as its content (Broughton
1981c). Given this social cognition approach, the normative
change in social location appears as an extrinsic factor
invading the developmental process. From the standpoint of
a theory of cognitive psychology, less ground would have to
be ceded to sociology if it were possible to locate in individual subjectivity the cognitive correlates, or, better,
prerequisites, of such extrinsic social transitions. This
would allow the changes associated with work socialization
to be imbedded in, or at least tied into, the genesis of
consciousness.

It therefore comes as no surprise that the cognitive-developmental theory of adolescence (Kohlberg & Gilligan 1971; Kohlberg 1973; Colby 1978; Gilligan & Murphy 1979, ref. note 7) has begun to focus upon the Eriksonian notion of identity formation.[6] In particular, these writers have implicated the so-called "identity crisis" (Erikson 1956) in the transition to cognitive adulthood. Empirically (Meilman 1977, Bourne 1978a), identity formation tends to be most intensely experienced during the college years, the years that are of such interest to Kohlberg. The suggestion of a relationship between moral and identity development would presumably be congenially received by Erikson, who has always maintained that moral ideology is a factor in identity development (e.g., Erikson 1963).

Marcia's Ego-identity Statuses.    Much of the recent thinking on identity has been stimulated by Marcia's (1966, 1976) work.  He has broken down Erikson's fifth stage of "identity versus role diffusion" into four ego-identity statuses, and operationalized them in terms of interview methods and analysis. The four statuses are: "diffusion", "foreclosure", "moratorium", and "identity achievement". The two referents employed for identifying identity status are "crisis" and "commitment" which are equivalent, respectively, to a period of making a decision between alternatives, and a firm choice among those alternatives. Deliberation and selection may take place in one of three areas: occupation, religion, and politics.[7] Diffusion is a state where commitment is absent, unless one counts the commitment to a lack of commitment (Marcia 1964). Foreclosure is a state of clear commitment despite never having experienced crisis. Moratorium is the state of crisis itself, an active struggle which temporarily precludes commitment. Identity achieved is the status in which a commitment is sustained following such a period of crisis.

Originally, Marcia envisaged this breakdown as a fourfold typology (Bourne 1978b, p. 381), but in a recent review (Marcia 1980), he has expressed a reasoned desire to rethink this in terms of developmental process. On the basis of his own findings, Matteson (1977) had already recommended such a course of action. Whether or not the ego-identity statuses comprise a single, unilinear sequence is still under debate. However, the data on age relatedness, and the kinds of transitional forms found, suggest that there is a general developmental tendency to move from the diffusion status toward the achieved status (Waterman, in press; Stark & Traxler 1974; Meilman 1977; Adams et.al. 1979b), at least for men (Marcia 1980). It is not yet clear that movement out of dif-fusion is always transitional to foreclosure, and only thereafter can moratorium be reached. On theoretical and empirical grounds, Adams et.al. (1979b) have indicated that there is support for the alternative route directly from diffusion to moratorium. Matteson (1977) has suggested that foreclosure would be expected to precede diffusion in sequence. However, the received view is more in line with Adams and Shea (1979a, p.88) who argue empirically that dif-fusion is the normal first step, an opinion that would appear to be more compatible with Erikson's perspective.

The only thorough longitudinal study for which the data are presented clearly is Marcia's six-year follow-up of 30 students. All 16 diffusion and foreclosure students did not depart from theoretical expectations, showing considerable stability. However, to everyone's embarrassment and con-fusion, 3 out of 7 identity-achieved subjects "regressed" to foreclosure, as did 2 out of 7 moratorium; 2 other moratori-um subjects "regressed" to diffusion status. Marcia admit-ted that some of these apparent anomalies might be attribu-table to measurement error. Afrifah's (1980) discussion of scoring conventions encourages us to consider that the mora-torium "regressors" might well reflect instances of measure-ment error, since, for pragmatic reasons, Marcia ascribed the moratorium status wherever there was any mixture of scores across occupation, religion, and politics domains. And a variety of studies have reported that foreclosure and achieved statuses are psychometrically close to each other, making it hard to discriminate between them (Schenkel 1975; Marcia & Friedman 1970; Toder & Marcia 1973; Afrifah & Broughton, ref. note 1). Last year, Marcia (personal commu-nication) suggested that a serious identity crisis may elude detection because of repression, causing an identity-achieved subject to be scored at the foreclosure status. However, and despite the small $n$, his final conclusion in the 1976 report of the study stated that, to a large extent, his results were the consequence not of error but of an ac-tual rigidification in his subjects' identity occasioned by expedient commitments to career advancement. The fact that his most recent reflection on these longitudinal results (Marcia 1980, p. 172) appears to minimize their incongruity suggests that he may have had second thoughts about the whole study and its implications.

The   Relationship  of  Identity  Formation  to  Moral  Development.    Remember  that  prior  to  1973  (e.g.,  Kohlberg  &  Gilligan  1971),  the  prevalent  conception  of  and  scoring   criteria  for  post-conventional  moral  judgment  were  consonant  with
the   expectation  that  it  would  most  frequently  appear  around
the  transition  from  adolescence  into  adulthood.  True    adulthood  was   still   thought   of   as  morally  principled,  a  view
which  Marcia  (1980)  maintains  to  this  day.   Within  the   context  of  these  assumptions  about  adulthood  post-conventionality,   an   archetypal   study  of  considerable  significance  was
carried  out  by  Podd  (1972).   Podd  related  Kohlbergian   moral
development   to  Marcia's  neo-Eriksonian  statuses,  in  a  study
of  112  male  college  juniors  and  seniors.   In  even  attempting
such  a  thing,  he  presupposed  that  the   statuses   represented
developmental  phases  in  the  resolution  of  identity  issues.

About   half   the  identity-achieved  subjects  reached  the
post-conventional  level  in  moral  judgment,   the   other   half
being  at  the  conventional  level.   Only  about  40%  of  conventional  subjects  attained  the  status  of  identity  achievement.
Only  15%  of  the  subjects   at   the   other   identity   statuses
reached   the  post-conventional  level.   In  comparison  to  chi-
square  expecteds,  subjects  in  the  foreclosure  status   showed
a   tendency  towards  conventional  moral  judgment,  while  those
in  the  diffusion  status  showed  a  tendency  to  be  transitional
(between  conventional  and  post-conventional).[8]  Moratorium
subjects  were  infrequent  and  were  indiscriminately  distributed   across  conventional,  transitional  and  post-conventional  levels,  a  phenomenon  which  may  have  to  do  with  the
peculiar  scoring  conventions  for  that  status  which  were  mentioned  in  the  last  section.

Podd   interpreted   his  results  as  evidence  of  a  one-to-
one  correspondence  between  Kohlberg  levels  and   Marcia   statuses.  In  so  concluding,  Podd  was  clearly  guided  by  his  concern  to  show   that  his  study  "supports  Erikson's  view  that
moral  ideology  is  a  factor  in  ego  identity"  (1972,   p.  505).
[9]  For  example,  he  interpreted  his  data  as  showing  that
"conventional  morality  is  representative  of  a   foreclosure-
type  process"  (p.  504).   However,  the  data  are  subject  to
another  interpretation  which  reverses  the  relationship,  making  ego  identity  a  factor  in  moral  development.   This  alternative  interpretation  is  the  one  chosen  by  Kohlberg  and  Gilligan  (1971,  pp.  1077f.),  and  later  adopted  by   others   such
as  Habermas  (1975)  and  Marcia  (1980).   They  interpreted  the
results  as  showing  that  the  achievement  of   identity   was   a

necessary but insufficient condition for the attainment of
principled moral judgment, and not vice versa. The fact
that about half the identity-achieved subjects were still
morally conventional appears to support the insufficiency
claim; but, barring the possibility of error, the existence
of 3 foreclosures and 3 diffusions at the post-conventional
level, not to mention the possible 9 foreclosures and 19
diffusions in the transitional phase, indicates that the
"necessity" involved may not be very strict in nature.

There is something puzzling about Kohlberg and Gilligan's interpretation of Podd's study. After adopting the
"necessary-but-not-sufficient" stance which we have all come
to know and love, they also claim that the study demonstrates that morally transitional subjects are in a transitional identity status. Here they seem to be shifting to a
"necessary-and-sufficient" stance, one closer to Podd's
"one-to-one correspondence" hypothesis. Kohlberg et.al.
(note 5) have since stated this revised assessment of the
relationship more boldly as a basic assumption of the Kohlberg approach, seeming to suggest that there is just a single phenomenon of transition which has two aspects: the moral and the identity sides. This position has been seconded
and clarified by Colby (1978).

It makes sense that Kohlberg would shift to this
revised position after the 1973 paper in which, as we have
seen, he becomes invested in construing the college years as
a tertium quid that merges cognitive-structural experiences
and transformations with ego-developmental ones. Nevertheless, the new position flies in the face of the fact that
half of Podd's identity-achieved students were still morally
conventional. In addition, only 4 out of the 34 subjects in
the moral transition phase were classified by Podd as moratorium. It would appear that more than twice as many were
still foreclosed in ego identity. And nearly five times as
many were still in an even more developmentally primitive
state of diffusion.

However, if one looks at the literature, one can redeem
Kohlberg by noting that there has been some support for the
notion of a more mature "post-crisis" diffusion. Podd
argues for such a distinction on the basis of his data.
Marcia himself had already suggested the distinction in his
dissertation (1964), in characterizing a type of student who
was "committed to a lack of commitment" (p. 196). Empirical

evidence in favor of Marcia's and Podd's view comes from Toder and Marcia (1973), Adams et.al. (1979), Schenkel (1975),
Raphael (1975), and Fannin (1979), all of whom report that
diffused and moratorium statuses are hard to separate out,
with a tendency for this to be especially difficult with
respect to women subjects. Thus, it is still a possibility
that there is a post-crisis diffusion, developmentally parallel to moratorium and comprising an alternative transition
from foreclosed to achieved statuses. If this turns out to
be so, then it may be possible to partially redeem Kohlberg's claim that there is a single process of transition
combining moral and identity aspects, since Podd's morally
transitional diffusion subjects could be located in a form
of identity transition rather than in a developmentally primitive state of role-confusion.


Morality and Identity: Other Research and Conclusions.
There are some data from other studies that are consistent
with Kohlberg's claim. Keniston (1968) found that his 14
New Left activists, who were all morally post-conventional,
had all experienced a late adolescent "crisis" of identity.
In Döbert and Nunner-Winkler's (1975) study of German
draft-resisters and army volunteers, the 3 "stage 4-1/2"
subjects were all caught up in an intense identity crisis,
and 9 draft resisters who were post-conventional all showed
clear evidence of having passed through an identity crisis.

However, there are more data that are inconsistent with
Kohlberg's claim. Some of the 15 army volunteers in the
Döbert and Nunner-Winkler study were post-conventional but
had experienced no crisis. Gilligan and Murphy (ref. note 7)
found that only 1 out of 10 subjects in transition to post-
conventional morality was experiencing identity crisis.
Moreover, 3 out of 4 individuals who were in crisis had
already reached Kohlberg's stage 5, and only 1 out of the 8
post-conventional subjects had achieved an identity. And
in Afrifah's (1980) study of Ghanaian youth, five identity-
achieved subjects were found to be at Kohlberg stage 3,
while 2 fell below that level; and only 1 scored as high as
3/4. These data seem to contradict not only Kohlberg's
claim, but also the earlier Kohlberg and Gilligan "necessary-but-not-sufficient" hypothesis based on Podd's results.
Finally, there is the finding (Afrifah & Broughton, ref.
note 1) that Piagetian logical stages show no systematic relation to the attainment of specific identity statuses. If

moral development is regularly contingent upon logical development in the way that Kohlberg and his school claim,
then some clear relation should obtain between stages of
logic and identity formation. This does not appear to be
the case.[10]

Of course, the above-mentioned changes in scoring criteria and the discrimination of "3-1/2" from "4-1/2" might
alter the pattern of findings (Colby 1978), especially in
the case of the Keniston and Döbert studies, which used an-
tiquated methods of scoring. However (Broughton 1981, note
3), a downward revision of moral scores promises to result
in the general finding that identity is already attained by
either of the two conventional stages of moral development.
This pattern would be damning to the later Kohlberg claim,
and only partially supportive of the earlier Kohlberg and
Gilligan hypothesis. Most important, it would indicate that
neither identity crisis nor identity achievement is a central part of the accession to moral post-conventionality.
This conclusion in turn leads us to a further suspension of
faith in the revised "structural-functional" theory by which
Kohlberg (1973) attempted to explain the end of adolescence
and the beginning of adulthood.

Finally, the equivocality of these various empirical
results leads us to wonder whether cognitive-developmental
theory is capable of giving a convincing interpretation of
major life-transitions in psychological terms. If not, the
sociological grounding of those transitions would appear to
place greater claims upon our attentions. Or, possibly, an
alternative psychology would have to be sought.

## THE DIAGNOSIS OF COGNITIVE-DEVELOPMENTALISM

Explanatory Devices. When one traces the vicissitudes
of Piagetian and neo-Piagetian psychology--for example, as
it pertains to adolescence--one is impressed with the fruit-
fulness of its central concepts. Its dense internal structure and its labile interface with empirical activity have
provided a framework for the elaboration of a rich matrix of
ideas and facts. As it has evolved in diverse directions, it
has developed and articulated old questions about the
adolescent process, and generated new ones. It has suggested the possible role of a range of forms of consciousness in shaping adolescent phenomena.

Thus, the <u>beginning</u> of adolescence has been postulated as the birth of formal logic, the advent of hypothetical thought, the emergence of idealization, the proliferation of possibilities, and the entertainment of counterfactuals. It has been redescribed in terms of the capacity to think about thinking. This, in turn, has been equated with the appearance of self-consciousness, or, in the Kohlberg school, the capacity for reciprocal role-taking and even the discovery of self itself. Table 1 summarizes some of the many constructions placed upon early adolescence within this tradition.

TABLE 1

Characteristics of Emerging Adolescent "Self"
According to Developmental Theories of Cognition

| | |
|---|---|
| G.S. Hall (1904) | "Self-feeling"; self-consciousness; selfishness and altruism; idealism and moral heroism; romantic love; religious enthusiasm, historical sense; love of nature. |
| J.M. Baldwin (1906-1911) | Psycho-physical dualism; differentiation of self from experience; differentiation of subject-self from object-self. |
| J. Piaget (1929) | Subjectivism (instruments of thought situated within ourselves); relativism (conceptions are relative to a point of view, and there are alternative points of view). |
| G.H. Mead (1936) | Self-conscious creation of own life-history ("romantic self"); an opposition to norms. |
| M. Debesse (1943) | "La prise de conscience de soi"; need for deviation from norm; self-isolation, and secretiveness; introspective monologue. |
| B. Inhelder & J. Piaget (1958) J. Piaget (1969) | Generation of all possible realities; propositional logic; formation of a life-plan; ideological egocentrism and Messi- |

(continued)

Table 1   (Continued)

anism;   epistemological   self,   thinking
about thinking.

| | |
|---|---|
| D. Elkind<br>(1967) | As for Inhelder & Piaget, plus: awareness of arbitrariness of own mental constructions; differentiation of cognition from perception; grandiose sense of imaginary audience; personal fabulation (of immortality, uniqueness, etc.). |
| P.A. Osterrieth<br>(1969) | As for Inhelder & Piaget, plus: discovery of interiority (freedom, originality, authenticity, and responsibility); withdrawal into imagination. |
| L. Kohlberg &<br>C. Gilligan<br>(1971)<br>C. Gilligan &<br>L. Kohlberg<br>(1978) | Self-consciousness; interpretation of truth and morality within subjective experience; solipsism and Cartesian "cogito"; romantic love; tension between real and ideal; transcendental religiosity, esthesis, and naturalism. |
| H. Gruber &<br>J. Vonèche<br>(1976) | As for Inhelder & Piaget, plus: formation of personal point of view, personal code; withdrawal from action; epistemological reflection on processes of knowing. |
| M.J. Chandler<br>(1975a, 1975b,<br>1978) | Epistemological relativism, perspectivism and solitude; construction of artificial consensus, suppression of individuality; religious and secular (e.g., scientific) conversion; prejudice and stereotypy; spiralling formalism; appreciation of irony, paradox, and contradiction. |

The end of adolescence has been a less popular subject of discussion. Most developmental psychologies have been content to be child psychologies, and have refrained from trying to characterize the adult mind. As a consequence, they have been unable to delineate that fine but significant line between late adolescence and early adulthood. Within the cognitive-developmental approach, both Piaget and Kohlberg have had difficulties on this score. Piaget has suggested that a flood of accommodation and an ebb of assimila-

tion must surely compensate in late adolescence for the ebb
of accommodation and flood of assimilation in early adolescence. Pressed to concretize this metaphor, he has fallen
back on his Baldwinian equation of excessive assimilation
with play and compensatory accommodation with imitation.
Thus, the end of adolescence comes with the imitative reproduction of the established social order.[11] The young person (man?) responds to the vocational summons, and assumes a
pre-structured work role by the grace of society's generos-
ity. Discipline descends upon the unruly imagination of the
pubescent youth, channeling its diffuse energies in a single
direction. This labor instructs the intellect and dictates
the form and occasion of its rationality. The self-construc-
tion and autonomy of reason appear to be compromised by the
limits of the developmental theory. And it is therefore not
a surprising finding that the attainment of formal thought
varies with social stratum, available forms of labor, and
degree of societal modernization (see, for example, Buck-
Morss 1975).

Kohlberg, on the other hand, has postulated a specific
cognitive transition to explain the end of adolescence. It
is the transit from conventional to post-conventional morality, founded upon the consolidation of formal logic, yet extending beyond it. As a consequence of psychometric innovations, empirical findings, and conceptual realizations, he
has withdrawn from this position and forged a theoretical
rapprochement with functional theories of ego development.
For the new Kohlberg, personality is no longer an after-
thought; it merely comes after thought.

At the nexus of thought and personality is the identity
crisis, in which self is formed. Thus, paradoxically, the
discovery of self serves a double explanatory function: it
initiates and it terminates adolescence. Of course, the
discovery is construed in two different, developmentally
disparate forms. Around puberty, the self discovered is the
Baldwinian "subject-self" (Broughton, 1981d), the mental
knower aware of its own subjective states. At college age,
the self discovered is the self-responsible person, transcending its social roles by exerting the power of choice
among them. The genesis of the individual is one from mind
to self, and only thence to society. Adolescence has, somewhat magically, transformed the contemplative, cognitive
self into an active, existential self. We have seen above
how difficult it is to understand this change from head to
heart in terms of the merely structural elaboration of for-

mal or conventional consciousness. Also, the unruly empirical findings fail to fall neatly into line with the predictions. As a result of this explanatory weakness in cognitive-developmental theory, its confidence in the generative
power of thought has been persistently undermined by the
feasibility of the rival sociological explanations that
account for development in terms of socialization processes.

Self and Society. In general, it seems fair to say
that cognitive-developmental theory has sound body parts,
but is weak at the joints. In themselves, the notions of
"preconventional" / "conventional" / "post-conventional" and
"concrete" / "formal" levels seem reasonable enough. However, the major turning-points between developmental phases
seem to present all kinds of problems. Crucial stage transitions seem poorly understood as it is. But to make matters
worse, they do not seem to coincide with each other across
domains (logic, morality, identity), nor do they appear to
coincide with major periods of the lifespan (childhood/ado-
lescence/adulthood) or the important life events (puberty,
going to college, taking a first job) that mark the boundaries of those periods.

Of course, one possible escape route from this dilemma
is to dismiss "life periods" and "life events" as superfi-
cial colloquialisms imposed upon the course of our experience, ones that conceal and distort the real underlying
genetic processes. However, this is easier said than done.
The conceptual and pragmatic segmentation of the life-span
and the articulation of its parts are the primary ways in
which the development of individuals becomes understandable
to society. Just as clearly, this societal self-understand-
ing tends to provide the foundation for the self-understand-
ing of the growing individuals themselves, and their capacity to orient to and participate in their own development.

The unfortunate truth of the matter is that cognitive-
developmental theory, despite all its credits, exhibits the
major debit of having effectively eliminated both self and
society from psychology. Certainly, self gets a mention,
but only as the object of knowledge or discovery (Broughton
1981c). It is never seen as comprised by or formed in terms
of its own subjectivity (Blasi, this volume). It is equally
true that society gets a mention, but only as the object of
knowledge (Broughton 1978c), or as a rather mysterious abstraction that the formal, post-conventional, or identity-

achieved individual comes to reflect on and construe as a
formal instantiation of one of a system of possibilities.
But society, like the self, is a concrete actuality. Both
self and society have their living histories, by virtue of
which they exist. A theory that eviscerates these two vital
poles of human reality will have to engage in some mighty
fancy footwork in order to retain a modicum of viability.

The Concept of Adolescence. Something similar could
be said of adolescence. It is an historically created phenomenon, not a natural category (cf. Keniston 1970). It has
its own course of development, including multiple phases in
each of which it has undergone a thorough reinterpretation
(Kett 1977; Gillieron 1979). Adolescence is not amenable to
a naturalist psychology. It cannot and will not sit still
long enough. Thus, even in the successive views of teen-age
presented by Inhelder and Piaget, then Kohlberg and Gilligan, and now the contemporary writers in the cognitive-
developmental tradition, we see reflected the historical
sequence of evanescent and disjunctive images of youth. The
upright, inner-directed, old middle-class, young Swiss Cal-
vinist man is transformed abruptly into the swinging,
other-directed, new middle-class Yanqui androgyne, who in
turn metamorphoses into a confused but responsible modern
middle-class careerist, intent upon prudent occupational
choice and secure citizenship as a refuge from post-Viet
Nam, post-Watergate, post-Keynesian cynicism. The shifting
series of authors carrying the cognitive-developmental baton
unwittingly bears witness to the shifting nature of the
adolescent that is the object of its theory. In doing this
so unwittingly, perhaps they even promote and potentiate the
tendency toward further shifts.

While even cognitive-developmentalists implicitly
acknowledge that the nature of adolescence is historically
constructed and transformed, they are committed to an evolutionary model rather than a historical one. This model
underlies their conception of individual, society, and
(especially) the individual-society relationship. Although
the model appropriated has been a form of naturalism somewhat subtler than G.S. Hall's crude social Darwinism, it has
maintained Hall's notion that adolescence holds the key to
man's transcendence.

If, like Piaget and Kohlberg, one focuses upon the increasing rationalization of society, progress appears as the

dominant feature of social history. If, in addition, one's world-view is based in liberalism (e.g., Hobhouse 1964), this progress must be gradual and reformist in caliber (Roberts 1977, Unger 1975). If, in addition, one holds to an underlying naturalism, then this gradual progress must reflect an "inheritance of acquired characteristics". If, in addition, one's naturalism is individualistic, then it must be the discoveries and creations of successive generations of individuals that contribute to the ever-renewed transcendence of previous social norms and achievements. Ontogeny then assumes a pivotal role in introducing new advances that sustain the general trend of social progress (Baldwin 1902; Broughton, 1981a). This ontogeny must have a form such that its normative course allows for the personal transcendence of established societal structure.

Psychology is certainly a promising discipline within which to formulate an explanation of the form of development, particularly since thought would appear to be a very viable way in which the individual in the full course of growth could pass beyond previous achievements by criticising the received view and constructing a more reasonable perspective. Of course, one would not want imagination to run riot; an excess of imagination threatens to amount to uncontrolled impulse as Adorno (1968) has ironically pointed out. Therefore, one construes imagination within the narrow scientific vein, as the formulation of hypotheticals to be tested against an ever-present and insistent reality. Reality, like cognition, is a structure; and so any particular creation or discovery is muted by the necessity that it be assimilated into the habitual, and that it restructure all the established relations of parts, before it can be said to have transformed the whole.

Piaget, Kohlberg, and Erikson all share this view of a constrained and therefore domesticated creativity. Their epistemology is a classical realism, in which mental constructions can never retain validity while running ahead of reality. "Transcendence" (formal thought, post-conventionality, generativity) aspires only to reach the real, not to transcend it. This kind of safe and homely reality principle dissembles and presents itself as transcendence only by virtue of the fact that the majority of people are sufficiently truncated in their development that reality is still beyond them.

For Piaget, the task of mental construction is not entirely lonely. The reality that confronts the thoughtful individual is social (i.e., interpersonal) as well as environmental. However, even the social world is natural, in the sense that laws of equilibrium apply to the ecology of inter-individual transactions as much as they do to that of intra-individual ones. Perspectives can be subjected to a calculus of reversible exchange just as well as operations can (Piaget 1967a; Kitchener, in press). Thus, interpersonal structures could share the same very general form that pertained to cognitive structures. In the social world, rather than the constraint of converging schemas, it is others' viewpoints that act as a criterion and corrective, sustaining a social-cognitive balance, and modulating any wild excesses of particular individuals. Thus the cognitive-developmental approach has tempered its individualism by incorporating the equivalent of symbolic interactionism, and with it, the notion that truth is, more or less, the prerogative of a logical community (Peirce 1931-58; Baldwin 1906-11; Habermas 1971).

It is not surprising that a theory compounded of these various conceptual layers comes to focus on adolescence. As Piaget has noted, modern adolescence is, par excellence, the time of peer relations. Several social theorists have pointed out how such a view was promulgated in advanced capitalist states as part of the transition from direct hierarchical authority to indirect lateral social control via ideology (Henry 1963; Seeley et.al. 1956; Bernstein 1975; Riesman 1950; Silver 1979, ref. note 8; Lasch 1977; cf. Broughton, 1981a). For Piaget, adolescent peer relatedness held the crucial promise of autonomy in the form of freedom from adult constraint (Youniss 1981). "It is because the child cannot establish a genuinely mutual contact with the adult that he remains shut up in his own ego....The adult takes advantage of his situation instead of seeking equality" (1932, pp. 61f.). This important sociological insight was gradually eclipsed in Piaget's later work, since, as we have seen, Piaget devoted himself increasingly to a purified bio-psychological theory of the development of logico-mathematical structure. The formality of thought came to be seen as the saving grace of development, and in the later, better-known, theory it was logical closure of cognitive relations, rather than escape from heteronomous social relations, that was seen as the harbinger of autonomy.

Kohlberg assisted the Piagetian eclipse of the role  of
authoritarian social relations in development.  In addition,
he  perceived  a  deeper  level  of  autonomy that had to be
attained in moral reasoning (Kohlberg 1981).   In  realizing
that  it  was  rights and not rules, ethical revision rather
than cooperative regulation, that guaranteed autonomy, Kohlberg delayed the age of majority for  freedom.   Development
was revealed as something much harder and longer than Piaget
had acknowledged.  The acquired characteristics necessary to
advance  the species as a whole were not as simply gained as
the Genevan school had thought.

However, maintaining the late capitalist image  of  the
adolescent  as  hero,  Kohlberg preserved a charismatic role
for the teenager at the fulcrum of development.  Rather than
a telos, adolescence became a transition, one with the  double whammy of self-discovery and social transcendence.  This
two-stage  rocket was meant to launch the child rapidly into
the orbit of the social critic.  The  propulsive  mechanism
was  to  be role-taking, the quintessentially other-directed
cognitive act that the social  theorists  like  Riesman  had
described as the modern form of social control.

The  Conventional Nature of Symbolic Interaction, Role-
Taking, and Identity.   Nevertheless, as we have  seen,  the
course  of  events  has  revealed that development is not as
hard as Kohlberg supposed; it is harder.  To echo Jacqueline
Susann, adolescence is not enough.  But since it  was  not
enough,  one  more difficulty was in turn posed.  If adolescence was the peer-oriented phase adapted to the development
of conventional morality via symbolic interaction and cognitive conflict, the insufficiency of adolescence for autonomy
implied the insufficiency of symbolic interaction and systematic  role-taking  for  autonomy.   Symbolic  interaction
threatened to be incapable of generating its own  self-tran-
scendence.  At  this  point, Kohlberg (1973) introduced the
alternative mechanism of identity formation,  and  swallowed
his pride enough to allow for "functional" Eriksonian theory
as  a  necessary  supplement to the ailing structural theory
that was promising to truncate development  at  the  conventional level.

Several  other  cognitive-developmental  theorists have
joined this flight to the lifeboats, looking to a  range  of
functional,  pragmatic, contextual, and existential alternatives as the basis for adult development (Perry 1968;  Gibbs

1977; Gilligan & Murphy 1979; Gilligan 1979; Philibert, in press). As Blasi (1975) has pointed out, cognitive role-taking or any other purely perspectival view of sociality (including Piaget's) is inherently incapable of generating a post-conventional perspective. This is why even "post-conventionality" has such a conventional air about it (Broughton, 1981b). In fact (Broughton, forthcoming), the very notion of perspective implies a particular kind of spatialized continuum that precludes the possibility of fundamental social critique. This weakness must be added to the abovementioned inadequacy: that cognitive conflict and exchange not only cannot account for biography, but they also systematically exclude the life-historical dimension from consideration (Broughton 1981c).

Can "identity" rescue developmental theory by providing a mode of individual transcendence? No. At the root of identity formation[12], as well, we find the convergence of perspectives in symbolic interaction:
The sense of ego identity, then, is the accrued confidence that one's ability to maintain inner sameness and continuity (one's ego in the psychological sense) is matched by the sameness and continuity of one's meaning for others. (Erikson 1980, p. 94)

The underlying concept is the organismic and ego-psychological one of mutual adaptation. The sensed match that Erikson describes can be negotiated by either adaptation of the self to the other or by adaptation of the other to the self (Josselson 1973, p. 19). Neither the augmenting of one's own confidence and self-esteem, nor persuading the other to view one the way one views oneself, can guarantee objectivity. In fact, such a definition of identity as Erikson gives seems to encourage the possibility of a "folie a deux." "Meaning" and contextual integration are here stressed at the expense of validity, as they are in related functional theories (e.g., Perry 1968, Fowler 1978, Gilligan 1979, Loevinger 1966, Levinson 1978). It is consensus that is vital, and the match that this consensus produces between self as object of self and self as object to other not only does not require a restructuring of social meanings, it almost seems to demand adherence to their traditional forms. Furthermore, since objectivity and subjectivity are correlative (Cassirer 1923), the sacrifice of objectivity in the understanding of self by both self and other amounts to an equally deleterious sacrifice of true subjectivity.

The Eriksonian Josselson (1973) has graphically illus-
trated the dilemma of functional theory in her thorough
case-study examination of the psychodynamic bases of ego-
identity statuses in women. She reported, not without surprise, that even her identity achieved subjects were remark-
ably conventional. Furthermore, she found that the transition out of moratorium into achievement entailed a loss of
what she considered fundamental human virtues (e.g., vital-
ity, openness, and sensitivity). This pattern is confirmed
by Loevinger's (Loevinger & Wessler 1970) studies of women's
ego development. As documented elsewhere in great detail
(Broughton & Zahaykevich 1980; Broughton, in press), even
her "autonomous" and "integrated" subjects are essentially
conventional, as witnessed by their voluntary subordination
to social tradition, their "renunciation of the unattainable", and their active subscription to standard stereotypes
of work and sex roles.

Levinson's (1978) and Vaillant's (1977) research confirms a similar pattern in men. They find that maturity
brings individuation and a clear identity for most of the
males studied. Yet this "growth" appears as little more
than an increasingly internalized acceptance of conventional
adulthood, defined by traditional criteria of competition,
possessive individualism, and successful domestic and public
life. The disquiet, dissatisfaction, and disillusionment of
these adult men amounts to an experience of self-doubt, of
disappointment that they have been unable to fulfill their
dreams, rather than a serious questioning of the societal
structures to which they have so willingly committed themselves, and which made possible the construction of illusory
goals in the first place (Broughton, in press). Unfortunately, it is this self-defeating matrix of male "identity
achievement" into which the modern professional woman is
unwittingly entering (Zahaykevich 1982).

The Transcendence of Developmental Theory. If developmental theory has so much difficulty in accounting for
transcendence, we seem to be left with two alternatives,
both of which require a radical change of course. On the
one hand, we can choose to transcend developmental theory
itself, i.e., look toward a critical developmental theory
that has a basis in some other, more philosophically, historically, and politically adequate set of presuppositions
about life and society. In this case, current developmental
theory would appear as a defense against acknowledging real

developmental processes. On the other hand, we can choose to accept and refine current cognitive-developmentalism while admitting that it accounts only for the adaptation of individuals to currently existing social structures (Broughton 1981e).

The problem with the second alternative is twofold: it could not prescribe any future ideals by virtue of which potential transcendence could be realized, guided, or even imagined; in this absence, it would be bound to continue to present and promote the act of equilibration or adaptation as the absolute definiens of developmental progress.

If we settle for the first alternative, even by default, there seem to be some potentially exciting directions to follow. Questioning the assumptions of cognitive-developmental theory immediately draws our attention to the fact that Cartesian and Kantian philosophy (in Kohlberg's case), Hobbesian individualism, evolutionary naturalism, and liberal pragmatism (in Erikson's and Kohlberg's cases) have all been subjected to major critical revisions and superseded by more socially sensitive frames of thought. Those that come to mind first are by Rousseau, Nietzsche, Hegel, and Marx. In the modern era, these critiques and alternative theories have been brought to bear on the sphere of subjectivity by critical psychoanalysis and phenomenology (as in Reich, Marcuse, Merleau-Ponty, and others). The gist of these critiques was to reintroduce the concepts of suppression, oppression, and mystification (see Broughton 1979a, 1979b).

The notion of suppression suggests, at the psychic level, the "suppressed": the need or its impulse, and the self that defines the character of desire. The impulsive dimension had been denied by Descartes, reduced to aesthetic sensuousness by Kant, reified as instinct by naturalism, and assimilated to reason by liberal structural or functional psychology. Subjectivity had been given only a formal dimension, or reduced to a mere source of bias, error, or caprice (Broughton & Riegel 1977).

The concept of oppression suggests that the very structure of social life maintains unfreedom, and that the liberation which autonomy presupposes has as its requirement a real and life-threatening struggle (Benjamin 1980). Structural unfreedom and struggle had been either reduced to a mere cognitive act, or justified in terms of the natural

need for power or the arbitrary biological competition for
survival.   What was forgotten was the historical domination
of man by man.

The notion of mystification has never really been formulated within the standard traditions mentioned.   There has
been a pervasive tendency in Western thought to reduce politics  to  morality  and  epistemology to science.  These, in
turn, tend to be reduced to instrumental cognition,  genetic
egoism/altruism, or biological equilibration.  Small wonder,
then, that there has been a systematic eclipse of any political  epistemology  that  could  identify and illuminate the
ways in which state ideology masks reality and  presents  in
its  stead  a  less  true  yet  more palatable truth (Jacoby
1975).

A critical developmental theory could thus  start  from
an  original  basis that recognizes subjectivity, historical
domination, and ideology.  According to  such  an  approach,
adolescence would take on a renewed significance.  Its character  would  not  be  defined by cognitive transformations.
These might have a place in the  overall  scheme,  but  they
would be subordinated to the full formation of the personality  as  a  whole.  They would comprise changed meanings for
self (Blasi, this volume) rather than self-sufficient cognitive structures tending  toward  logical  closure.  Rather,
adolescence  would  appear as a historically motivated phase
of socialization in which either the possibility or impossi-
bility of transcendence is prepared.  Thus adolescence would
no longer be explicable in terms of  an  internal  dialectic
purely at the intellectual level.  Instead, it would have to
be  explained in terms of the integration of the theoretical
component of consciousness with a practical one.  Since  the
construction  of reason in adolescence would have a concrete
personal significance, it would be tied throughout to active
choice,  interpersonal  conflict,  and  socio-political
demeanor.

Of  course, anomie and alienation would tend to encourage the degeneration of this link between theory and  practice.  This simply implies that the regrettable processes of
systematically induced deterioration (e.g., the splitting of
theory from practice), would need to be included in the purview  of  a  new  adolescent  psychology.  Thus, central to
research concerns would be the tension between the internalization of suppression/oppression/mystification and the  ex-

ternalization of the political subject through acts of individual and collective emancipation.

Instead of erecting purely internal sub-domains of
cognition and personality or thought and action, and then
studying their relations, an account of the differentiated
subjective totality would be arrived at by considering the
organized sociopolitical institutions and agencies (e.g.,
military, industrial, pedagogic, medical, scientific) which
selectively form, draw out, recruit, and co-opt areas of
subjectivity.[13] Developmental stages, domains, and transitions lend themselves to and are rapidly subjected to a
formative appropriation by societal agencies interested in
selective promotion and allocation of power and resources
(cf. Bourdieu & Passeron 1977; Vonèche, forthcoming).

Following this line of thought, a new adolescent psychology would not split "occupational identity" from "formal
operations" and then hand on responsibility to empirical
procedures as a way of relating them again. Instead, it
would focus upon their common origin within the depersonal-
ized-hierarchical structure of bureaucratic rationality and
instrumental exchange (Broughton, in press; Buck-Morss
1975). The concepts of "occupation", "vocation", and
"career choice", inherent in the classist notion of an occupational identity are dependent upon the division of labor
and organization of production/reproduction specific to certain monopoly phases of capitalist states. Formal thought
is the preferred mode of regulating and safeguarding such an
organizational activity. To be effectively self-reproduc-
ing, these categories and modes of social control must have
subjective psychological representations (Horkheimer 1972;
Adorno 1974).

However, the empirical connections between formally
similar phases of these internalized representations need
not be simple correspondences, or relations of implication
and necessity. On the contrary, we should expect that the
contradictory nature of liberal "democratic" social organization would appear as clearly in its psychological form as
in its sociological structure. A society deeply divided
against itself implies a severely fragmented psyche. Even
identity itself would need to be internally inconsistent.
This is, in fact, precisely what is found empirically: there
is characteristically inconsistency between occupational,

religious, and political sub-domains of identity formation (Afrifah 1980).

Different societies in different historical phases require more or less internal consistency (Parsons 1964). They may even need to interrupt cultural transmission altogether, in order to permit new forms of stratification and control to emerge (Bernstein 1975). In the modern era, such cultural interruption would naturally be mediated by adolescents, who are in the course of achieving potency, yet still sufficiently labile to accommodate new values and loyalties dispersed through media-linked peer networks, and sufficiently confident that they could even re-socialize their own tradition-bound parents (Lasch 1977).

What we do not yet know is the extent of "adaptive disjunction" between cognitive development and the development of other aspects of consciousness such as the psychodynamic. It still seems possible that cognitive socialization pursues a course that can be characterized in terms that are not unrecognizable to Piagetian psychologists, while the real basis of adolescence is a dynamic psycho-social one which is carried out more or less regardless of specific cognitive phases. Such arrangements of disjunction may be less a reflection of the "natural structure of consciousness" than an expression of the specific needs of a certain type of society at a particular point in its history.

In terms of empirical psychological predictions, a social analysis of adolescent cognitive development might lead to a reversal of the typical expectations. Physical cognition may turn out to be dependent on social cognition for its true form (Broughton 1978c); and logic may derive its normativity from the form of moral judgment, as Kant's original analysis suggests (Macmurray 1957). Moral development, in turn, might be expected to be dependent upon political development (cf. Adelson 1971; Broughton, 1981b). As suggested above, since all the forms of adolescent consciousness previously construed as structures need to be reinterpreted as interdependent meanings to an active self-directing self, each cognitive-developmental life-event may enter into the interpretation of every other event regardless of domain. Logical, moral, and political may infuse each other in a variety of subjectively meaningful ways, each of which may be more or less under the control of the particular subject involved depending upon the grasp that

the individual has taken on his or her own biographical formation.

While it is not possible at the present to be more than
speculative about a "new" adolescent psychology, the reexamination of the foundations of cognitive-developmentalism
seems bound to transform the approach. This chapter has been
offered as a necessary, although insufficient, condition of
that transformation.


## ACKNOWLEDGEMENTS

I would like to thank the following colleagues, conver-
sation with whom has been instrumental in the development of
the ideas presented here: Afrifah Adjepong, Augusto Blasi,
Anne Colby, John Gibbs, Christiane Gillieron, James Marcia,
Paul Philibert, Jacques Vonèche, Ronna Weiss, Tom Wren, and
Marta Zahaykevich.


## ENDNOTES

1.  Of course, any philosopher worth her salt would be
    reduced to a high-frequency shudder by this passage.
    It compounds an erroneous interpretation of epistemol-
    ogy with a conflation of solipsism and the cogito, both
    of which are in turn misunderstood. To add insult to
    injury, on the same page, all the above are confounded
    with the "romantic" self and the "transcendental" self!
    There is a peculiar indecisiveness in this neo-Piaget-
    ian account which passes enigmatically from Cartesian
    to Kantian and Hegelian positions, and back again. A
    more detailed dissection of these errors has been pre-
    sented elsewhere (Broughton 1979b), as has a clarifica-
    tion of the historically different meanings of these
    various terms (Broughton 1980).

2.  Kohlberg (1981) has even explicitly opposed the disap-
    pearance of the self in Piaget's account.

3.  Some writers (e.g., Marcia 1980; Philibert, in press;
    Döbert, Habermas, & Nunner-Winkler, forthcoming) still
    maintain this erstwhile claim.

4.  It is not entirely clear why Kohlberg and Gilligan pur-
    sue the issue of how adolescents' moral judgment is
    related to their logical reasoning, since they give the
    impression that it is the <u>epistemology</u> of formal opera-
    tions, more than its logic, that is significant.

5.  One can see here why the original (discredited) notion
    arose that transition to principled thought was depen-
    dent on the availability of formal operational reason-
    ing.  Like the latter, relativism requires the genera-
    tion of a universe of hypothetical alternatives (Döbert
    et.al., forthcoming).

6.  Kohlberg, Colby, et al. are apparently less inclined
    to appropriate the rest of Erikson's theory. They are
    especially reticent about the four stages prior to that
    identity. Feasibly, this could have something to do
    with the fact that Erikson's earlier (childhood) stages
    parallel the Freudian psychosexual stages that Kohlberg
    (1966, 1969) had already rejected in no uncertain
    terms. In addition, as we have seen, Kohlberg wants to
    argue that childhood and early adolescence are periods
    of development best characterized as cognitive-struc-
    tural, rather than in terms of life tasks, functional
    choices, and the expanding "social radius" (Erikson
    1959) determined by socialization processes. Despite
    this, the Kohlbergian research of Belenky et al. (1979)
    has demonstrated the interrelation of moral and psycho-
    social development at ages 9–10 and 17.

7.  In some recent research, additional domains, such as
    attitudes to sex, have been included (Marcia 1980).

8.  Podd reports 17 "preconventional" subjects who must
    actually have been transitional "stage 4-1/2s" (cf.
    Broughton 1978b). For the purposes of the present
    discussion, it is assumed that they were, and the data
    are reported accordingly.

9.  Cauble's (1976) study appeared to refute Erikson's
    view, since she found no empirical relationship between
    an Eriksonian measure and a Kohlbergian one. However,
    the latter was Rest's "D.I.T." which is only loosely
    related to Kohlberg's own structural measure, and the
    former was the Constantinople measure, which does not

discriminate the four phases of identity concerns. Therefore the comparability of Cauble's findings to Podd's is severely limited. It should be noted that others besides Kohlberg (e.g. Ries, 1978) have argued in favor of one-to-one correspondence between phases of moral and identity development.

10. A small study by Rowe (1978) claimed to find a relationship; but out of his 26 subjects, only 3 were at the identity-achieved status, making any generalizations difficult.

11. In fact, in Baldwin, this order is the reverse: one imitates (Baldwin's "projective" phase), then plays (Baldwin's "ejective" phase). What this suggests is that the initiation of societal roles in late adolescence is followed by the "play" of fully mastered role behavior in adulthood, which then becomes the context for the imitative activities of the next generation, who are, so to speak, observing their elders at play (cf. Baldwin 1897). However, the circular, inter-generational possibilities of Baldwin's scheme have never been appreciated by Piaget.

12. Ego-identity, as construed by Erikson, is a complex concept. It comprises a variety of dimensions (Leites 1971). However, the symbolic-interactional feature is the one most prominent in Erikson's accounts.

13. A compatible perspective is presented by Döbert and Nunner-Winkler (1975).

REFERENCE NOTES

1. Afrifah, A. and Broughton, J.M. "The Relation of Cognitive Development and Identity Development of Identity in Ghanaian Youth". In preparation.

2. Colby, A. and Kohlberg, L. 1976. "The Relation of Cognitive to Moral Development". Unpublished paper, Harvard Graduate School of Education.

256 J. M. BROUGHTON

3. Broughton, J.M. 1981. "Adolescent Self, Logic, Morality and Identity". Unpublished manuscript, Columbia
University.

4. Broughton, J.M. 1969. "'College Regression': Analysis
of Facts and Theories". Unpublished manuscript, Harvard
University.

5. Kohlberg, L. et al. 1975. "Moral Development in Adulthood and Aging". Unpublished research proposal, Harvard
University.

6. Broughton, J.M. 1981. "Functional and Structural Developmental Theories: Are They Compatible? Unpublished
manuscript, Columbia University.

7. Gilligan, C. and Murphy, M.J. 1979. "The Dilemma of
the Fact: Development in Late Adolescence and Adulthood". Unpublished manuscript, Harvard University.

8. Silver, A. 1979. "Small Worlds and the Great Society:
The Social Production of Moral Order". Paper presented
at the American Sociological Association, Boston.

REFERENCES

Adams, G.R. & J.A. Shea 1979a. "The Relationship Between
Identity Status, Locus of Control, and Ego Devel-
opment". Journal of Youth and Adolescence, VIII,
2:81-9.
Adams, G.R., J.A. Shea, & S.A. Fitch 1979b. "Toward the
Development of an Objective Assessment of Ego-
Identity Status". Journal of Youth and Adoles-
cence, VIII, 2:223-37.
Adelson, J. 1971. "The Political Imagination of the Young
Adolescent". Daedalus, C, 4:1013-50.
Adorno, T.W. 1968. "Sociology and Psychology II". New
Left Review, XLVII:90-115.
Adorno, T.W. 1974. "The Stars Down to Earth". Telos, XIX:
30-90.
Afrifah, A. 1980. "The Relationship Between Piaget Cogni-
tive Stages and Erikson Ego-Identity Statuses".

Unpublished doctoral dissertation, Teachers College, Columbia University.

Baldwin, J.M.   1897.   Social and Ethical Interpretations in
Mental Development.   New York: Macmillan.

Baldwin, J.M.   1902.   Development and Evolution.   New York:
Macmillan.

Baldwin, J.M.   1906-11.   Thought and Things, 3 vols.   London: Swan Sonnenschein.

Belenky, M.F., J.M. Tarule, & A.M. Landa.   1979.   Education
and Development: The Role of Schools in Ego-
Identity and Moral Development of Normal and
Troubled Adolescents.   The Goddard/Teacher
Corps Institute in Developmental Education,
Plainfield/Burlington, Vermont.

Benjamin, J.   1980.   "The Bonds of Love: Rational Violence
and Erotic Domination".   Feminist Studies, VI,
1:144-74.

Berkowitz, M.W., J.C. Gibbs, & J.M. Broughton.   1980.   "The
Relation of Moral Judgment Stage Disparity to
Developmental Effects of Peer Dialogues".
Merrill-Palmer Quarterly, XXVI, 4:341-57.

Bernstein, B.   1975.   "Class and Pedagogies: Visible and
Invisible".   In Class, Codes and Control, III
(ed. by Bernstein).   London: Routledge and Kegan
Paul.

Blasi, A.   1975.   "Role-Taking and the Development of Social
Cognition".   Paper presented at the American
Psychological Association, Chicago.

Blasi, A.   & E.C. Hoeffel.   1974.   "Adolescence & Formal
Operations".   Human Development, XVII, 5:344-63.

Bourdieu, P.   & J.C. Passeron.   1977.   Reproduction: In
Education, Society, and Culture.   London: Sage
Publications.

Bourne, E.   1978a.   "The State of Research on Ego Identity:
A Review and Appraisal, Part I".   Journal of
Youth and Adolescence, VII, 3:223-51.

Bourne, E.   1978b.   "The State of Research on Ego Identity:
A Review and Appraisal, Part II".   Journal of
Youth and Adolescence, VII, 4:371-91.

Boyd, D.R.   1976.   "Education Toward Principled Moral
Judgment".   Unpublished doctoral dissertaion,
Harvard Graduate School of Education.

Boyd, D.R.   1980.   "The Condition of Sophomoritis and Its
Cure".   Journal of Moral Education, X, 1:24-39.

Broughton, J.M.   1978a.   "The Cognitive-Developmental
Approach to Morality: A Reply to Kurtines and

Greif". _Journal of Moral Education_, VII, 2:
81-96.

Broughton, J.M. 1978b. "Criticism and Moral Development
Theory". _Journal Supplement Abstract Service_,
VIII:1-38.

Broughton, J.M. 1978c. "The Development of Concepts of
Self, Mind, Reality, and Knowledge". In _New
Directions for Child Development: Social Cogni-
tion_ (ed. by W. Damon), San Francisco: Jossey-
Bass.

Broughton, J.M. 1979a. "Developmental Structuralism: Without Self, Without History". In _Recent Approaches
to the Social Sciences_ (ed. by H.K. Betz), Win-
nipeg: Hignell Printing Co.

Broughton, J.M. 1979b. "Self and Identity in Adolescent
Development". Paper presented at the Conference
on Recent Approaches to the Self, Center for
Psychosocial Studies, Chicago, October.

Broughton, J.M. 1979c. "The Limits of Formal Thought". In
_Adolescents' Development and Education_ (ed. by
R.A. Mosher), Berkeley: McCutchan.

Broughton, J.M. 1980. "Psychology and the History of the
Self: From Substance to Function". In _Psychol-
ogy: Theoretical-Historical Perspectives_ (ed. by
R.W. Rieber and K.W. Salzinger), New York: Academic Press.

Broughton, J.M. 1981a. "The Genetic Psychology of James
Mark Baldwin". _American Psychologist_, XXXVI,
4:396-407.

Broughton, J.M. 1981b. "Hegel's Stepchild: The Degeneration of Developmental Theory". In _Human
Development: Loyola Symposium on the Person_ (ed.
by T. Wren), Evanston: New University Press.

Broughton, J.M. 1981c. "Piaget's Structural Developmental
Psychology, IV: Knowledge Without a Self and
Without History". _Human Development_, XXIV,
5:325-50.

Broughton, J.M. 1981d. "Genetic Logic and the Developmental Psychology of Philosophical Concepts". In
_The Cognitive-Developmental Psychology of James
Mark Baldwin_ (ed. by J.M. Broughton & D.J. Freeman-Moir), Norwoood, N.J.: Ablex Pub. Co.

Broughton, J.M. 1981e. "Piaget's Structural Developmental
Psychology, V: Ideology-Critique and the Possi-
bility of a Critical Developmental Psychology".
_Human Development_, XXIV, 6:390-411.

Broughton, J.M. (in press). "The Psychology and Ideology of the Self". In Psychology and Ideology (ed. by K. Larsen), Norwood, New Jersey: Ablex Publishing Company.

Broughton, J.M. (Ed.) (forthcoming). Critical Developmental Theory. New York: Plenum Press.

Broughton, J.M. & K.F. Riegel. (1977). "Developmental Psychology and the Self". Annals of the New York Academy of the Sciences, 291:149-67.

Broughton, J.M. & M.K. Zahaykevich. 1980. "Personality and Ideology in Ego Development". In La Dialectique dans les Sciences Sociales (ed. by J. Gabel & V. Trinh van Thao), Paris: Anthropos.

Buck-Morss, S. 1975. "Socio-Economic Bias in Piaget's Theory, and Its Implications for Cross-Cultural Studies". Human Development, XVIII:35-49.

Cassirer, E. 1923. Substance and Function. New York: Dover Press.

Cauble, M.A. 1976. "Formal Operations, Ego Identity, and Principled Morality: Are They Related? Developmental Psychology, XII:363-64.

Chandler, M.J. 1975a. "Relativism and the Problem of Epistemological Loneliness". Human Development, XVIII:171-80.

Chandler, M.J. 1975b. "Irony and the Dialectics of Self-Other Differentiation". Paper presented at the American Psychological Association, Chicago.

Chandler, M.J. 1978. "Adolescence, Egocentrism, and Epistemological Loneliness". In Topics in Cognitive Development, II. New York: Plenum Press.

Colby, A. 1973. Logical Operational Limitations on the Development of Moral Judgment. Unpublished doctoral dissertation, Columbia University.

Colby, A. 1978. "Evolution of a Moral-Developmental Theory". In New Directions for Child Development (ed. by W. Damon). San Francisco: Jossey-Bass.

Debesse, M. 1943. L'Adolescence. Paris: Presses Universitaires de France.

Döbert, R. and Nunner-Winkler, G. 1975. Adoleszenskrise und Identitätsbildung. Frankfurt: Suhrkamp Verlag.

Döbert, R., J. Habermas, & G. Nunner-Winkler. (forthcoming). "The Development of the Self". In Critical Developmental Theory (ed. by J.M. Broughton), New York: Plenum Press.

Elkind, D. 1967. "Egocentrism in Adolescence". Child
     Development, XXXVIII:1025-34.
Erikson, E. 1956. "The Problem of Ego Identity". Journal
     of the American Psychoanalytic Association, IV:
     56-121.
Erikson, E. 1959. "Identity and the Life Cycle". Psy-
     chological Issues, I, 1.
Erikson, E. 1963. Childhood and Society, (2nd ed.). New
     York: W.W. Norton.
Erikson, E. 1968. Identity: Youth and Crisis. New York:
     W.W. Norton.
Fannin, P. 1979. "The Relation Between Ego-Identity Status
     and Sex-Role Attitude, Work-Role Salience, Atyp-
     icality of Major, and Self-Esteem in College
     Women". Journal of Vocational Behavior, XIV:
     12-22.
Fowler, J. 1978. "Life/Faith Patterns: Structures of Trust
     and Loyalty". In Life Maps (ed. by J. Fowler &
     S. Keen), Waco, Texas: Word Press.
Fowler, J. 1980. "Faith and the Structuring of Meaning".
     In Toward Moral and Religious Maturity (ed. by
     C. Brusselmans & J. Fowler), Morristown, N.J.:
     Silver Burdett Co.
Gibbs, J.C. 1977. "Kohlberg's Stages of Moral Judgment: A
     Constructive Critique". Harvard Educational
     Review, XLVII:43-61.
Gibbs, J.C. 1979. "Kohlberg's Moral Stage Theory: A
     Piagetian Revision". Human Development,
     XXII:89-112.
Gillieron, C. 1979. "La pensée de l'adolescent". Totus
     Homo, XI, 9:11-24.
Gillieron, C. 1980. "The Epistemic Subject is not the
     Competent Subject". Paper presented at Teachers
     College, Columbia University, October.
Gilligan, C. 1979. "Woman's Place in Man's Life Cycle".
     Harvard Educational Review, XLIX, 4:431-46.
Gilligan, C. & L. Kohlberg. 1978. "From Adolescence to
     Adulthood: The Rediscovery of Reality in a
     Postconventional World". In Topics in Cognitive
     Development (ed. by K. Appel & B. Presseissen),
     New York: Plenum Press.
Gilligan, C. & M. Murphy. 1979. "Development From Adoles-
     cence to Adulthood: The Philosopher and the
     Dilemma of the Fact". In New Directions for
     Child Development: Intellectual Development

Beyond Childhood (ed. by D. Kuhn), San Francisco:
Jossey-Bass.

Gobar, A.   1968.   Philosophic Foundations of Genetic
Psychology and Gestalt Psychology.   The Hague:
Martinus Nijhoff.

Gruber, H.E.   & J.J. Vonèche.   1976.   "Reflexions sur les
operations formelles de la pensée".   Archives de
Psychologie, XLIV:45-55.

Gruber, H.E.   & J.J. Vonèche.   1977.   The Essential Piaget.
New York: Basic Books.

Habermas, J.   1971.   Knowledge and Human Interest.   Boston:
Beacon Press.

Habermas, J.   1975.   "Moral Development and Ego Identity".
Telos, XXIV:41-55.

Hall, G.S.   1904.   Adolescence, 2 vols.   New York: Appleton.

Henry, J.   1963.   Culture Against Man.   New York: Vintage.

Hobhouse, L.T.   1964.   Liberalism.   New York: Oxford
University Press.

Holstein, C.B.   1976.   "Development of Moral Judgment: A
Longitudinal Study of Males and Females".   Child
Development, XLVIII:51-61.

Horkheimer, M.   1972.   Critical Theory.   New York: Herder
and Herder.

Inhelder, B.   & J. Piaget.   1958.   The Growth of Logical
Thinking from Childhood to Adolescence.   New
York: Basic Books.

Jacoby, R.   1975.   Social Amnesia.   Boston: Beacon Press.

Josselson, R.L.   1973.   "Psychodynamic Aspects of Identity
Formation in College Women".   Journal of Youth
and Adolescence, II, 1:3-52.

Keniston, K.   1968.   Young Radicals.   New York: Harcourt,
Brace and World.

Keniston, K.   1970.   "Youth: A "New" Stage of Life".
American Scholar, XXXIX:631-54.

Kett, J.F.   1977.   Rites of Passage.   New York: Harper and
Row.

Kitchener, R.F.   (in press).   "Piaget's Social Psychology".
Journal of the Theory of Social Behavior.

Kohlberg, L.   1958.   "The Development of Modes of Moral
Thinking and Choice in the Years Ten to Sixteen".
Unpublished doctoral dissertation, University of
Chicago.

Kohlberg, L.   1966.   "A Cognitive-Developmental Analysis of
Sex-Role Concepts and Attitudes".   In The Devel-
opment of Sex Differences (ed. by E. Maccoby),
Stanford: Stanford University Press.

Kohlberg, L.  1968.  "Early Education: A Cognitive-Developmental Approach".  Child Development, XXXIX: 1013-62.

Kohlberg, L.  1969.  "Stage and Sequence".  In Handbook of Socialization Theory and Research (ed. by D. Goslin), Chicago: Rand McNally.

Kohlberg, L.  1973.  "Continuities and Discontinuities in Childhood and Adult Moral Development Revisited".  In Lifespan Developmental Psychology, III (ed. by P.B. Baltes & K.W. Schaie), New York: Academic Press.

Kohlberg, L.  1978.  "The Meaning and Measurement of Moral Development".  Paper presented at the American Psychological Association, Toronto.

Kohlberg, L.  1981.  "Moral Development".  In The Cognitive-Developmental Psychology of James Mark Baldwin (ed. by J.M. Broughton & D.J. Freeman-Moir), Norwood, N.J.: Ablex Pub. Co.

Kohlberg, L.  & R. Kramer.  1969.  "Continuities and Discontinuities in Childhood and Adult Moral Development".  Human Development, XII:93-120.

Kohlberg, L.  & C. Gilligan.  1971.  "The Adolescent as Philosopher: The Discovery of Self in a Postconventional World".  Daedalus, C, 4:1028-61.

Kohlberg, L.  & R. de Vries.  1971.  "Relations Between Piaget and Psychometric Assessments of Intelligence".  In The Natural Curriculum (ed. by C. Lavatelli), Urbana, Illinois: ERIC.

Kramer, R.  1968.  "Moral Development in Young Adulthood".  Unpublished doctoral dissertation, University of Chicago.

Kuhn, D.  1976.  "Short-term Longitudinal Evidence for the Sequentiality of Kohlberg's Early Stages of Moral Development".  Developmental Psychology, XXII, 2:162-66.

Kuhn, D., J. Langer, L. Kohlberg, & N. Haan.  1977.  "The Development of Formal Operations in Logical and Moral Judgment".  Genetic Psychology Monographs, XCV:97-188.

Lasch, C.  1977.  Haven in a Heartless World.  New York: Harper and Row.

Leites, N.  1971.  The New Ego.  New York: Science House.

Levinson, D.  1978.  Seasons of a Man's Life.  New York: Ballantine.

Loevinger, J.   1966.   "The Meaning and Measurement of Ego
        Development".  American Psychologist, XXI:195-
        206.
Loevinger, J.  & R. Wessler.  1970.  Measuring Ego Develop-
        ment, 2 vols.  San Francisco: Jossey-Bass.
MacMurray, J.  1957.  Self as Agent.  London: Faber and
        Faber.
Marcia, J.E.  1964.  "Determination and Construct Validity
        of Ego Identity Status".  Unpublished doctoral
        dissertation, Ohio State University.
Marcia, J.E.  1966.  "Development and Validation of Ego
        Identity Status".  Journal of Personality and
        Social Psychology, III:551-58.
Marcia, J.E.  1976.  "Identity Six Years After: A Follow-up
        Study".  Journal of Youth and Adolescence,
        V:145-60.
Marcia, J.E.  1980.  "Identity in Adolescence".  In Handbook
        of Adolescence (ed. by J. Adelson), New York:
        J. Wiley.
Marcia, J.E. and Friedman, M.L.  1970.  "Ego Identity Status
        in College Women".  Journal of Personality,
        XXXVIII:249-63.
Matteson, D.R.  1977.  "Exploration and Commitment: Sex
        Differences and Methodological Problems in the
        Use of Identity Status Categories".  Journal of
        Youth and Adolescence, VI, 4:349-70.
Mead, G.H.  1936.  Movements of Thought in the Nineteenth
        Century.  Chicago: University of Chicago Press.
Meilman, P.W.  1977.  "Crisis and Commitment in Adolescence:
        A Developmental Study of Ego Identity Status".
        Unpublished doctoral dissertation, University of
        North Carolina.
Murphy, M.J.  & C. Gilligan.  1980.  "Moral Development in
        Late Adolescence and Adulthood: A Critique and
        Reconstruction of Kohlberg's Theory".  Human
        Development, XXIII:77-104.
Osterrieth, P.A.  1969.  "Adolescence: Some Psychological
        Aspects".  In Adolescence: Psychological Per-
        spectives (ed. by G. Caplan & S. Lebovici), New
        York: Basic Books.
Parsons, T.  1964.  Social Structure and Personality.  New
        York: Free Press.
Peirce, C.S.  1931-1958.  Collected Papers, I-VIII.  Cam-
        bridge: Harvard University Press.

Perry, W.   1968.   Intellectual and Ethical Development in
            the College Years.   New York: Holt, Rinehart and
            Winston.
Philibert, P.  (in press).   "The Motors of Morality:
            Religion and Relation".   In Moral Development
            Foundations: Theological Alternatives (ed. by
            D. Joy), Nashville: Abingdon Press.
Piaget, J.   1929.   The Child's Conception of the World.
            London: Routledge and Kegan Paul.
Piaget, J.   (1932).   The Moral Judgment of the Child.   New
            York: Free Press, 1965.
Piaget, J.   1967a.   "The Mental Development of the Child".
            In Six Psychological Studies (ed. by J. Piaget),
            New York: Random House.
Piaget, J.   1967b.   Etudes Sociologiques.   Geneva: Librairie
            Droz.
Piaget, J.   1969.   "The Psychological Development of the
            Adolescent".   In Adolescence: Psychological
            Perspectives (ed. by G. Caplan & S. Lebovici),
            New York: Basic Books.
Piaget, J.   1972.   "Intellectual Evolution from Adolescence
            to Adulthood".   Human Development, XV:1-12.
Piaget, J.   & B. Inhelder.   1969.   The Psychology of the
            Child.   New York: Basic Books.
Podd, M.H.   1972.   "Ego Identity Status and Morality: The
            Relationship between Two Developmental Con-
            structs".   Developmental Psychology, VI, 3:497-
            507.
Raphael, D.   1975.   "An Investigation into Aspects of Iden-
            tity Status of High-School Females".   Unpublished
            doctoral dissertation, University of Toronto.
Ries, S.   1978.   "The Psychological Phenomenon of Moral
            Relativism and its Relationship to Identity".
            Unpublished doctoral dissertation, Harvard
            Graduate School of Education.
Riesman, D., N. Glazer, & R. Denney.   1950.   The Lonely
            Crowd.   New Haven: Yale University Press.
Roberts, B.   1977.   "George Herbert Mead: The Theory and
            Practice of His Social Philosophy".   Ideology and
            Consciousness, II:81-106.
Rowe, I.   1978.   "Ego Identity Status, Cognitive Develop-
            ment, and Levels of Moral Reasoning".   Unpub-
            lished masters thesis, Simon Fraser University.
Schenkel, S.   1975.   "Relationship Among Ego Identity
            Status, Field Independence and Traditional

Femininity". Journal of Youth and Adolescence, IV:73-82.

Seeley, J., R.A. Sim, & E.W. Loosley. 1956. Crestwood Heights. New York: Basic Books.

Selman, R.L. 1980. The Growth of Interpersonal Understanding. New York: Academic Press.

Stark, P.A. & A.J. Traxler. 1974. "Empirical Validation of Erikson's Theory of Identity Crises in Late Adolescence". The Journal of Psychology, LXXXVI: 25-33.

Toder, N. & J.E. Marcia. 1973. "Ego Identity Status and Response to Conformity Pressure in College Women". Journal of Personality and Social Psychology, XXVI, 2:287-94.

Turiel, E. 1966. "An Experimental Test of the Sequentiality of Developmental Stages in the Child's Moral Judgments". Journal of Personality and Social Psychology, III:611-18.

Turiel, E. 1972. "Stage Transition in Moral Development". In Second Handbook of Research on Teaching (ed. by R.M. Travers), Chicago: Rand McNally.

Turiel, E. 1974. "Conflict and Transition in Adolescent Moral Development". Child Development, XLV: 14-29.

Turiel, E. 1980. "The Development of Social-Conventional and Moral Concepts". In Moral Development and Socialization (ed. by M. Windmiller, N. Lambert, & E. Turiel), Boston: Allyn and Bacon.

Unger, R.M. 1975. Knowledge and Politics. New York: Free Press.

Vaillant, G. 1977. Adaptation to Life. Boston: Little Brown.

Vonèche, J.J. (forthcoming). "Jacqueline, Laurent, Lucienne et al.". In Critical Developmental Theory (ed. by J.M. Broughton), New York: Plenum Press.

Walker, L.J. 1980. "Cognitive and Perspective-Taking Prerequisites for Moral Development". Child Development, LI:131-39.

Walker, L.J. & B.S. Richards. 1979. "Stimulating Transitions in Moral Reasoning as a Function of Stage of Cognitive Development". Developmental Psychology, XV, 2:95-103.

Waterman, A.S. (in press). From Adolescence to Adult. Boston: Houghton Mifflin.

Young, R.M.  1970.  <u>Mind, Brain, and Adaptation</u>.  Oxford:
          Clarendon Press.
Youniss, J.  1981.  "A Revised Interpretation of Piaget".
          In <u>Recent Piagetian Psychology</u> (ed. by I. Sigel).
          Hillsdale, N.J.: Lawrence Erlbaum.
Zahaykevich, M.K.  1980.  "Critical Perspectives on Adult
          Women's Development".  <u>Psych Critique</u>, III.

# A NEO-PIAGETIAN APPROACH TO OBJECT RELATIONS

Robert Kegan

Harvard University

## INTRODUCTION

Terry is fifteen years old. She is admitted to the psychiatric ward of a general hospital by her mother, who can no longer deal with her. She is a chronic truant, will obey none of her mother's rules, steals money. According to Terry her mother is "stern", "stubborn", "nagging", "unwilling to compromise", and "headstrong". Terry experiences her mother, and a lot of other people, as "trying to invade my life". Terry worries that she is "going to have to submerge my personality". She feels that she is "no longer whole". "I feel like others are being woven into me."

Diane is twenty. She took forty-two pills, "not to die, but to show him how much he hurt me". She had lived with him nearly a year and now he was seeing another woman. Her hopes for "an exclusive relationship" were destroyed. She was furious with him, but she was afraid that if she expressed this she would lose him completely. She felt she could not continue without him. As her anxiety and depression increased, she had terrifying nightmares with recurring images of death. She experienced several episodes of rage (over minor irritations with her boyfriend) during which she pictured herself sitting on the mouth of a large plastic head resembling her own, staring at the back of the inside of the structure. She had several experiences of waking up abruptly and relating to her boyfriend "as if he were my father and I was about four years old"; then she would fall back to sleep and wake with no memory of these events. As she began to make plans to separate from her boyfriend, she became extremely upset and took an overdose of drugs. This has happened to her before, she says. She finds a man, finds herself becoming increasingly dependent upon him, finds herself merging into him; he eventually finds the relationship too

267

burdening and moves to end it or reduce its intensity; she
feels then "I can't live with him and I can't live without
him". On two other occasions when things got to this state
she tried to kill herself.

Rebecca is thirty-six. She is in therapy because she is
afraid "I am going to come apart". A hard-won sense of control feels like it is threatened. "I know I have very defined boundaries and I protect them very carefully. I don't
want to give up the slightest control. In any relationship I
want to decide who gets in, how far and when." "How exhausting it has become holding it all together." "I feel like I
am slipping, and that it will all come apart."

Whatever else might be said about Terry, Diane, and
Rebecca, they seem to share a common concern about personal
boundaries. Their sense of what is "self" and what is "others" seems either to be shaky or to have become shaky. Difficulties of this sort--especially the inability to maintain
the differentiation between self and other--are now widely
understood to reflect on the vicissitudes of earliest life,
particularly the infant's separation and individuation from
its undifferentiated state at birth (Mahler, Pine, & Bergman
1975). From this point of view, the phenomena of infancy
take on the quality of the very context in which object
relations throughout the life-span are considered. Recurring issues of differentiation and integration throughout
life come to be understood as the consequences, reflections,
or offspring of this earliest period. The recognition of
this crucial era prior even to the Oedipal years has led in
effect to a restatement of Freud's dictum: now it is the
infant who is father to the man.

This chapter, which develops not out of a neo-psychoanalytic framework but a neo-Piagetian one, takes a somewhat
different view--both of the phenomena of infancy, and of
their representations in the later living of a Terry, a
Diane, or a Rebecca. Its argument, in effect, is a shift of
figure and ground. It suggests that, rather than understanding issues of differentiation and integration in the
context of infancy, the phenomena of infancy are better
understood in the context of the psychological meaning of
evolution, a lifetime activity of differentiating and integrating what is taken for "self" and what is taken for
"other". The consequences of doing so, as the chapter tries
to demonstrate, are not only a somewhat different conception

of infancy, but the possibility of understanding--in some
way which would not otherwise be available--that Terry,
Diane, and Rebecca are <u>not</u> any longer eighteen months old.
Everyone has heard the story of the man who was searching
for something under a streetlight. "You lost it around
here?" he is asked. "No, over there", he says, pointing to
a dark corner some distance away. "Well, why are you looking for it here?" "Because this is where the light is good"
comes the reply. This chapter wonders whether a neo-Piagetian approach to "object relations" might light up some of
these dark corners so that we are not forced, by default, to
understand all evolutionary phenomena in terms of the one
transformation we understand well.

## WHAT IS AN "OBJECT"?

What is an "object"? People frequently find the term
"object relations" strange or distasteful, since what we are
most of all speaking about, they say, is other human beings,
and the notion of persons as things seems unfortunate. And
yet there is a meaning to the word "object" that must not be
lost and that no other word conveys. We can start by looking for this in its very etymology. The root <u>ject</u> speaks
first of all to a <u>motion</u>, an activity rather than a thing--
more particularly, to <u>throwing</u>. Taken with the prefix, the
word suggests the motion or consequence of "thrown from" or
"thrown away from". "Object" speaks to that which some motion has made separate or distinct from, or to the motion
itself. "Object-relations", by this line of reasoning,
might be expected to have to do with our relations to that
which some motion has made separate or distinct from us, our
relations to that which has been thrown from us, or the experience of this throwing itself. Now I know this preliminary definition sounds peculiar, but it does have more in its
favor than a Latin pedigree. It is, I suggest, the underlying conception of object relations to be found in neo-Piagetian theory (Kegan 1979, 1982).

Central to that theory is an understanding of a <u>motion</u>
as the prior context of personality. Simply put, this is
the motion of evolution; less simply, it is evolution as a
meaning-constitutive activity. As the prior context of personality (I mean, of course, philosophically prior; not tem-
porally), it is argued to be not only the unifying <u>but the</u>
<u>generating</u> context for both (1) thought and feeling (about

which more later) and (2) "subject" and "object" or "self"
and "other" (about which more now). Evolutionary activity
involves the very creating of the object (a process of differentiation) and our relating to it (a process of integration). By such a conception, "object relations" (really,
"subject-object relations") are not something that go on in
the "space" between a worldless person and a personless
world; rather, they bring into being the very distinction in
the first place. Subject-object relations emerge out of a
lifelong process of development, the essence of which is a
succession of qualitative differentiations of the self from
the world, thereby creating each time a qualitatively more
extensive object with which to be in relation, a natural
history of qualitatively better guarantees to the world of
its distinct integrity, successive triumphs of "relationship
to" rather than "embeddedness in". By such a conception the
term "object-relations" is an acceptable, even welcome, term
(rather than something more human-sounding) because, properly understood, the term does not relate persons to things
but creates a more general category, recognizing that any
given person may differ for us not only by his distinctness
from other persons but by the differing ways in which we
ourselves make sense of him, of which differences none may
be so important as the extent to which we distinguish him
from ourselves.

    Psychoanalytic object-relations theory looks to the
events of the first years of life for its basic themes and
categories. While early infancy has great importance from a
neo-Piagetian view, it is not, in its most fundamental respect, qualitatively different from any other moment in the
life-span. What is taken as fundamental is the activity of
meaning-constitutive evolution. Inasmuch as infancy marks
the beginning in the history of this activity, establishes
themes that can be traced through the lifespan, originates a
disposition on the part of the person toward this activity,
the first years of life have great salience. But it is not
a salience sui generis. The distinctive features of infancy
are to be understood in the context of that same activity
which is the person's fate throughout his/her life. The recurrence of these distinctive features in new forms later on
in development are understood not as later manifestations of
infancy issues, but as contemporary manifestations of meaning-making, just as the issues of infancy are, at that time,
contemporary manifestations of meaning-making. What does it
mean to look at the psychological phenomena of infancy in

the context of meaning-making, rather than to look at  meaning-making in the context of infancy?

## INFANCY AS AN INSTANCE OF A LIFELONG RHYTHM

Piagetian and psychoanalytic psychologies share a quite
similar conception of  the newborn's state.  Both consider
the newborn to live in an objectless world, a world in which
everything sensed is taken to be an extension of  me,  where
out-of-sight (or  touch  or  taste or hearing or smell) can
mean out of existence. Freud (1911) considered "mental functioning" eventually to be under the sway  of  "the  pleasure
principle" and "the reality principle", but in the newborn
only under the first.  Piaget (1952) considers mental  functioning  eventually  to  be under the sway of "assimilation"
(fitting one's experience to one's present means of organizing reality) and "accomodation" (reorganizing one's  way  of
making  meaning  to  take account of experience), but in the
newborn only under the first.  Taken at a broad enough level
of generality, the notion of "orality"  is  consistent  with
the Piagetian conception of the all-assimilative, incorporative newborn.  Both perspectives see the central psychological  achievement of the first eighteen months in terms of an
end to this objectless world and the dawning of object relations.

From a psychoanalytic point of view, the baby's binding
energy is directed away from itself toward another  or  some
part  of another.  The infant's natural narcissism, or self-
absorption, gradually comes to an end as  it  withdraws  an
attachment  to itself in favor of a new "object choice" outside itself.  The notion of object relations as  an  energy
redirection  or  an object choice can be contrasted with the
neo-Piagetian notion of object creation.  By  this  understanding,  the consequence of the organism's gradual "emergence from embeddedness" (Schachtel 1959) is the dawn  of  an
object  world.  By differentiating itself from the world and
the world from it the organism brings into being that  which
is  independent  of  its own sensing and moving. As Piaget
himself writes, such an understanding makes early life "narcissistic" only in a very special sense of the word:
> One could not describe it as a focus of emotion  on  the
> activity itself,  as a self contemplation of self, pre-
> cisely because the self has not yet developed.   Narcis-
> sism  is  nothing other than emotion associated with the

non-differentiation between the self and the non-self
(the adulatory stage of Baldwin, or emotional symbiosis
of Wallon). The primary narcissism of nursing is really
a narcissism without Narcissus. (1964, p. 35)

From the neo-Piagetian view, the transformation in the
first eighteen months of life--giving birth to object rela-
tions--is but the first instance of that basic evolutionary
activity taken as the fundamental ground of personality development. The infant's "moving and sensing", as the basic
structure of its personal organization (the reflexes), get
"thrown from"; they become an object of attention, the "content" of a newly evolved structure: rather than "being" my
reflexes, I now "have" them and "I" am something else. "I"
am that which coordinates, or mediates, the reflexes (i.e.,
impulses and perceptions)--the new subjectivity. This creates a world, for the very first time, separate from me--the
first qualitative transformation in the history of guaranteeing the world its distinct integrity, of having it to-re-
late-to rather than to-be-embedded-in. But this transformation does not take place over a weekend; and it does not
take place without cost to the organism, which must suffer
what amounts to the loss of itself in the process. The la-
borious gradualness and personal cost of this transformation
can be considered in the context of the two best researched
phenomena of this period--the construction of the permanence
of the object, and infant protest upon separation from the
primary caretaker. From a neo-Piagetian view, both phenomena are easily misunderstood.

In a film describing their scales for measuring object
permanence, Uzguris and Hunt (1968) show us the same infants
from the first months of their lives until after their second birthdays. The experimenter interests the child in some
small object and then, right under the child's eyes, con-
ceals it in some way--a bead necklace is covered by a small
blanket, or a ball is rolled under a chair. Before the object is covered, the child of 4 or 5 months is able to involve himself with the toy (pursuing it with eyes and hands,
holding it, bringing it to his mouth); but if the object is
covered, all involvement with it ceases. This cannot be
attributed to the child's loss of interest in the toy, for
the experimenter has only to lift the cover and the child
lights up, vocalizes, reaches again for the object. It
seems not that the child loses interest in the covered object, but that the covered object loses its existence for

the child.  Beginning somewhere around 8-10 months, children
make some tentative efforts to retrieve the covered object--
though if it is covered by more than one screen or displaced
from one screen to another, their exploration often comes to
an early halt.  But by the time they are two years old, they
usually have no problem retrieving the hidden object despite
the experimenter's multiple displacements.

These experiments, widely known, are not so widely understood. When I show graduate students the Uzguris and Hunt
film for the first time, they usually become bored and rest-
less halfway through.  It is clear that the children are behaving "as they should", that as they get older they get
"better at finding the object". The theme is set  early  on,
and  the  predictability  of the infants' behavior makes the
film tiresome.

And yet I have seen students watch the film  again  and
again  once they understood what they were seeing, so fascinating does it become; and I find that I do not tire of seeing it repeatedly.  For the film amounts to time-lapse  photography  of the most difficult sort--the evolution of a relationship.  It is not a  rose unfolding, nor is it a single
organism of any kind exactly (not  the  physically  growing
child)  that  is  most interesting.  It is rather time-lapse
photography of the relationship between an organism and  its
environment; if one turns an eye to it, he can see the child
being "hatched  out" (to  use  the term of Mahler, Pine, &
Bergman 1975) of a world in which  it  was  embedded. It is
difficult  for anyone, unless forewarned, to resist the perception that the little necklace or the rubber ball  is  remaining  the  same  throughout the film while only the infant
is changing, that the two-year drama  contains  two  charac-
ters--an  infant and an object--whose entitivity remains the
same.  The film takes on a whole new life if it  is  considered  that  one  is  watching  a  single  dynamic organism,
"baby-and-ball", which is gradually undergoing a process  of
transformation.

One  can  see,  over  the  period  roughly from 9 to 21
months, that although the baby-and-ball becomes no longer  a
single entity, neither does it quite constitute, as yet, two
distinct  entities.  The hidden object is not given up immediately; at the same time, its pursuit is  easily  defeated.
One has the sense of a differentiation so fragile, so tentative,  that it can very easily merge back into oneness. Even

the experimenter's taking of the object out of the hands of
the infant becomes a fascinating picture of a scientist
probing the integrity of a specimen in a critical period in
its evolution.  Having given the object to the infant to enhance his interest in it, the experimenter then routinely
pulls it gently away in order to cover it.   In the early
months the child gives it up without protest of any kind. He
does not--it seems clear, <u>have</u> it--in the sense of its being
something  apart from, something to be <u>bound up with</u>.  As he
gets older, it seems that it is not only his <u>physical</u>  grasp
that  intensifies  and articulates (from gross to fine motor
coordination, e.g.), but a <u>psychological</u> one as   well.   All
in all, the film, then, is capturing a motion--the motion of
"throwing from",  of  differentiation  (which   creates   the
object) and the motion of  integration  (which   creates   the
object relation).

    But what can be found can also be lost.   The process of
differentiation,   creating   the possibility of integration,
brings into being the lifelong theme of  finding-and-losing,
which  before now could not have existed.   The universal infant reaction of protest upon separation  from  the  primary
caretaker,  a  great  number  of  researchers tend to agree,
first appears around 10 months,   peaks  at  12  months,  and
ceases at about 21 months (Kagan 1971).

    From  the neo-Piagetian perspective, these simultaneous
phenomena--the gradually developing capacity  to  orient  to
the  object even when it is absent, and the definable course
of protest upon separation--are the cognitive and  affective
dimensions  of  a  single, more basic phenomenon: the evolutionary transition from an  undifferentiated  state  to  the
first  equilibrium (what will appear in Table 1 as the transition from subject-object relation "zero" to subject-object
relation "1").   In arguing for evolutionary activity as  the
fundamental  ground  in  personalty, neo-Piagetian theory is
not choosing between "affect" or "cognition" as  the  master
of  development--a  choice  which has limited both classical
Freudian and Piagetian theory--but is putting forth a candidate for a ground of consideration prior to, and  generative
of, cognition <u>and</u> affect. When we view this evolutionary activity with respect to any particular subject-object differentiation  which is being maintained, we are considering its
implications for <u>knowing</u>.  The events of the first  eighteen
months  culminate  with  the creation of the object and make
evolutionary activity henceforth an activity  of  equilibra-

tion, of preserving or renegotiating the balance between what is taken as subject or self and what is taken as object or other.

Neo-Piagetian research makes clear that human development involves a succession of renegotiated balances, or "biologics", which come to "know" or organize the experience of the individual in qualitatively different ways. In this sense, evolutionary activity is intrinsically cognitive—but it is no less affective, for we are this activity and we experience it. Affect is essentially phenomenological, the felt experience of a motion (hence, "e-motion"). In identifying evolutionary activity as the fundamental ground of personality, the neo-Piagetian perspective is suggesting that the source of our emotions is the phenomenological experience of evolving—of defending, surrendering, and reconstructing a center. The universally recognized anxiety between nine and twenty-one months is understood, in the neo-Piagetian perspective, as that distress which attends every qualitative decentration—which from the point of view of the developing organism amounts to the loss of its very organization.

"Separation anxiety", by this account, is not well understood as anxiety about the loss of an object or another, of the mother or the comforts "she" provides. It seems more correct to imagine that, from the perspective of the infant (the only perspective that counts when attempting to understand an infant's anxiety), the distress is not about the loss of an object—for an object does not yet wholly exist, and when it does wholly exist (around 21 months) that is exactly when the anxious behavior comes to an end! Central to the experiences of qualitative change or decentration (phenomenologically, the loss of my center) are the affects of loss: anxiety and depression. Infant distress understood as the felt experience of an evolutionary transformation—an emergence from embeddedness—seems to be not so much a matter of separation from the object as separation from myself, from what is gradually becoming the old me, the old subject-object balance from which I am not yet sufficiently differentiated to integrate as other. The extraordinary vulnerability of the infant to an actual prolonged separation from the primary caretaker—which Spitz (1973) and others (e.g., Bowlby 1973) document seems not to exist before six months or after two years—seems to be due to the misfortunate combination of an actual disappearance of what was part of me at the very time I am psychobiologically beginning to separate myself from it.

Anxiety and depression may be the affective experience of the wrenching activity of differentiation in its first phases; but sooner or later the balance as to which self is "me" begins to shift, and the old equilibrium can be "reflected" upon from the new emerged position. This experience—which begins the process of integration, of taking the old equilibrium as "object" in the new balance—is often affectively a matter of anger and repudiation. From a neo-Piagetian perspective, the familiar sequence from depression to anger is not so much a matter of redirecting an emotion from self to other—i.e., from one target to another—as it is the moving of the target of the anger itself from self to other. Emergence from embeddedness involves a kind of repudiation, an evolutionary "re-cognition" that what before was me is now not-me. This move we see, too, on a universal basis in the first years of life. What is the possibility that the "terrible twos", with its rampant negativism and declarations of "No!" are a communication more to the old self, now gradually becoming object, than to those exasperated parents who feel they are being defied as distinct and separate people? When the new balance becomes more secure, the infant will have less of a need to "protest too much" and the parents will become more "others" than "not-me's".

Hence, if I started by discussing how "object creating" must mean "subject losing", I have come round to showing how "subject-losing" can lead to "object finding". This is a rhythm central to the underlying motion of personality. Its discovery during difficult periods in the lives of a Terry, Diane, or Rebecca may not be so much a matter of revisiting an infancy rhythm as experiencing the contemporary manifestation of a lifelong activity which begins in infancy.

Let me summarize some of the implications of this section. It should be clear that the neo-Piagetian conception neither subsumes affectivity to the cognitive realm, as traditional Piagetians tend to do (Schaefer & Emerson 1964), nor does it make intellectual life the offspring servant of affect, as psychoanalysis tends to do. "Thinking", from such a perspective, does not have to wait upon the discovered insufficiencies of "primary process", as classic psychoanalysis would suggest. It begins in its own form at birth, with the moving hand and the sensing eye—the newborn's body

being its mind, the prehensile grasp a preabstracted fore-
runner to the grasp of apprehension. As Piaget himself has
said (despite the inability of his own work to fully realize
it): "There are not two developments, one cognitive and the
other affective, two separate psychic functions, nor are
there two kinds of objects: all objects are simultaneously
cognitive and affective" (1964, p. 39). This is because all
objects are themselves the elaboration of an activity which
is simultaneously cognitive and affective.

Also, it should be clear how such a model speaks directly to the ambition of modern psychoanalytic theorists to
recognize an intrinsic motivation for object relations,
rather than just positing one (Winnicott 1965, Fairbairn
1962, Guntrip 1971). Though the person is seen now to
"think" from birth, it is not "thinking" which motivates its
growth. Though it "feels" from birth, it is not "feeling"
or drive-states or energetics which motivates its growth.
In identifying a context "prior" to thought and feeling, the
neo-Piagetian perspective posits a motive which subsumes
thought and feeling, and makes object-relations intrinsic
(or better put, makes intrinsic that which brings them into
being). Psychoanalytic theory views the individual as primarily motivated by the desire to reduce or eliminate un-
pleasurable affect. By this reasoning, the individual turns
away from himself to the object (whether it be the object's
representation as in primary process, or the world of real
objects as in secondary "cognition") because his own system
of warding off noxious experience has broken down. Object
relations are thus formed extrinsically, a kind of necessary
inconvenience never put better than in Freud's depiction of
ego formation and reality orientation as an unavoidable "detour" the psychic system must make to secure for itself the
same peace it has desired since inception and which was much
more efficiently obtained in utero.

While the neo-Piagetian perspective shares the perception that it is the newborn's inability to satisfy himself
that brings on the birth of object relations, this perception is couched in a conception of human motivation which is
not tied to affect alone. It is the greater coherence of
its organization which is the presumed motive (White 1959),
a transorganic motive shared by all living things. A more
cognitive-sounding translation of the motive is to say that
the organism is moved to make meaning or to resolve discrepancy, but this would not be different than to say that it is

moved to preserve and enhance its integrity. In any case, the neo-Piagetian view concurs that it is the infant's inability to satisfy himself that prompts his development--but for a rather different reason.

The in utero state represents a kind of nirvana to many ways of thinking. One of its most impressive features is that the needs of the foetus are perfectly met: the host organism nourishes it, breathes for it (through the host's blood), and so on. It is most appealing, in light of the neo-Piagetian understanding of growth as a process of emergence from embeddedness, to consider the experience of birth itself as the beginning of the transition out of the first evolutionary balance. What must be most dramatic about this new world for the infant is an end to this harmony. Innate reflexes may cause the eyes to close automatically before bright light, but the stomach contractions brought on by hunger do not cause food to enter the system. These discrepant experiences the organism is prompted to resolve--however not (the neo-Piagetian perspective would understand) in order to return to the homeostatic state of the foetus, but to bring its organization into a coherence that can take account of the greater complexity with which it is faced; not to return to an old reality, but to establish a meaning for, or make sense out of (yes, even at one year of age) its present reality.

The further elaboration this occasions--which always brings about a qualitatively new object relation is hence neither an effort to recreate the foetal state, nor an extrinsic detour. Object relations, from a neo-Piagetian view, are thus oriented to the present reality, and are brought into being for their own intrinsic value. While growth, from a neo-Piagetian perspective, is no merry ride, neither is each qualitative change regarded as a greater defeat, or further indebtedness (an ever more complex and less elegant way of keeping the system free of stimulation). Rather, each qualitative change--hard won--is a response to the complexity of the world, the essence of which response is a recognition of its distinct integrity and the possibility of my relation to it, rather than fusion with it.

## THE EVOLUTION OF OBJECT RELATIONS

I am going to suggest in this section something like a life history of qualitative reconstructions of self and other, each one emerging out of a process of differentiation and integration. The framework--its logic and process-- seems to provide an integrating context for an enormous amount of data (the bulk of which cannot be dealt with here, but especially the research of Piaget-oriented stage theorists: see, e.g., Piaget 1954, Kohlberg 1976, Loevinger 1976, Perry 1970, Selman 1980, Broughton 1975, Fowler 1981, Gilligan 1978). Some of these theorists (e.g., Piaget, Broughton, Kohlberg, and Selman) feel they are only studying forms of cognition; but I believe they are tapping in to something deeper, that the phenomena they have discovered on the protocol page are a consequence of transformations in the evolutionary activity of the person. Others among these theorists (and most clearly, Loevinger) are trying to attend to this "something deeper", to the master motion in personality; the implication, for her work, of the framework I am about to sketch is that it may provide the missing logic and process that underlie her ego stages.

Table 1 summarizes an argument for six different levels of subject-object relations throughout the lifespan. The table differs from any number of similar comparisons of theories in its suggestion, at each stage, of the underlying structure in which the various theories might be rooted.

By consulting the table, the reader will see that what we traced in the previous section on infancy was the transition from the Incorporative stage to the Impulsive stage. This transformation was accomplished through a process which we shall see repeated. It has been called a process of decentration (Piaget 1952), an emergence from embeddedness (Schachtel 1959), and the recurring triumph over egocentrism (Elkind 1976). It has been referred to as a process in which the whole becomes a part to a new whole, in which what was structure becomes content on behalf of a new structure, in which what was ultimate becomes preliminary on behalf of a new ultimacy, in which what was immediate gets mediated by a new immediacy. All these descriptions speak to the same process, which is essentially that of adaptation--a differentiation from that which was the very <u>subject</u> of my personal organization and which becomes thereby the <u>object</u> of a new organization on behalf of a new subjectivity which coordinates it. In Mahler's terms we are "hatched out"--but over and over again. And, as we shall see, we are vulnerable each time to a qualitatively new kind of "separation anxiety".

In disembedding itself from its reflexes, the two-year-old comes to "have" them rather than "be" them, and the new

self is embedded in that which coordinates the reflexes
(namely, the "perceptions" and the "impulses"). The tremen-
dous lability, cognitively and emotionally, of the preschool
child is suggested to be a function of this new embeddedness. The child is able to recognize objects separate from
itself, but those objects are subject to the child's perception of them (this is, I suggest, the underlying structure
of Piaget's pre-operational stage); i.e., if the child's
perception of an object changes, the object itself has
changed in the child's experience, (he is unable to hold his
perception of the liquid in one container with his perception of the liquid in the taller, thinner container precisely because he cannot separate himself from his perceptions).

   The same is true of the structurally equivalent psychological category, the "impulse". The pre-schooler has poor
"impulse control", it is suggested, not because he lacks
some quantitative countering force, but because his "bio-
logic" (the living logic which he is) is composed in a qualitatively different way. "Impulse control" requires their
mediation, but the impulses are immediate to this subject-
object balance. When I am subject to my impulses their
non-expression raises an ultimate threat; they risk who I
am. Similarly the preschooler's inability to hold two perceptions together (which is what gives the object world its
concreteness, à la Piaget) is paralleled in the preschooler's inability to hold two feelings about a single thing to-
gether--either the same feelings about a thing over time,
which creates the "enduring disposition" (the psychological
structure which I call "needs, interests, wishes" in the
table) or competing feelings at the same time. This latter
suggests why the preschooler lacks the capacity for ambivalence, and it understands the tantrum--the classic expression of distress in this era--as an example of a system
overwhelmed by internal conflict because there is no self
yet which can serve as a context upon which the competing
impulses can play themselves out; the impulses are the self,
are themselves the context.

   The extremely varied assortment of phenomena suggesting

a "five-to-seven shift" finds a unifying context in the consideration that the underlying shift is exactly that of the
next transformation in object relations, from the "Impulsive" balance to the "Imperial" one. Consider findings as
apparently disparate as these: (1) children on the younger
side of the five-to-seven shift seem to need rewards that
are fairly immediate, sensual, and communicating of praise;
children on the older side seem to feel more rewarded by the
information that they have been correct; (2) children who
lose a limb or become blind before they are through the
shift tend not to have phantom-limb responses or memories of
sight; children on the other side of the shift do have them
(Gardner 1978). The capacity to take my impulses and perceptions as an object of my meaning-making not only brings an
end to the lability of the earlier subject-object relation,
it brings into being a new subject-object relation which
creates a more endurable self--a system of its own which
does its own praising, so to speak, but needs the information that it is correct, as confirmation of its system; this
system which can store memories, feelings, and perceptions
(rather than immediately being them) so that a feeling arm
or a seeing eye lives on in some way. The examples mean, I
think, that the evolving context is more than cognition or
affect.

One way of characterizing the new subject-object relation (stage two: the Imperial balance) is in terms of the
construction of the role. This is true whether we are speaking of the social-cognitive capacity to assume the role of
another person, or the affective differentiation within the
impulse life of the family, permitting me to take my appropriate role as a "child" in relation to a "parent" rather
than "being" my impulse life bound up with another. A distinguishing feature of this new subject-object relation is
that the child seems to "seal up", in a sense: there is a
self-containment appears that was not there before; the
adult no longer finds himself engaged in the middle of
conversations he has begun all by himself; he no longer
lives with the sense that the parent can read his private
feelings. He has a private world, which he did not have
before.

TABLE 1

Balances of Self and Other as the Common
Ground of Several Developmental Theories

| THE EVOLVING | PIAGET | KOHLBERG |
|---|---|---|
| INCORPORATE<br><br>(Stage Zero) | Sensorimotor | |
| Underlying Structure $\longrightarrow$ | | SELF -- OTHER: |
| IMPULSIVE<br><br>(Stage One) | Pre-operational | Punishment &<br>Obedience<br>Orientation |
| Underlying Structure $\longrightarrow$ | | SELF -- OTHER: |
| IMPERIAL<br><br>(Stage Two) | Concrete<br>Operational | Instrumental<br>Orientation |
| Underlying Structure $\longrightarrow$ | | SELF -- OTHER: |
| INTERPERSONAL<br><br>(Stage Three) | Early Formal<br>Operations | Interpersonal<br>Concordance<br>Orientation |
| Underlying Structure $\longrightarrow$ | | SELF -- OTHER: |
| INSTITUTIONAL<br><br>(Stage Four) | Full Formal<br>Operations | Societal<br>Orientation |
| Underlying Structure $\longrightarrow$ | | SELF -- OTHER: |
| INTER-INDIVIDUAL<br><br>(Stage Five) | Post-Formal?<br>Dialectical? | Principled<br>Orientation |
| Underlying Structure $\longrightarrow$ | | SELF -- OTHER: |

| LOEVINGER | MASLOW | McCLELLAND/ MURRAY | ERIKSON |
|---|---|---|---|
| Pre-Social | Physiological Survival Orientation | | |

Reflexes (Sensing, Moving) -- NONE

| | | | |
|---|---|---|---|
| Impulsive | Physiological Satisfaction Orientation | | Initiative vs. Guilt |

Impulses, Perception -- Reflexes (Sensing-Moving)

| | | | |
|---|---|---|---|
| Opportun- istic | Safety Orientation | Power Orientation | Industry vs. Inferiority |

Needs, Interests, Wishes -- Impulses, Perceptions

| | | | |
|---|---|---|---|
| Conformist | Love, Affection Belongingness Orientation | Affiliation Orientation | Affiliation vs. Abandonment |

The Interpersonal, Mutuality -- Needs, Interests, Wishes

| | | | |
|---|---|---|---|
| Conscien- tious | Esteem and Self-Esteem Orientation | Achievement Orientation | Identity vs. Identity Diffusion |

Authorship, Identity, Psychic __ The Interpersonal,
Administration, Ideology      Mutuality

| | | | |
|---|---|---|---|
| Autonomous | Self- Actualization | Intimacy Orientation? | |

Individuality, Interpenetrability__Authorship, Identity,
    of Self Systems            Psychic Administration,
                               Ideology

It is not only the physical world which is being con-
served, but internal experience, too, with the constitution
of the "enduring disposition" (which I call, for shorthand
purposes, the "needs"--but it should be clear that I am not
talking about "need" as a content), there comes as well the
emergence of a self-concept--a more or less consistent notion of a me (what I am), as opposed to the earlier sense of
self (that I am) and the later sense of self (who I am).

With the capacity to take command of one's impulses (to
have them, rather than be them) can come a new sense of
freedom, power, independence--above all, agency. Things no
longer "just happen" in the world; with the capacity to see
behind the shadows, to "come in" with the data of experience, I now have something to do with "what happens". The
end of Kohlberg's first moral stage, where authority is all
powerful and right by virture of its being authority, is
probably brought on by the construction of one's own authority.

As is the case with every new development, the new liberation carries new risks and vulnerabilities. If I now have
to do with what happens in the world, then whether things go
badly or well for me is a question of what I can do. Looming
over a system whose hallmark is newly won stability,
freedom, and control, is the threat of the old lability,
loss of control, and what now appears as the old subjugation
from without. How much of the control and manipulation we
experience when we are the object of this meaning-making
balance is a matter of the person's efforts to save itself
from an old world which carries with it the threat--real or
imagined--of ungovernable and overwhelming impulse life?

Every new balance is a triumph over the constraints of
the past evolutionary truce, but a limit with respect to the
truce which might follow. What are the limits of stage two,
the "Imperial" I? I cannot, for example, feel feelings which
might arise out of a simultaneous consideration of my own
impulse-coordinating and another's. To do so would require
me to take impulse-coordinating as an object, which is precisely the limit of this evolutionary balance. This means,
for example, that if I betray a confidence because it suits
my needs to do so, I do not experience whatever it is one
experiences when one simultaneously considers one's own impulse-coordinating with another's (often called "guilt").
What I may experience is concern about whether the person I

have betrayed will find out, and what the consequences of their finding out will be. I am certainly prepared for their dissatisfaction with my deed, as I am able to see that they, like me, have needs and interests. I am able to understand how they might feel about being betrayed, but how they will feel is not a part of the very source of my own feeling or meaning-making. For it to be so would require me to be able to integrate one needs-perspective with another, which would be not just an additive but a qualitative construction of the balance in which I hang. Such a reconstruction entails not just a new level of social perspective, but a new organization and experience of interior life as well. When one's own needs and another's are not integrated, I am unable to hold the other imaginatively and so must seek to hold him in some other way. This is because, being unable to hold him imaginatively, I am left having to await or anticipate the actual movements or happenings of others in order to keep my world coherent.

The development of conscience, or the creation of guilt, may seem a terrible burden and a terrible loss. And, of course, in some way, it is; but it is also quite liberating, as it frees one of having to exercise so much control over an otherwise unfathomable world. Guilt frees one of the distrust of a world which I am radically separate from. Without the internalization of the other's voice in one's very construction of self, how one feels is much more a matter of how external others will react, and others will feel the universal effort to preserve one's integrity as an effort to control or manipulate. When you are the object of the stage 2 balance, you are subject to its projecting onto you its own embeddedness in its needs. I constitute you as that by which I either do or do not meet my needs, fulfill my wishes, pursue my interests, confirm my self-competence. Instead of seeing my needs, I see through my needs. You may experience this as manipulation, or being imperialized, because in order for me to "keep my balance" I have to actually control (or at least predict) the behavior "out there" of people who, in carrying around their own agendas separate from me, make it impossible (unless I can exercise such control) for me to gauge reality--the essence of which is, at this point, knowing the consequences of my actions. What makes the balance "imperial" is our sense of the absence of a shared reality. The absence of that shared reality names the structural limits of the second stage.

Gradually a new evolutionary truce is struck--a qualitatively new object relation--with the emergence from an embeddedness in one's own needs. "I" no longer _am_ my needs, rather I _have_ them. In having them, I can now coordinate (or integrate) one need-system with another, and in so doing, I bring into being that need-mediating reality which we refer to when we speak of the interpersonal or mutuality.

Now the feelings the self gives rise to are a priori shared; somebody else is "in there" from the beginning. The self becomes conversational. To say that the self is located in the interpersonal matrix is to say that it embodies a plurality of voices. Its strength lies in its capacity to _be_ conversational, freeing itself of the prior balance's frenzy-making constant charge to find out what the voice will say on the other end. But its limit lies in its inability to consult itself about that shared reality. It cannot, because it _is_ that shared reality.

Thus, stage three's ambivalences or personal conflicts are not really conflicts between what I want and what someone else wants. When looked into, they regularly turn out to be conflicts between what I want to do to meet the expectations of _this_ shared reality and what I want to do to meet the expectations of _that_ shared reality. To ask someone in this evolutionary balance to resolve such a conflict by bringing both shared realities _before itself_ is to name precisely the limits of this way of making meaning. "Bringing before itself" _means_ not being subject to it, being able to take it as an object--just what this balance cannot do.

With no coordinating of its shared psychological space, "pieced out" in a variety of mutualities, this balance lacks the self-coherence from space to space that is taken as the hallmark of "identity". From some perspectives, this more-public coherence is what is meant by ego itself, but in my view it would be wrong to say that an ego is _lacking_ at stage three--just as it would be wrong to say that at stage three there is a _weaker_ ego. What there is is a qualitatively, not a quantitatively, _different_ ego--a different way of making the self cohere. When I live in this balance as an adult, I am the prime candidate, for example, for the assertiveness-trainer, who may tell me that I need to learn how to stand up for myself, be more "selfish", less pliable--and so on--as if these were mere skills to be _added on_ to whomever else I am. The popular literature will talk about me as "lacking self-esteem", or being a pushover because I "want other people to like me". This does not quite address me in my predicament. It is more that there is no self independent of the context of "other people liking". It is not as if this self, which is supposedly not highly esteemed, is the same self as the one that can stand up for itself independent of the interpersonal context; it is rather a wholly different self, differently constructed. The difference is not just an affective matter—how much I like myself, how much self-confidence I have; the difference goes to that fundamental ground which is itself the source of affect and thought: the evolution of meaning.

This balance is "interpersonal" although it is not "intimate", because what might appear to be intimacy here is the self's source rather than its aim. There is no self to share with another; instead, the other is required to bring the self into being. Fusion is not intimacy. If one can feel manipulated by the Imperial balance, one can feel devoured by the Interpersonal one.

Stage three is not good with anger and may in fact not even be angry in any number of situations which might be expected to make a person angry. Anger owned and expressed is a risk to the interpersonal fabric, which for this balance is the holy cloth. Getting angry amounts to a declaration of a sense of self separate from the relational context—a declaration that I still exist, that I am a person too, that I have my own feelings, which I would continue to own apart from this relationship. It is, as well, a declaration that you are a separate person and can be, that you can survive my being angry, that it is not an ultimate matter for you. If my meaning-making will not permit me to know myself this way, it will surely not permit me to guarantee this kind of distinctness to you either. There are a myriad of reasons why people may find it hard to express anger when they feel it. But it appears that persons in this balance undergo experiences such as being taken advantage of or victimized, yet do not get angry at all because they cannot know themselves separate from the interpersonal context; instead they are more likely to feel sad, wounded, or incomplete.

Thus, if the Interpersonal balance is able to bring inside to itself the other half of a conversation the Imperial balance always had to listen for in the external world, the

Interpersonal balance suffers the vicissitudes of its own externalities. It cannot bring onto itself the obligations, expectations, satisfactions, purposes, or influences of interpersonalism; they cannot be reviewed, reflected upon, mediated--and so they rule.

Each new balance sees you (the object) more fully as you--guarantees, in a qualitatively new way, your distinct integrity. Put another way, each new balance corrects a too-subjective view of you; in this sense, each new balance represents a qualitative reduction of what another psychology might call "projected ambivalence". In the Imperial balance (stage 2), "you" are an instrument by which I satisfy my needs and work my will. You are the other half of what, from the next balance, I "recognize" as my own projected ambivalence. In the move to the new evolutionary grammar of stage three, I claim both sides of this ambivalence and become internally "interpersonal". But stage three brings on a new "projected ambivalence". You are the other by whom I complete myself, the other whom I need to create the context out of which I define and know myself and the world. At stage four I "recognize" this as well, and again claim both sides as my own--bringing them onto the self. What does this mean for my inner life?

In separating itself from the context of interpersonalism, meaning-evolution authors a self which maintains a coherence across a shared psychological space and so achieves an identity. This authority, sense of self, self-dependence, self-ownership is its hallmark. In moving from "I _am_ my relationships", to "I _have_ relationships", there is now somebody who is doing this having--the new I, who, in coordinating or reflecting upon mutuality, brings into being a kind of psychic _institution_ (in + _statuere_: to set up; _statutum_: law, regulation--as in "statute" and "state").

As stage 3, in appropriating a wider other, was able to bring onto itself the other half of a conversation stage 2 had always to be listening for in the external world, stage 4's wider appropriation internalizes those conflicts between shared spaces which were formerly externalized. This makes stage 4's emotional life a matter, for example, of holding both sides of a feeling simultaneously--where stage 3, for example, tends to experience its ambivalences one side at a time. But what is more central to the interior change between 3 and 4 is, perhaps, the way 4 is regulative of its feelings. Having moved the shared context over from subject to object, the feelings which arise out of interpersonalism do not reflect the structure of my equilibrative knowing and being, but are in fact reflected upon by that structure. The feelings which depend on mutuality for their origin and their renewal remain important but are relativized by that context which is ultimate: the psychic institution and the time-bound constructions of role, norm, self-concept, auto-regulation (which maintain that institution). The socio-moral implications of this balance are of course the construction of the legal, societal, normative system. But what I am suggesting is that these social constructions are reflective of that deeper structure which constructs the self itself as a system, and makes ultimate (as does every balance) the maintenance of its integrity.

Talk of "transcending the interpersonal" often makes people uneasy who want to point out—and rightly, I would think—that "other people" remain or should remain what is important to us throughout our lives. But others are not lost by an emergence from embeddedness in the interpersonal. (On the contrary; in a sense, they are found.) The question always is how "other people" are known. The institutional balance does not leave one bereft of interpersonal relationships, but it does appropriate them to the new context of their place in the maintenance of a personal self-system.

A strength of this is the person's new capacity for independence—to own oneself, rather than having all the pieces of oneself owned by various shared contexts. The sympathies which arise out of one's shared space are no longer determinative of the "self", but are taken as preliminary, mediated by the self-system. But in this very strength lies a limit. The "self" is identified with the organization. The "self" at ego stage 4 is an "administrator" in the narrow sense of the word—one whose meanings are derived out of the organization, rather than deriving the organization out of one's meaning/principles/purposes. It has no "self", no "source", no "truth" before which it can bring the operational constraints of the organization, because its "self", its "source", its "truth" is invested within these operational constraints. In this sense, ego stage 4 is inevitably ideological (as Erikson recognized must be the case for identity formation; 1968)—a truth for a faction, a class, a group. And it probably requires the recognition of a group (or persons as representatives of groups) to come

into being—either the tacit ideological support of American
institutional life (which is most supportive to the "institutional" evolution of middle-class white males), or the
more explicit ideologies in support of disenfranchised
social class, gender, race.

The immediacy of mutuality-generated feeling (stage 3)
is thus replaced by a mediacy of the regulation of the mutuality-generated feeling; a regulation which is not itself
mediated by anything. This makes stage 4's emotional life
more internally controlled. Where for stage 3 it is those
events that risk the integrity of the shared context which
mobilize the "self's" defensive operations for stage 4 it is
those events that threaten chaos for the interior polity.
The question is not, as it was earlier, "Do you still like
me?", but "Does my government still stand?". A variety of
feelings and doubts, especially erotic or affiliative feelings and doubts around performance and duty-discharge, come
to be viewed as potential dissidents which must be subjected
to the psychic civil polity. Stage 4's delicate balance is
that in self-government it has rescued the "self" from its
captivity by the shared realities, but, in having no "self"
before which it can bring the demands of that government it
risks the excesses of control that may obtain to any government not subject to a wider context in which it roots and
justifies its laws.

The last object relation for which we have evidence is
characterized by a rebalancing that separates the self from
the "institution" and thus creates the "individual", that
self who can reflect upon or take "as object" the regula-
tions and purposes of a psychic administration which formerly was the subject of one's attentions. "Moving over" the
"institutional" from subject to object frees the self from
that displacement of value whereby the maintenance of the
institution has become the end in itself; there is now a
self who runs the organization. Where previously there was
a self who was the organization, there is now a source before which the institutional can be brought, by which it is
directed; before, the institution was the source.

Every subject-object relationship amounts to a kind of
"theory" of the prior stage; this is another way of speaking
about subject moving to object, or structure becoming content. Stage 2 is a "theory" of impulse; the impulses get organized or ordered by the needs, wishes, or interests. Stage

3 is a "theory" of needs; they get ordered by that which is
taken as prior to them, the interpersonal relationships.
Stage 4 is a kind of theory of interpersonal relationships;
they get rooted in and reckoned by the institutional.  Stage
5 is a theory of the institutional; the institutional is
ordered by that new self which is taken as prior to the institutional.

     Kohlberg's (1976) moral stage 5, for example, requires
a "prior-to-society perspective", by which he refers to that
dislodging by which the self is no longer subject to the societal; this is accomplished at the transitional disequilibration between his stages 4 and 5.  To be "stage 5" in Kohlberg's framework, the person must have, in addition to this
"prior-to-society" perspective, a kind of theory that roots
the legal institution in principles which give rise to it,
to which conflicts in the law can be appealed, and by which
the rights that the legal institution protects might be hi-
erarchized.  What this amounts to, more than disequilibrial
transition out of the stage-4 balance, is that re-equilibration by which the legal institution has been recovered or
recollected as object in a new balance.  No longer is "the
just" derived from the legal; now the legal is derived from
a broader conception of the just.  And no longer does the
person make no reference to the former institutional context
("'Should' is no longer in my vocabulary.").  The hallmark of
every rebalancing is that the past, which may during transition be repudiated, is not finally rejected but reappropri-
ated (integration).

     But what is a kind of theory of the legal institution
may be an expression, in the moral realm,of that deep structure which is a theory of the self as institutional.  And
that which constructs this theory--the new subjectivity--is
the next self-other balance.

     What happens to the community of identification in this
object relation?  The capacity to coordinate the institutional permits one to now join others, not as fellow-instru-
mentalists (stage 2) nor as partners in fusion (stage 3) nor
as loyalists (stage 4) but as individuals--people who are
known ultimately in relation to their actual or potential
recognition of themselves and others as value-originating,
system-generating, history-making individuals. The community
is for the first time a "universal" one in that all persons
are eligible for membership by virtue of their being

persons; the group which this equilibrium's self knows as "its own" is not a pseudo-species, but the species. The self is no longer identified with the coordination of the interpersonal which, in constructing the institution that handles this task, brings the interpersonal "into" the "self"; rather, one's "self" is coordinating one's own and others' "institutions", which brings the self back into the interpersonal. The great difference between this and stage 3 is that there now is a "self" to be brought to, rather than derived from, others; where stage 3 is interpersonal (a fused commingling), stage 5 is interindividual (a commingling which guarantees distinct identities).

This new locating of the self, not in the structure of my psychic institution but in the coordinating of the institutional, brings about a revolution in Freud's favorite domains, "love" and "work". If one no longer is one's institution, neither is one any longer the duties, performances, work-roles, career which institutionality gives rise to. One has a career and no longer is a career. The self is no longer so vulnerable to the kind of ultimate humiliation which the threat of performance-failure holds out, for the performance is no longer ultimate. The functioning of the organization is no longer an end in itself; and one is interested in the way it serves the aims of the new self, whose community stretches beyond that particular organization. The self seems available to "hear" negative reports about its activities; previously, it was those activities, and therefore literally "irritable" facing those reports. (Every balance's irritability is simultaneously testimony to its capacity to grow and its propensity to preserve itself.) Every new balance represents a capacity to listen to what previously one could hear only irritably, and the capacity to hear irritably what previously one could hear not at all.

But the increased capacity of the stage 5 balance to hear (and to seek out) information which might cause the self to alter its behavior, or share in a negative judgment of that behavior, is but a part of that wider transformation which makes stage 5 capable--as was no previous balance--of intimacy. At stage 4, one's feelings seem often to be regarded as a kind of recurring administrative problem which the successful ego-administrator resolves without damage to the smooth functioning of the organization. When the self is located not in the institutional but in the coordinating of

the institutional--one's own and others--the interior life
gets "freed up" (or "broken open") within oneself, and with
others. This new dynamism, flow, or play results from the
capacity of the new self to move back and forth between psychic systems within itself. Emotional conflict seems to become both recognizable and tolerable to the self. At stage
3, emotional conflict cannot yet be recognized by the self;
one can feel torn between the demands from one interpersonal
space and those from another, but the conflict is taken as
"out there" (it is the ground and I am the figure upon it).
At stage 4, this conflict comes inside. The dawn of the
"self-as-a-self" (the institutional self) creates the self
as the ground for conflict, and the competing poles are figures upon it. Emotional conflict is recognized but not tolerable; i.e., it is ultimately costly to the self. The self
at stage 4 was brought into being in order to resolve such
conflict, and its inability to do so jeopardizes its balance. Stage 5, which recognizes pluralities of institutional selves within the (inter-individual) self, is thereby
open to emotional conflict as an interior conversation.

One way of speaking of stage 5's capacity for intimacy,
then, is to say that it springs from the capacity to break
open its institutionality to itself--i.e., to be intimate
with oneself. Locating itself now in the coordination of
psychic institutions, the self surrenders its counter-depen-
dent independence for an inter-dependence. Having a self,
which is the hallmark of stage 5's advance over stage 4, it
now has a self to share. This sharing of the self at the
level of intimacy permits the emotions and impulses to live
in the intersection of systems, to be "re-solved" between
one self-system and another. Rather than the attempt to be
both close and auto-regulative, "individuality" permits one
to "give oneself up" to another--to find oneself in what
Erikson has called "a counterpointing of identities" which
at once shares experiencing and guarantees each partner's
distinctness, which permits persons (again Erikson's words;
1968) "to regulate with one another the cycles of work, pro-
creation, and recreation". Every re-equilibration is a qualitative victory over isolation.


THE EVOLUTION OF SEPARATION AFFECTS

I have tried to indicate how the processes so salient

to the first two years of life--differentiation, the creation of the object relation--recur throughout life, each new
object relation serving as a culture of embeddedness out of
which the evolving personality might emerge. What I would
like to do now is indicate that the affects so salient to
these processes in infancy also recur in qualitatively new
ways. Mahler understands her work to suggest that the person is not born only once, as it were, but is psychologically born a second time as it emerges from an extra-uterine
symbiosis. My reading of the structural-developmental research, in concert with my own clinical work and clinically
oriented research, suggests that we are psychologically born
again and again. With this bigger picture in mind it becomes less satisfactory, I believe, to see the transformations of later living in terms of the earliest transformation, and more satisfactory to see all the transformations
in the context of that which subtends them: meaning-constitutive evolution.

Evolutionary activity gives rise to constructions of
balance, to truces. But what is at stake in these truces?
Viewed from the outside, it is what we shall take for "self"
and what we shall take for "other"; will I, in other words,
continue to know? Hence, equilibrative activity is naturally epistemological. But what is at stake in maintaining the
balance, when viewed from the "inside", is whether I (this
constitution of "I") shall continue to be; hence equilibrative activity is also naturally ontological. Viewed with
respect to the experience of being one who equilibrates, we
open ourselves to the study of feeling, from which the traditional Piagetian orientation to stages must necessarily
exclude us. For we are not our stages; we are not the self
who hangs in the balance at this moment in our evolution. We
are the activity of this evolution. We compose our stages,
and we experience this composing. Out of this evolutionary
motion which we are, we experience emotion; this is what the
word means (ex + motion: out of, or from, motion). Any theory of emotion must begin by naming that motion it regards
as the source. I have named my candidate. Feeling may be
the sensation of evolution--more complexly, the phenomenology of personality in its predicament as self-constituting
meaning-making.

Nowhere is that predicament more painful than those
times in our lives when the specter of loss of balance is

looming over the system. These are the moments when I experience, fleetingly or protractedly, that disjunction between
who I am and the self I have created--the moments when I
face the possibility of losing myself, the moments which
Erikson hauntingly refers to as "ego chill". The chill comes
from the experience that "I am not myself", the experience
of a distinction between who I am and the self I have
created. My colleagues (especially Laura Rogers and Donald
Quinlan) and I interviewed 39 persons, Terry and Diane among
them, during their stay on a psychiatric ward. We used
Kohlberg dilemmas and analyzed them for level of self/other
differentiation. We also had their permission to acquaint
ourselves with their intake interviews, the report of what
life had been like prior to admission, and the events of
their hospital stay. From the files, we thought we detected
three very heterogeneous but nevertheless qualitatively different kinds of depression:

Type A. The fundamental concern seemed to be the loss
of my own needs, or an unhappiness at the increasing personal cost of trying to satisfy my own needs. When the orientation is outward, I may feel constrained, deprived, cur-
tailed, controlled, or interfered with in the effort to satisfy my needs, wishes, desires. When the orientation is
self-directed, there is a tension between feeling unavoid-
ably irresponsible, a slave to my own interests, insensi-
tive, on the one hand; or feeling compromised away, or that
I have lost my own distinct personality, on the other. The
overall foreboding feeling seems to be that, with the possible loss of the satisfaction of my wants, I may no longer
be. At stake, directly or lurkingly, seems to be the very
meaningfulness of an ultimate orientation to "my needs" in
the first place.

Type B. The fundamental concern seemed to be the loss
of, or damage to, an interpersonal relationship. When the
orientation is outward, I may feel unbearably lonely, de-
serted, betrayed, abandoned, stained. When the orientation
is self-directed, there is a tension between feeling vulnerable to incorporation, fusing, the loss of myself as my own
person, on the one hand; and on the other, feeling selfish,
heartless, coldly prideful, uncaring as a result of beginning to "put myself first". The overall foreboding feeling
seems to be that, with the rupture of my interpersonal con-

text, I may no longer be. At stake, directly or lurkingly, seems to be the very meaningfulness of an ultimate orientation to the interpersonal context in the first place.

Type C. The fundamental concern seemed to be a blow to self-concept, a failure to meet my own standard, perform or control myself as I expect to. When the orientation is outward, I may feel humiliated, empty, out of control, closed in on--that the world is unfair, that life is somehow meaningless. When the orientation is self-directed, there is a tension betweeen feeling vulnerable to crippling self-attack, identifying with my performance, isolated in self-containment, on the one hand; or feeling weak, ineffective, out of control, evil, decadent, without boundaries, on the other. The overall foreboding feeling seems to be that, with the disruption to my self-conscious psychic organization or authorship, I may no longer be. At stake, directly or lurkingly, seems to be the very meaningfulness of an ultimate orientation to the maintenance of this psychic administration in the first place.

Each subject was assigned to a depression type without knowledge of his/her subject-object level. (The Kohlberg scores--grouped as 2 or 2-3 transitional, 3 or 3-4 transitional, 4 or 4-5 transitional--were identical between two raters in 94% of the cases; depression scores were identical in 84% of the cases.) When the subject-object levels are compared with depression type, an extremely strong relation emerges between persons at stage 2 or 2-3 and Type A depression, between persons at stage 3 or 3-4 and Type B depression, and between persons at stage 4 or 4-5 and Type C depression (Kendall's tau = .815, < .001; Chi square = 46.63, 8 d.f., < .001). This correlation is impressive, but inconclusive by itself. For one thing, to even confirm the phenomenon, it needs to be replicated on a broader scale; for another, even confirmed, the phenomenon itself only suggests a relation, not a cause. My colleagues and I are presently at work not only to confirm the phenomenon, but, by longitudinal investigation, to test the hypothesis that the depression is caused by the disequilibrium, or threat to the balance, of a given evolutionary truce. Still, even the present study has much to tell us beyond these numbers and correlations. However in need of elaboration and confirmation it may be, a clear picture does come through of successive transformation processes and their costs to the evolving person.

Before we take a look at that picture, it might be useful to point out that the types of depression I think I see are not wholly unconfirmed. After I completed this study, I learned of the work of Blatt and his colleagues (1974), who have described two types of depression which are quite consistent with the latter two just described. Blatt calls the first of these "anaclitic" or "dependent" depression, which "is characterized by feelings of helplessness, weakness, and depletion,...intense fears of abandonment and desperate struggles to maintain direct physical contact with the need-gratifying object"; the second he calls "introjective" or "guilty" depression, which "is characterized by feelings of worthlessness, guilt, and a sense of having failed to live up to expectations and standards,...intense fears of a loss of approval, recognition, and love from the object".

Blatt, D'Afflitti, and Quinlan (1976) have constructed a Depression Experience Questionnaire, deriving items from clinical assessments of aspects of patient-reported depressive experience. The items consist not of symptoms but of thoughts, feelings, and experience of self, which subjects rate on a seven-point scale from "strongly like my experience" to "strongly unlike my experience". By factor analysis, three major factors emerged—their stability confirmed by further investigation. One of these was characterized by positive, self-confident, efficacious feelings about the self. The two others seemed to the authors to be characterized on the one hand by dependency ("items primarily related to interpersonal relationships and concerns about helplessness, loneliness, and dependency,...concerns about losing someone, about being rejected, hurting or offending people, of being very dependent and lonely, and having difficulty in managing anger and aggression for fear of losing someone") and on the other hand by self-evaluation ("items about feeling guilty, empty, hopeless, unable to meet responsibilities, having failed to meet expectations and standards, disappointing to others, unsatisfied, insecure, threatened by change, and tending to blame and feel critical toward oneself"). These researchers at once provide further evidence of the phenomenological reality of kinds of depression differing qualitatively in the ways first suggested, and, at the same time, make it difficult to claim that I am involved in a fancy tautology (i.e., that I looked at depression in just the sort of way that would naturally make it correlate with constructive-developmental stages); I am looking at the same phenomenon and describing it in ways that are similar, but from a radically different perspective.

There is nothing in factor analysis that even <u>suggests</u> an explanation of the underlying nature of a given depression. Blatt, et al., (1976) are willing to put forth such an explanation, but would be equally willing, I am sure, to admit that there is nothing in factor analytic instrumentation to support their explanation over any other. Their explanation appears de novo from psychoanalytic theory; and it is thus wholly dependent for its validity on the validity that theory has been able to amass in its long, if troubled, research history. They call the findings of their factor-analytic study consistent with Blatt's earlier formulation of two types of depression (it certainly is that), but then go on to "describe" those types as follows: "an anaclitic, dependent type of depression which evolves at a lower level of development, primarily around the individual's dependent relationship with the mother; and an introjective or guilt depression which evolves during the phallic and oedipal stages of development where the child is establishing his identification and competition with the parent of the same sex and is struggling to win attention, approval, and affection of the parent of the opposite sex"!

While their study supports the contention of phenomenologically discrete types of depression--discrete along lines Blatt had earlier suggested--it is equally clear that nothing about their study points either to or away from the psychoanalytic understanding the authors <u>attach to</u> the discrete types. The explanation of differing depressions that I am suggesting can be said to derive directly from an understanding of what the empirical measure <u>means</u> in relation to a given type of depression with which it is strongly correlated.

Whatever might be said about the <u>early</u> object relations of Terry, Diane, or Rebecca, their present experience seems most understandable in terms of their <u>present</u> object relations. Each subject-object level is a new resolution of the tension between integration and differentiation (to put it in stark, biological terms), or the tension between the lifelong human yearnings to feel included, related, "a part of", on the one hand, and on the other, to feel my own distinct agency integrity (to put the same thing more "psychobiologically"). One of the promising features of the frame-

work I am presenting here, it seems to me, is that it offers
a corrective to all present developmental frameworks which
univocally define growth in terms of differentiation,
separation, and increasing autonomy--losing sight of the
fact that adaptation is equally about integration, attachment, inclusion. The net effect of this myopia, as feminist
psychologists (Gilligan 1978, Low 1978) are now pointing
out, is that differentiation (the stereotypically male overemphasis in this most human ambivalence) is favored with the
language of growth and development, while integration (the
stereotypically female overemphasis) gets spoken of in terms
of dependency and immaturity. The history of subject-object
levels, I am suggesting,indicate how each is a level of differentiation and integration (beginning with "level zero",
which is not only "undifferentiated"--as everyone says--but
also "unintegrated"). What is more, the levels trace a history of oscillating over-balancing in the lifelong solving
and resolving of this eternal tension. The echoes between
stages 2 and 4 (the Imperial and the Institutional) and
between 3 and 5 (the Interpersonal and the Inter-individual)
are a feature of the fact that both 2 and 4 are "over-dif-
ferentiated" and 3 and 5 are "over-integrated". We know the
balances of any subject-object level are not perfect; this,
in fact, is why we grow. And the framework suggests a way
of better understanding the nature of our "vulnerability to
growth" at each level.

If the reader reconsiders the brief glimpses of Terry,
Diane, and Rebecca at this chapter's beginning, it may be
clear that while boundary issues seem salient for them,there
are some clear distinctions to be made. The youngest and
oldest of the three express their concerns in terms of preserving a boundary which feels like it is giving way.
Diane, on the other hand, expresses her concerns in terms of
an inability to preserve the lack of a boundary. Terry and
Rebecca are experiencing the problems of over-differentia-
tion. More particularly, as I will elaborate in a moment, I
think they are experiencing a "hatching" out of the Imperial
or Institutional levels, repectively. Diane is experiencing
the problems of over-integration, and is, I will suggest
more particularly, at the limits of the Interpersonal
subject-object relation.

Terry did not have a good experience in the hospital,
where group therapy and participation in the life of a ward
community were the primary media of treatment (milieu thera-

py). She could be seen in group meetings struggling unsuc-
cessfully to talk about herself in a way the staff would ap-
prove of. When she would say, in many different ways, that
"My problems are not mental; they have to do with getting
along with my family", the staff, however gently and indirectly, was not satisfied with the formulation and took it as
resistance to deal with herself, or her own feelings and responsibilities. The staff could see that when she did speak
the language of internal reflexivity, she was doing it by
triangulation or imitation, that she was not really speaking
her own voice; this, too, they took as a kind of slipperi-
ness or dishonesty. Their exasperation with her grew; and
when it was discovered that she both spoke to people outside
the ward about people inside the ward (a cardinal violation
of patient-patient bonds) and had used "unprescribed drugs"
during a weekend pass (a violation of patient-staff bonds),
the staff threw her out of the community. They understood
her behavior as hostility and acting-out; and they justified
their own behavior on the grounds that she "would not do the
work of the ward", "was argumentive", "was disruptive", "was
inciting to other patients", and "was a staff-splitter". It
seems they were a touch angry with Terry.

What is the possibility that Terry was not helped because she was not well understood—that if the nature of her
present subject-object relations were better understood, her
behavior would be less infuriating and the means of treatment would have been more appropriate?

From Terry's Kohlberg-interview, researchers who had no
familiarity with her pre-hospital history or her hospital
course could discern that she was just emerging from the Imperial balance. If she is just beginning that activity which
will eventually lead to the recovery of "my needs" as an object of attention, it may now feel to her as if they are
being ripped from her without hope of return. She seems to
speak to this kind of experience in her sense that she is
going to have to "submerge my personality", her sense that
she is "no longer whole", and—most poignantly accurate of
the 2-3 navigation—that, strangely, "others are woven into
me". It is this weaving, which now feels so alien, that constructs the shared psychological context that is the hallmark of the next balance.

In the absence of that Interpersonal balance and this
"sharing", the subject's relation to her emotional life—and

her emotional life itself—does not allow the kind of psy-
chologizing of one's interior states that make one reachable
through a therapy that depends on such reflectivity. Without the internalization of the others' voices in one's very
construction of "self", how one feels is much more a matter
of how actual external others will react, what they will or
will not do in consequence of what I do. One's emotional
situation is thus ever-changing, depending upon the actions
and reactions of others who are not easily decipherable
since their interior life cannot be constructed by the subject any more complexly than she can construct her own. She
is, thus, left looking "out there" in her construction of
her problems, which are therefore "not psychological but
have to do with getting along with my family".

What we must ask then is whether her posture on the
ward is necessarily the recalcitrance it is seen to be by a
frustrated staff. Is her inability to speak about her problems in group in the acceptable fashion, which is viewed by
the staff as strategic manipulation to find out the right
thing to say, some kind of resistance? Well, yes—but not
in the sense that a "truer self" is hiding there under a defensive shield; rather, her "resistance" or "defense" be-
speaks the integrity of that which makes a system a system.
It seems more accurate to say that when she describes her
problems as "out there", or as really "social problems", she
is locating the arena of her "mental" or "psychological"
life as accurately as another patient who may talk about
guilt. The question is whether this is because she is "so-
ciopathic" (as diagnosed by the hospital) or whether, subject to her own needs and unable to locate the self in the
coordinating of needs that constructs the mutual and the
"shared", she is left at the mercy of "whether her mother
knows she lied" in order to know what her own "feeling" can
be. This makes the "socio" itself the psychological, suggesting that her inability to keep the bargains of a
milieu-treatment approach—which requires as a baseline the
very balance (stage 3) that is her growing edge—is due not
to any psychiatric "illness" but to the limits of her subject-object relating.

In the hospital, Diane, whose subject-object interview
suggested she was just moving out of the Interpersonal level, was at first afraid to talk about her "problems" because
"others may not like me". She eventually discovered a parallelism about her feelings toward her father and her boy-

friend: in both cases, despite her subjective experience of being rejected by the man, she persisted in trying to please him and shelter him from what she perceived to be his fragility in the face of any direct confrontation. She began dealing with the conflict she felt in relation to her family: a desire to be loved and cared for and approved of versus a desire for independence. She felt a similar conflict in relation to her boyfriend, complicated by an emerging fear of sexual intimacy: it felt to her as though there was no way to hold him and be with him without losing herself to him in the relationship. She began to feel—and deal with—her feelings of anger toward her father, anger toward the hospital staff, anger toward her boyfriend. Toward the end of hospitalization, she had put controls on the pattern of her father's visits and had prepared for discharge to a living-and-working situation which would put into effect her decision to separate from her boyfriend and to continue her career. She expressed her apprehension on leaving in terms of being afraid of whether she "can accept responsibility for myself".

I wonder to what extent the disruption, pain, and restoration in Diane's life is not a function of the loss of and emerging reconstruction of her subject-object relating. I wonder if her story is not that of the gradual and harrowing disidentification by her evolutionary activity from that invented self founded on the interpersonal. Prior to hospitalization the loss or threatened loss of the relationship seemed to raise the question of the loss of her very being; by the time of discharge, a newly emerging self takes into its own hands the future of that relationship. Taking a relationship into one's hands means "moving over" the very structure of "relationship" from subject to object. But that "going over" (uebergang) phenomenologically amounts to, first, the relativizing of what was taken for ultimate—a loss of the greatest proportion—and second, a period of not-knowing, of delicate balance between what can feel, on the one hand, like being devoured in the boundarilessness of the old construction, and on the other, the selfishness, loneliness, or coldness of being without "the interpersonal".

The quality of the loss and the desperate coldness and isolation before recovery are both reflected in an early recurrent dream: "I am in water as red as blood. There is an island on the water. On the island there are people whom I

have known in the past. I cry for help but feel myself
drowning." Drowning in water as red as blood, my gaze is
upon the departed: solid and familiar ground, peopled with
those I knew before. Yet, viewed from the perspective of
the new transitional self (the self "at sea"), even the familiar world, as Diane's later hospital course suggests, can
feel like a place to drown in. The newly emerging Institutional self constructs the self-conscious self-system, an
intactness that "brings inside" the conflicts which earlier
were located between oneself and another (since between oneself and another was where the self resided), a kind of
self-sufficiency reflected in the expression "In order to
get myself together, I first had to get myself apart". This
new self moves from being a fragile, easily reabsorbed pres-
ence--as in the powerful and terrifying image of the tiny
self struggling against ingestion by the giant plastic
head--to a more reliable context which places limits on a
relationship the termination of which had earlier raised the
specter of annihilation. Diane's hospital diagnosis, of
course, is "borderline". To what extent is this a matter of
her earliest object relations, and to what extent a matter
of her present ones--a matter of walking the fearful border
between two ways of making the world cohere, one left behind
but the other not yet clearly evolved?

     I met Rebecca as a therapist, not a researcher. I have
no intersubjective means of supporting my sense that she
gives us a picture into the separation affect of the last
transformation in the framework. But it seems to me that
the very self-sufficiency Diane desperately needed when she
entered the hospital had long been familiar to Rebecca. It
seems to have become now too familiar, to have worn out its
welcome. But because "it" is how Rebecca is herself composed, she is herself worn out. I think one can hear in her
words (1) a fleeting glance back to the interpersonalist
balance, long ago transcended, (2) the personal authority
and self-conscious integrity of the institutional self, (3)
the fatigue of experiencing its limits:
     I know I have very defined boundaries and I protect them
     very carefully. I won't give up the slightest control.
     In any relationship I decide who gets in, how far, and
     when.
     What am I afraid of? I used to think I was afraid people
     would find out who I really was and then not like me.
     But I don't think that's it any more. What I feel now
     is--"That's me. That's mine. It's what makes me. And

I'm powerful. It's my negative side, maybe, but it's also my positive stuff--and there's a lot of that. What it is is me, it's myself--and if I let people in maybe they'll take it, maybe they'll use it--and I'll be gone." Respect, above all, is the most important thing to me. You don't have to like me. You don't have to care about me, even; but you do have to respect me.
This "self"--if I had to represent it, I think of two things: either a steel rod that runs through everything, a kind of solid fiber; or sort of like a ball at the center that is all together. What you just really can't be is weak. How exhausting it's becoming holding all this together. And until recently I didn't even realize I was doing it.

Rebecca shares a dream which seems to convey powerfully the experience of separation from an embeddedness in the Institutional subject-object relation:
I am in a crowded subway, rushing to get a train. I am standing. There is a woman next to me. The woman makes motions I interpret as a request for money. I open my purse, but it turns out the woman is asking me to identify myself. So I begin taking out all my cards--my social security card, my license, my work identification card, my health insurance--and I show them to the woman. All the while, it is on my mind that I am late and might miss the train. While I am showing my cards the subway gets into the train station. I grab up all my cards and get off the subway. I run for the train. I come to a sort of revolving door and it closes on me with my arm outside, clutching all my cards, and the rest of me looking out toward the train. I feel just completely stuck. I know that if I could just let go of all these cards, just let them drop, I could get through and catch the train. But I am just completely panicked. And I am panicked both at the idea of having to give up all these cards and at the idea of missing the train.

The neo-Piagetian framework, as I have sketched it here, makes the process of development--not the stages of development--its ultimate loyalty in its attempt to engage the odyssey of the changing person in time. An odyssey is known not only by describing the various ports of call, but by some recognition of the costs and courage involved in leaving any port for unfamiliar sea. Nothing is so central to a psychodynamic psychology as its caring for that pain which is part of growth itself--the pain of grief, of separation, of loss. The neo-Piagetian framework understands this pain not as a re-creation of experience in infancy (any more than the emergence of the inter-personal is a re-creation of the emergence of the enduring disposition) but as the phenomenology of evolution itself.

## CONCLUSION

The first way to care for that pain which is part of growth itself--the first way to care for anything, perhaps--is to try to understand it. That has been the focus of this chapter. But is there not some contribution to the attempt to be of help--even a theory of therapy, perhaps--that grows out of this framework? I believe there is. It is not a subject to be taken up here, but in the same spirit in which I began (of seeking to join the conversation so well begun by Margaret Mahler and her colleagues), I would like to close by referring to the way she herself has answered this question (Mahler 1968).

In seeking to help an autistic child by joining what I will call the child's "culture of embeddedness" and then seeking to foster the mother's ability to provide such a culture, Mahler is, I believe, acting out of the single most important principle developmental psychology can offer the clinical world. The principle amounts to this: if you want to know how to be therapeutic, study the instances and intricacies of unselfconscious therapy found over and over in nature. The mother acts as the person's first therapist--a natural therapist--by performing, to speak grossly, two crucial functions: (1) she provides a "culture of embeddedness" which recognizes and holds the child (all development is social; it arises out of a communal context) and (2) she assists in the termination of this embeddedness (she facilitates the infant's emergence by serving as a bridge, or transitional object, upon which further evolution can take place). I would suggest that "unnatural" therapists will find their greatest resource in the wisdom of nature; that the communal culture of embeddedness is required again and again but in different forms appropriate to each level of subject-object relations. If the professional helper would seek to provide that culture, does he not need to know what "emergency" he is seeking to support?

REFERENCES

Blatt, S.M. 1974. "Levels of Object Representation in Ana-
          clitic and Introjective Depression". In The Psy-
          choanalytic Study of the Child, (Yale University
          Press) XXIX:107–57.
Blatt, S.M., D. D'Afflitti & D. Quinlan. 1979. "Experi-
          ences of Depression in Normal Young Adults", Jour-
          nal of Abnormal Psychology, LXXXV(4):383–89.
Bowlby, John. 1973. Separation: Anxiety and Anger. New
          York: Basic Books.
Broughton, John. 1975. The Development of Natural Epistem-
          ology in Adolescence and Early Adulthood. Unpub-
          lished doctoral dissertation, Harvard University.
Elkind, David. 1968. Editor's introduction. In Six Psy-
          chological Studies (by Jean Piaget, New York:
          Vintage).
Elkind, David. 1968. Child Development and Education. New
          York: Oxford University Press.
Erikson, E.H. 1968. Identity: Youth and Crisis. New York:
          W.W. Norton.
Fairbairn, W.R.D. 1962. An Object Relations Theory of Per-
          sonality. New York: Basic Books.
Fowler, J. 1981. Stages of Faith. New York: Harper & Row.
Freud, S. 1911. "Formulations Regarding the Two Principles
          in Mental Functioning". In A General Selection
          from the Works of Sigmund Freud (ed. by J. Rick-
          man). New York: Doubleday, 1957.
Gardner, H. 1978. Developmental Psychology. Boston:
          Little Brown.
Gilligan, C. 1978. "In a Different Voice: Women's Concep-
          tions of the Self and of Morality". Harvard Edu-
          cational Review, XLVII, 4:481–517.
Guntrip, H. 1971. Psychoanalytic Theory, Therapy and the
          Self. New York: Basic Books.
Kagan, J. 1971. Change and Continuity in Infancy. New
          York: John Wiley & Sons.
Kegan, R. 1979. "The Evolving Self: A Process Conception
          for Ego Psychology". Counseling Psychologist,
          VIII, 2:5–34.
Kegan, R. 1982. The Evolving Self. Cambridge: Harvard
          University Press.
Kohlberg, L. 1976. Collected Papers on Moral Development
          and Moral Education. Cambridge: Center for Moral
          Education, Harvard University.

Loevinger, J.  1976.  Ego Development.  San Francisco:
     Jossey-Bass.
Low, N.S.  1978.  "The Mother-Daughter Relationship in
     Adulthood".  Paper presented to the Massachusetts
     Psychological Association.
Mahler, M.S.  1968.  On Human Symbiosis and the Vicissitudes
     of Individuation.  Vol.  I: Infantile Psychosis.
     New York: International Universities Press.
Mahler, M.S., F. Pine & A. Bergman.  1975.  The Psycholog-
     ical Birth of the Human Infant.  New York: Basic
     Books.
Perry, W.G., Jr.  1970.  Forms of Intellectual and Ethical
     Development in the College Years.  New York: Holt,
     Rinehart and Winston.
Piaget, J.  1952.  The Origins of Intelligence in Children.
     New York: International University Press.
     (Originally published in 1936)
Piaget, J.  1954.  The Construction of Reality in the Child.
     New York: Basic Books.  (Originally published in
     1937)
Piaget, J.  1964.  "Relations between Affectivity and Intel-
     ligence in the Mental Development of the Child".
     In Sorbonne Courses.  Paris: University Documenta-
     tion Center.
Schachtel, E.  1959.  Metamorphosis.  New York: Basic Books.
Schaefer, E.S. & P. Emerson.  1964.  "The Development of
     Social Attachments in Infancy".  Monographs of the
     Society for Research in Child Development.
Selman, R.L.  1980.  The Growth of Interpersonal Understand-
     ing: Developmental and Clinical Analyses.  New
     York: Academic Press.
Spitz, R.A.  1950.  "Anxiety in Infancy: A Study of its Man-
     ifestations in the First Year of Life".  Interna-
     tional Journal of Psychoanalysis, XXXI:138-43.
Uzguris, I.  & J. Hunt.  1968.  "Object Permanence", from
     Ordinal Scales of Infant Psychological Develop-
     ment--A Series.  Film: 40 min.  Champaign: Univer-
     sity of Illinois Visual Aids Service.
White, R.W.  1959.  "Motivation Reconsidered: The Concept of
     Competence".  Psychological Review, LXVI:297-333.
Winnicott, D.W.  1965.  The Maturational Processes and the
     Facilitating Environment.  New York: International
     Universities Press.

# TOWARDS A VYGOTSKIAN THEORY OF THE SELF

Benjamin Lee, James V. Wertsch, and
Addison Stone

Center for Psychosocial Studies and
Northwestern University

## INTRODUCTION

Vygotsky proposed his theoretical framework in an attempt to deal with what he saw as the "crisis in psychology". Although he was speaking of a crisis that existed in the early part of the twentieth century, his critique and proposals have a great deal of relevance for today's Western social science. The crisis Vygotsky saw was that there was no broad philosophical foundation or metapsychological theory upon which to build an integrated explanation of hypotheses and empirical findings. Quoting Brentano, Vygotsky (1956) wrote that "there exist many psychologies, but there does not exist a single psychology" (p. 57). As Vygotsky went on to point out, the implications of this are quite serious.

...The absence of a single scientific system that would embrace and combine all of our contemporary knowledge in psychology results in a situation in which every new factual discovery...that is more than a simple accumulation of details is forced to create its own special theory and explanatory system. In order to understand facts and relationships investigators are forced to create their own psychology--one of many psychologies. (1956, p. 57-58)

Vygotsky proposed to deal with this fragmentation and with the age-old dualism between idealism and materialsim that often accompanied it by creating a general theoretical framework based on Marxist foundations. His proposal for such a framework was one of the most ingenious treatments in Soviet social science of the Marxist issue of the relationship between a materialist reality and human consciousness.

A basic postulate of this framework is that unlike the case with animals, man's representation and control of his activity is based on more than "inherited experience" (i.e., habits and patterns of behavior learned through direct, individual experience). Vygotsky argued that human consciousness is also based on "historical experience" and "social experience". In connection with the former he wrote:

Man makes use not just of physically inherited experience: throughout his life, his labor and his behavior draw broadly on the experience of former generations, which is not transmitted at birth from father to son. We may call this historical experience. (1979, p. 13)

In connection with social experience he wrote:

Ranked alongside this historical experience is social experience, the experience of other people, which constitutes a very important component in human behavior. I do not possess only those connections that have been formed in my personal experience... I also have at my disposal a multitude of associations and connections formed in the experience of other people. If I know Berlin or Mars although I have never looked into a telescope, the origin of this experience is obviously linked to the experience of other people who have traveled to Berlin and have looked into a telescope. It is just as obvious that animals usually do not have such experience. Let us call this the social component of our behavior. (1979, p. 13)

Historical and social experience were so essential to Vygotsky's theoretical framework that Soviet authors (e.g., Smirnov, 1975) refer to it as the "socio-historical" or "cultural-historical" approach to the study of mind.

Vygotsky, following Marx's lead, believed that new levels of the organization of consciousness followed different principles of development, and these principles depend upon differentiating between the social nature of human practical activity and the "natural" activity of animals. Marx insisted that what distinguished men from animals was that human productive labor is necessarily social while animal production and consumption are only contingently social. Principles of Darwinian evolution were adequate to explain the development of animal social organization, but the structure of human productive labor requires the introduction of new principles of development, namely those of dialectical materialism. The social nature of man requires a new mode of dialectical analysis which shows how psychological development emerges from more general social processes.

Vygotsky applied this Marxist thesis to child development. He distinguished between a natural line of development, which includes the tool activity of primates and probably corresponds to Piaget's sensori-motor period, and a
social line of development which is heavily dependent upon
the child's acquisition of language. Vygotsky believed that
the processes underlying the natural line of development are
subject primarily to physiological laws of development along
with some relatively simple principles of learning. The
child's social line of development follows principles based
upon the structure of communication. It is at this point
that Marx's and Vygotsky's approaches converge. As the
child moves from a "natural" line of development to the "social" line, he also becomes part of a social system whose
evolution is governed by the principles of dialectical materialism.

It is my belief, based upon a dialectical materialist
approach to the analysis of human history, that human
behavior differs qualitatively from animal behavior to
the same extent that the adaptability and historical development of humans differ from the adaptability and development of animals. (1978, p. 60)

This Marxist-inspired formulation of the role of social
and historical experience provided only the first step in
Vygotsky's socio-historical account of human consciousness.
His most original and important contribution was to explicate the mechanisms that made the formation and transfer of
historical and social experience possible. It was in this
connection that he carried out his analysis of the role of
sign systems in social and individual activity. The importance of this issue in his thinking can be seen in one of the
entries he made in his personal notebooks near the end of
his life. In 1932 he wrote that the analysis of signs
(semicheskii analiz) is "the only adequate method for analyzing human consciousness". (Vygotsky, 1977, p. 94)


BASIC THEMES

In order to understand Vygotsky's claims about the role
of sign systems in mediating human activity, it is necessary
to understand how they fit into the general framework of his
socio-historical theory. Therefore, before turning to a
more detailed analysis of semiotic mechanisms, we will pro-

vide a brief overview of Vygotsky's general approach.  It is
possible to examine this approach in terms of three  general
themes:  (1)  the  use  of genetic analysis to examine human
activity, (2) the claim that the origins  of  higher  mental
processes  are  to  be found in social activity, and (3) the
claim that higher mental processes are mediated by sign systems found in social activity.

     "Genetic analysis" is the term we will use to refer  to
Vygotsky's  practice  of  examining  social  and  individual
activities by tracing them from their  origins  through  the
various transitions and dialectical negations that lead to a
later  state.   In  outlining his ideas about this method he
wrote:

> ...we need to conncentrate not on the <u>product</u> of develop-
> ment but on the very <u>process</u> by which  higher  forms  are
> established....to  encompass in research the process of a
> given thing's development in all its phases and changes--
> from birth to death--fundamentally means to discover  its
> nature,  its  essence, for "it is only in movement that a
> body shows what it is".  Thus, the historical [i.e.,  in
> the broadest sense of "history"] study of behavior is not
> an  auxiliary  aspect  of  theoretical  study, but rather
> forms its very base.  (1978, pp. 64-65)

     Vygotsky carried out most of his empirical research  on
ontogenetic  change,  but  following  the  lead  of Marx and
Engels, he also examined phylogenetic  and  socio-historical
change.   Furthermore,  some  of  his  arguments  tapped yet
another area of development or "genetic  domain".   We  will
term this last genetic domain "microgenesis".  The two types
of  microgenetic  transitions that he examined were: (1) the
development of a single psychological  act  as  it  unfolds,
usually over a span of milliseconds, and (2) the development
of  an  individual  or social activity as it is repeated and
perfected during a single training session.   All  of  these
genetic  domains  are  united  in Vygotsky's approach by the
fact that they could be used in genetic analysis.  They  all
allow  the  investigator to operate in accordance with Blon-
sky's dictum so often quoted by Vygotsky that "behavior  can
only be understood as the history of behavior".

     The  second  theme  in Vygotsky's writings is the claim
that higher psychological functions, the  human  personality
(lichnost'),  and human consciousness all derive from social

interaction. Vygotsky added a great deal of power to this
general claim by specifying the means whereby social ("interpsychological") and individual ("intrapsychological")
processes are linked. Rather than simply claiming that individual processes somehow derive from social functioning, he
made the much stronger claim that these individual processes
emerge by taking over and internalizing the patterns, means,
and structures found in social interaction. At its most general level this claim is seen in the following quote. As is
the case with most of Vygotsky's statements on this second
theme, the issue of genetic analysis also appears here.

To paraphrase a well-known position of Marx's, we could
say that humans' psychological nature represents the
aggregate of internalized social relations that have become functions for the individual and forms of his structure....Formerly, psychologists tried to derive social
behavior from individual behavior. They investigated individual responses observed in the laboratory and then
studied them in the collective....Posing the problem in
such a way is, of course, quite legitimate; but genetically speaking, it deals with the second level in behavioral development. The first problem is to show how the
individual response emerges from the forms of collective
life. (1981, pp. 164-165)

Vygotsky's understanding of how this line of reasoning
applied to ontogenesis is reflected in his formulation of
the "general law of cultural development".

Any function in the child's cultural development appears
twice, or on two planes. First it appears on the social
plane, and then on the psychological plane. First it
appears between people as an interpsychological category,
and then within the child as an intrapsychological category. This is equally true with regard to voluntary
attention, logical memory, the formation of concepts and
the development of volition. (1981, p. 163)

For Vygotsky, a corollary of the general law of cultural development was that the structures and mediational means
in the intrapsychological product reflected their interpsychological origins. As we noted earlier this is what gives
Vygotsky's second theme a great deal of its power. This corollary was implied in the preceding quote about this second
theme, but it is perhaps nowhere asserted more forcefully
than in the following statement by Vygotsky.

The very mechanism underlying higher mental functions is a copy from social interaction; all higher mental functions are internalized social relationships....Their composition, genetic structure, and means of action--in a word, their whole nature--is social. Even when we turn to mental processes, their nature remains quasi-social. In their own private sphere, human beings retain the functions of social interaction. (1981, p. 164)

Thus we see that the second theme in Vygotsky's approach involves two related notions. First, the origins of certain forms of individual functioning are to be found in social functioning; and second, this process involves the transferal of the means and structure of interpsychological functioning to the intrapsychological plane, thereby making individual functioning quasi-social.

The third theme in Vygotsky's approach operates within the framework supplied by the first two. It relies on genetic analysis and identifies the specific mechanisms that Vygotsky proposed to link the interpsychological and intrapsychological planes of activity. This third theme is that higher mental functions are mediated by the same sign systems that operate in human social activity. This is Vygotsky's semiotic theme. As is evident in the following statement, the ideas of the French psychiatrist Janet (1926-1927, 1928) played an important role in Vygotsky's understanding of this theme.

The history of signs...brings us to a much more general law governing the development of behavior. Janet calls it the fundamental law of psychology. The essence of this law is that in the process of development, children begin to use the same forms of behavior in relation to themselves that others initially used in relation to them.... With regard to our area of interest, we could say that the validity of this law is nowhere more obvious than in the use of the sign. A sign is always originally a means used for social purposes, a means of influencing others, and only later becomes a means of influencing oneself. (1981, p. 157)

Vygotsky uses the principle of the semiotic constitution of the higher mental functions in order to add a socio-psychological dimension to Marx's characterization of the difference between human and animal society. For Marx, the major differences between men and animals lie in the nature of human consciousness and labor. First, people are

aware of their relations with their environment and others,
while animals are not.

> My relations to my surroundings is my consciousness...for
> the animal, its relation to others does not exist as a
> relation. Consciousness is therefore, from the very
> beginning a social product and remains so as long as men
> exist at all. (in Avineri 1968, p. 71)

Second, human labor is self-sustaining.

> [Men] themselves begin to distinguish themselves from
> animals as soon as they begin to produce their means of
> subsistence, a step which is conditioned by their physi-
> cal organization. By producing their means of subsist-
> ence men are indirectly producing their actual material
> life... (ibid, p. 73)

These two theses lead to the position that human production
differs from animal production in presupposing a different
level of awareness. Human productive labor includes subjects who are aware of their relationships to others and
their activities, and it is this awareness which guides
their production.

> Labour is, in the first place, a process in which both
> man and Nature participate, and in which man of his own
> accord states, regulates and controls the material re-
> actions between himself and Nature...By thus acting on
> the external world and changing it, he at the same time
> changes his own nature. He develops his slumbering pow-
> ers and compels them to act in obedience to his sway. We
> are not now dealing with those primitive instinctive
> forms of labour that remind us of the mere animal...We
> presuppose labour in a form that stamps it as exclusively
> human. A spider conducts operations that resemble those
> of a weaver, and a bee puts to shame many an architect on
> the construction of her cells. But what distinguishes
> the worst architect from the best of bees is this, that
> the architect raises his structure in imagination before
> he erects it in reality. At the end of the labour-proc-
> ess, we get a result that already existed in the imagina-
> tion of the labourer at its commencement... (In Avineri
> 1968, p. 81)

Marx, as his dedication of <u>Kapital</u> to Darwin indicates, believed that the evolution of animal "society" could be
explained by Darwinian principles. Human society, on the
other hand, evolves through the dialectical interplay of
productive forces and relations, both of which presuppose a
level of social awareness not shared with animals.

### THE SEMIOTIC THESIS

Vygotsky's great contribution to the formulation of a Marxist psychology is his proposal that it is the semiotic mediation of tool use that creates the truly human forms of labor activity.

Although practical intelligence and sign use can operate independently of each other in young children, the dialectical unity of these systems in the human adult is the very essence of complex human behavior. Our analysis accords symbolic activity a specific organizing function that penetrates the process of tool use and produces fundamentally new forms of behavior. (Vygotsky 1978, p. 24)

The semiotic mediation of practical activity, primarily through speech, transforms man and creates the possibility of human society. Human labor differs from animal tool use because man is aware of and plans his actions using historically transmitted and socially created means of production. This awareness and planning ability is a form of generalization made possible only through speech.

...The most significant moment in the course of intellectual development, which gives birth to the purely human forms of practical and abstract intelligence, occurs when speech and practical activity, two previously completely independent lines of development, converge. Although children's use of tools during their preverbal period is comparable to that of apes, as soon as speech and the use of signs are incorporated into any action, the action becomes transformed and organized along entirely new lines. The specifically human use of tools is thus realized, going beyond the more limited use of tools possible among the higher animals.

Prior to mastering his own behavior, the child begins to master his surroundings with the help of speech. This produces new relations with the environment in addition to the new organization of behavior itself. The creation of these uniquely human forms of behavior later produce the intellect and become the basis of productive work: the specifically human form of the use of tools. (idem)

Language, as a historically determined social institution, is the means through which society converts the principles of cognitive development from biological to social dialectical.

We can now formulate the main conclusions to be drawn from our analysis. If we compare the early development of speech and of intellect--which we have seen, develop

along separate lines in animals and in very young chil-
dren--with the development of inner speech and of verbal
thought, we must conclude that the later stage is not a
simple continuation of the earlier. The nature of the
development itself changes, from biological to sociohis-
torical. (ibid, p. 51)
Vygotsky saw the incorporation of speech into human consciousness as the fundamental mechanism that transforms
cognitive development and organizes it along a completely
new line. Earlier development is of the type Piaget would
later call "sensori-motor", where the development of thought
is governed primarily by biological factors and simple reflex learning. When the child learns to speak, however, he
is acquiring a system of signs, which, like any social institution, develops according to socio-historical principles
of dialectical materialism. Vygotsky maintained that the
planning aspect of human labor activity, which is essential
to Marx's conception of human nature, is created through
man's acquisition of a social system of signs which itself
shares the same dialectical foundation as does the organization of human productive forces.

Verbal thought is not an innate, natural form of behavior
but is determined by a historical cultural process and
has specific properties and laws that cannot be found in
the natural forms of thought and speech. Once we acknowledge the historical character of verbal thought, we must
consider it subject to all the premises of historical
materialism, which are valid for any historical phenomenon in human society. It is only to be expected that on
this level the development of behavior will be governed
essentially by the general laws of the historical development of society. (idem)

Vygotsky began with Marx's praxis-interactionist thesis: consciousness develops through the organism's interaction with the world. The nature of practical activity determines consciousness. In particular, the nature of the
means in a goal directed activity transforms its user.
Vygotsky thus introduced the category of "externally" mediated activity--actions that involve the use of some external
means to reach some goal. There are two major types of 'external' mediators--tools and signs. Tools (i.e., as used by
infants or Kohler's apes) and signs differ fundamentally in
their organization. A tool is externally oriented towards
the goal, a mere instrument in the hands of its user who
controls it. Signs, however, are inherently 'reversible'--

they feed back upon or control their users. A favorite example of Vygotsky's is a knot used as mnemonic device.    An
external sign is used to control its user—to help him
remember.

In his first major paper which was given at the Second
All-Union Congress of Psychoneurologists in Leningrad in
1924, Vygotsky pointed out the role language plays in the
development of consciousness, particularly through its property of 'reversibility'.    Although the discussion is couched
in terms of reflexes, partly because of the nature of the
audience, Vygotsky's message is clear: speech and consciousness and society are intertwined.

There is one group of easily distinguishable reflexes in
humans that one could correctly call reversible reflexes.
These reflexes are elicited by stimuli that are, in turn,
humanly produced.    A heard word is the stimulus, and a
word pronounced is a reflex producing the same stimulus.
Accordingly, the reflex is reversible, since a stimulus
can become a response, and vice versa.    These reversible
reflexes, which constitute the foundation for social
behavior, serve to coordinate behavior on a collective
basis.    There is one group of stimuli that for me stands
out from among all others: namely, social stimuli, i.e.,
stimuli that originate in other human beings.    What distinguishes them is that I myself can reproduce them and
that they become reversible for me very early, and hence
determine my behavior in a fundamentally different way
from all others.    They liken me to others and make my actions identical with one another.    Indeed, in the broad
sense, we can say that speech is the source of social
behavior and consciousness.    (Vygotsky 1979, pp. 28-29)

This implies that "the mechanism of social behavior and the
mechanism of consciousness are the same" (idem). Self-aware-
ness also depends upon this interplay between speech, reversibility, and social consciousness. Vygotsky specifically rejected the hypothesis that "we know others because we
know ourselves" (idem), and instead, reversed the argument.
We know ourselves because we are aware of others, and we are
aware of others because our awareness of ourselves is derived from the awareness others have of us.

I am aware of myself only to the extent that I am as
another for myself, i.e., only to the extent that I can
perceive anew my own responses as new stimuli.    Between
the fact that I can repeat a word uttered by another
there is no essential difference, nor is there any fun-

damental distinction in their mechanisms:  both  are  reversible reflexes or, conversely, stimuli.

    Accordingly, a direct consequence of this hypothesis
will  be  the "sociologization" of all consciousness, the
recognition that the individual dimension  of  consciousness is derivative and secondary, based on the social and
construed in its likeness.  (ibid, p. 30)

    Speech  is  unique among semiotic systems because it is
both a system for communication and a system  for  mediating
and  reflecting other systems.  In particular, speech establishes an identity between the  mechanism  of  consciousness
and  the  mechanism  of social contact, thereby allowing the
incorporation of a historical dimension into  human  experience.
Historical  and  social  experience are not in themselves
different entities, psychologically speaking, since  they
cannot  be  separated  in experience and are always given
together.  We can link them with a plus sign.  As  I  have
tried  to  show,  their mechanisms are exactly the same as
the mechanism of consciousness, since consciousness  must
be  regarded  as  a particular case of social experience.
Hence, both these components may be readily  referred  to
by the same label of repeated experience.  (ibid, p. 30)


## SOCIAL, EGOCENTRIC, AND INNER SPEECH

    Up to this point we have not touched on what may be the
most  original and well-known aspect of Vygotsky's work: his
account of social, egocentric, and inner speech.   It  provides the best set of constructs with which to integrate all
aspects  of  his approach.  In fact in several instances the
only detailed explication of a theme  or  general  claim  in
Vygotsky's  work  is  to  be  found  in connection with his
account of these speech forms.  Therefore, in  addition  to
outlining  the  notions social, egocentric, and inner speech
we will be concerned in this section with how these  notions
explicate  the  general  theoretical  framework we have outlined.

    As we have already mentioned, the ultimate concern that
motivates Vygotsky's semiotic analysis is how  sign  systems
mediate  and  control  human  activity or,  in the words of
Ivanov (1977), how "The history of culture can be  described

to a great extent as the transmission in time of sign systems serving to control behavior" (p. 29). This general
claim has of course been at the foundation of many theories,
but Vygotsky's contribution is original both because of his
unique semiotic analysis and because of his concrete proposals for how semiotic mediation is instantiated psychologically. Vygotsky approached the problem of semiotic mediation by outlining a genetic analysis of semiotic functioning
that integrated the biological with the social, the social
with the individual, and the external with the internal.
What made this integration possible in Vygotsky's account
was the fact that various aspects of the <u>same sign systems</u>,
above all, speech, were involved throughout development.

In brief, Vygotsky's proposal was that inner speech
plays a crucial role in making it possible for humans to
plan and regulate their activity and that this inner speech
derives from previous participation in verbal social interaction. Egocentric speech, in his account, is "a speech
form found in the transition from external to inner speech"
(1956, p. 87). It is because egocentric speech provides insight into the development, and therefore the very nature of
inner speech that it has "such enormous theoretical interest" (1956, p. 87). The appearance of egocentric speech,
roughly at the age of three, reflects the emergence of a new
self-regulative function similar to that of inner speech.
Its external form is a reflection of the fact that the child
has not fully differentiated this new speech function from
the function of social contact and social interaction. Thus
Vygotsky described egocentric speech as "inner speech in its
psychological function and external speech in its physiological form" (1956, p. 87). Roughly at the age of seven, egocentric speech disappears, a fact that Vygotsky attributed
to its "going underground" to form inner speech. The close
genetic relationship that Vygotsky saw among the forms and
functions of social, egocentric, and inner speech is reflected in the following passages from the second chapter of
<u>Thinking and Speech</u>:

> Thus the first thing that links the inner speech of the
> adult with the egocentric speech of the preschooler is
> similarity of function: both are speech for oneself as
> opposed to social speech which carries out the task of
> communication and social relation with surrounding peo-
> ple....The second thing that links the inner speech of
> the adult with the egocentric speech of the child is
> their structural properties...the structural changes [in

egocentric speech] tend toward the structure of inner speech, namely abbreviation....according to this hypothesis egocentric speech grows out of its social foundations by means of transferring social,collaborative forms of behavior to the sphere of the individual's psychological functioning....Thus the overall scheme takes on the following form: social speech—egocentric speech—inner speech. (1956, pp. 82-88)

An essential point in Vygotsky's account of verbal regulation that is often not appreciated is that during early phases of development the speech that regulates the child's action is <u>social</u> speech. Because of the great interest in Vygotsky's ideas about egocentric and inner speech, this early interpsychological functioning is often not included in accounts of his notions on verbal regulation. This is clearly not what he intended as is evidenced by statements such as the following:

We know that the general sequence of the child's cultural development consists of the following: At first, other people act on the child. Then, he emerges or enters into interaction with those around him. Finally, he begins to act on others, and only at the end begins to act on himself. This is how the development of speech, thought, and all other higher behavior processes proceed. (1980, p.95)

Of course Vygotsky's insistence on using genetic analysis would alone indicate the general need to study the interpsychological precursors of egocentric and inner speech. What is more interesting in the case of these speech forms, however, is that specific semiotic properties of verbal regulation on the intrapsychological plane reflect the requirements placed on speech in social interaction. We will illustrate this point by briefly turning to the issue of dialogue.

The analysis of Vygotsky's writings has led us and others (e.g., Ivanov, 1977; Bibler 1975) to argue that he viewed inner speech as dialogic. The reasons for this dialogicality becomes obvious when we remember the social interactional nature of the origins of egocentric and inner speech. Vygotsky's recognition of this point is reflected in his claim that "egocentric speech grows out of its social foundations by means of transferring social, <u>collaborative</u> forms of behavior to the sphere of the individual's psychological functioning" (1956, p.87, italics ours). He expanded on this general comment in the course of analyzing Piaget's early work on collaboration among young children.    In this connection he wrote:

> The tendency of the child to put into practice in relation to himself the same forms of behavior that were earlier social forms of behavior is well known to Piaget and is used by him...in his explanation of the emergence of reflection from argumentation. Piaget showed how children's reflection emerges after the emergence in the child collective of argumentation in the strict sense of the word. It is in argumentation, in discussion, that the functional properties appear that will give rise to the development of reflection. In our opinion something similar happens when the child begins to converse with himself exactly as he had earlier conversed with others, when he begins to think aloud by conversing with himself when the situation calls for it.  (1956, p. 87)

On the basis of such passages one would expect that the regulative speech found on the intrapsychological plane should reflect the inherent dialogicality of that found on the interpsychological plane. A more concrete indication that Vygotsky understood this point can be found in his use of the notion of abbreviation.

Abbreviation is one of the properties of dialogue noted by Vygotsky. Like Yakubinskii (1923), he viewed the tendency toward abbreviation as a natural result of the "apperceptual mass" shared by interlocutors in dialogue.  In his own work, however, the main use he made of the notion of abbreviation was in characterizing egocentric and inner speech.  That is, he applied a construct specifically designed to account for propoerties of social dialogue in his analysis of egocentric and inner speech.  He certainly did not do this simply in order to emphasize the dialogic properties of these two speech forms. Rather, abbreviation turned out to be a useful notion in analyzing egocentric and inner speech because the requirements and resulting properties of interpsychological semiotic functioning are reflected in intrapsychological functioning.  An analysis of the structure and function of later forms of regulative speech must take into consideration the characteristics of earlier forms and the communicative reguirements that gave rise to these characteristics. The fact that most analyses have not begun with social speech has meant that many aspects of egocentric and inner speech have been overlooked or have remained a mystery.

With this in mind let us turn to Vygotsky's comments on the emergence and development of egocentric speech. His most complete treatment of egocentric speech can be found in a 54-page introductory essay he wrote for the Russian translation of two of Piaget's early books. This translation was published in 1932, and Vygotsky's essay later became the second chapter in his volume Thinking and Speech (originally published in 1934). Piaget's account of egocentric speech clearly played an important role in Vygotsky's thinking. As is the case with other authors, this does not mean that Vygotsky agreed with these ideas. Quite the contrary. However, Piaget was the investigator who identified and examined the speech form that was to play such an important role in Vygotsky's theoretical and empirical investigation of verbal regulation.

In his review of Piaget's account of egocentric speech Vygotsky noted that Piaget separated children's speech into two functionally defined categories: egocentric speech and socialized speech. Vygotsky summarized Piaget's views as follows:
"...speech is egocentric," writes Piaget, "because the child speaks only about himself, because he does not attempt to place himself at the point of view of the listener" (Piaget 1932, p. 74). The child is not interested in whether others listen, he does not expect an answer, he does not wish to influence others or in fact communicate anything to them. This is a monologue...the essence of which can be expressed in a single formula: "The child talks to himself as though he were thinking aloud. He does not address anyone" (Piaget 1932, p. 73). When he is doing lessons the child accompanies his action with various utterances. Piaget distinguishes this verbal accompaniment to the child's activity (egocentric speech) from the child's socialized speech, the function of which is quite different. In this case the child actually exchanges thoughts with others. He requests, orders, threatens, communicates, criticizes, questions. (1956, p. 25)

Before turning to Vygotsky's criticism of Piaget's conception of egocentric speech it is worthwhile to focus briefly on his comments about Piaget's notion of socialized speech. These comments provide some important insights into how these two authors disagreed, not only about the role of speech in development but about the development of human cognition in general. In contrast to his own approach, which argues for the priority of social forces, Vygotsky wrote that Piaget's approach argued for the priority of intrapsychological functioning. He wrote that for Piaget "The social lies at the end of develoment, even social speech does not precede egocentric speech but follows it in the history of development" (1956, p. 86). Vygotsky contrasted this with his own claim that:

> The initial function of speech is the function of communication, social connection, influencing others (children influencing adults as well as being influenced by them). Thus the initial speech of the child is purely social. To call it socialized would be incorrect since this word is tied to the presupposition of something that is initially nonsocial, something that becomes social only in the process of its change and development. (1956, p. 86)

In truth, Vygotsky was either engaging in a bit of polemics here or did not recognize an important difference in his use of "social" and Piaget's use of "socialized". It is not really a simple matter of what comes first--social (or socialized) or individual. Rather, it is a matter of what is meant by social and socialized and how these terms fit into a general theoretical framework.

The term "socialized" as used by Piaget properly belongs to the realm of the psychological characterization of the individual. It is a term that in fact could have been used by Vygotsky in his account of intrapsychological functioning. For example, in his account of the development of generalization and word meaning it would have been appropriate to use the term "socialized speech" to refer to individuals' speech behavior that reflects social norms, i.e., speech behavior that reflects socialization into the speech community. In this sense we see that Vygotsky's analysis certainly dealt with socialized, as opposed to unsocialized speech.

In contrast, the term "social speech" as used by Vygotsky refers to a phenomenon that is not analyzable within the realm of the psychology of the individual. He was reaching outside the realm of individual psychology into the realm of sociology or social interaction theory when he wrote of social speech. This is an instantiation of his general argument about the transition from social to individual processes. This fundamental assumption that we must go outside of psychology if we wish to carry out a complete genetic analysis of the psychology of the individual is a crucial issue over which Vygotsky's and Piaget's approaches in fact do disagree. Hence the confusion, intentional or real, over Piaget's assertion that "the social lies at the end of development". In truth, the socialized individual lies at the end of development in both Vygotsky's and Piaget's account. The difference between the two theorists is that Vygotsky searched for the beginning of development outside of psychology, whereas Piaget did not.

When we compare Vygotsky's and Piaget's accounts of egocentric speech, we find even more important differences. While recognizing the "indisputable and enormous credit" that was due Piaget for outlining the phenomenon of egocentric speech, Vygotsky disputed his interpretation of its function and fate. He argued that a natural consequence of Piaget's assumption that egocentric speech "fulfills no objectively useful or necessary function in the child's behavior" (1956, p. 78) is that he would conclude that it simply disappears or dies away with progressive socialization. In contrast, Vygotsky argued that egocentric speech plays an important role in the regulation of action and that it continues to be a part of the child's psychological functioning, eventually on the internal plane.

Vygotsky's claims about egocentric speech were based on a series of studies that he and his students (Leont'ev, Levina, Luria) conducted in the late '20's and early '30's. In these studies they examined children's use of egocentric speech in various settings. By altering these settings in certain ways they hoped to identify the cognitive and social factors that influence the use of egocentric speech. The studies carried out by these investigators can be divided into two groups, corresponding to two main issues over which Vygotsky disagreed with Piaget's interpretation of egocentric speech. These were the issues of the function and the fate (or more broadly the origin and fate) of egocentric speech.

Vygotsky's contention that the function of egocentric speech is to plan and regulate human action led him to argue that for children an increase in the cognitive difficulty of a task should result in an increase in the incidence of egocentric speech. In order to test this hypothesis he used a technique in which an experimenter surreptitiously intro-

duced an impediment into the flow of a child's action in
order to observe the effect this would have on the child's
use of egocentric speech. Vygotsky reported that:
> Our research showed that the coefficient of children's
> egocentric speech, calculated for those points of in-
> creased difficulty, was almost twice the normal coeffi-
> cient established by Piaget and the coefficient calcul-
> ated for these same children in situations devoid of
> difficulties. (1956, p. 79)

This finding has received corraborative support in a recent
study by Kohlberg, Yaeger, and Hjertholm (1968). These
researchers compared the amount of overt self-regulative
speech that children (4,6-5,0) used in four task settings
(bead stringing, easy jigsaw puzzle, building a block tower,
hard jigsaw puzzle) and reported a statistically significant
increase in the mean number of egocentric speech comments
with an increase in task difficulty.

On the basis of his findings about the relationship between task difficulty and incidence of egocentric speech
Vygotsky concluded that "egocentric speech very early...be-
comes a means of the child's realistic thinking" (1956, p.
84). He argued that his data clearly contradict Piaget's
claim that egocentric speech reflects egocentric thinking
(and hence serves no useful function) and support his own
claim about its role in planning and regulating action.
> The child's egocentric speech not only is not an expres-
> sion of egocentric thinking it fulfills a function that
> is diametrically opposed to egocentric thinking, the
> function of realistic thinking. This thinking does not
> approximate the logic of daydreaming and dreaming rather,
> it approximates the logic of intelligent, purposeful ac-
> tion and thinking. (1956, p. 84)

If Vygotsky's first criticism of Piaget's account of
egocentric speech concerned the function of this speech
form, his second criticism concerned the related issue of
its origin and fate. It was in this connection that he and
his colleagues carried out a second set of studies. Vygotsky
reasoned that if egocentric speech is "a speech form found
in the transition from external to inner speech" (1956, p.
87), one should be able first of all to document a close
relationship between social speech and early forms of egocentric speech. Then, with the development of egocentric
speech one should be able to document an increasing diverg-
ence between it and social speech and an increasing approxi-

mation of inner speech. These hypotheses about the origin
and fate of egocentric speech contrast sharply with those
generated by Piaget's position since according to him egocentric speech reflects egocentric thinking and should
therefore lose its egocentric quality and disappear with
progressive socialization.

The specific semiotic mechanism that Vygotsky posited
as giving rise to egocentric speech is the differentiation
of speech functions.

> In the process of growth the child's social speech, which
> is multifunctional, develops in accordance with the prin-
> ciple of the differentiation of separate functions, and
> at a certain age it is quite sharply differentiated into
> egocentric and communicative speech. (1956, pp. 86-87)

On the basis of this claim about the differentiation of
speech functions Vygotsky outlined the development of egocentric speech in the following terms:

> The structural and functional properties of egocentric
> speech grow with the child's development. At three years
> of age the distinction between this speech and the
> child's communicative speech is almost zero. At seven
> years of age we see a form of speech that is almost 100%
> different from the social speech of the three-year-old in
> its functional and structural properties. It is in this
> fact that we find the expression of the progressive dif-
> ferentiation of two speech functions and the separation
> of speech for oneself and speech for others out of a gen-
> eral, undifferentiated speech function which, during the
> early years of ontogenesis, fulfills both assignments
> with virtually identical means. (1956, p. 346)

This understanding of functional differentiation during
the period when egocentric speech is used provided the
foundation for specific empirical hypotheses. The first
issue Vygotsky addressed concerned the relationship that one
should find between social and egocentric speech. According
to his line of reasoning one should find a lack of differentiation or even a thorough confusion between social and egocentric speech in young children's verbal behavior. In this
connection Vygotsky carried out three "critical experiments"
that were designed to determine the degree to which children's use of egocentric speech is affected by phenomena
that affect the use of social speech (e.g., the presence of
absence of an interlocutor). His argument was that if the
use of egocentric speech is sensitive to the same factors

that affect the use of social speech it is not egocentric in
the sense that Piaget had in mind.    Furthermore, it would
reflect a close connection between egocentric speech and its
genetic precursor, social speech.

Each of the three studies conducted by Vygotsky focused
on a requirement of egocentric speech that reflected its
lack of differentiation from social speech. The three requirements examined were: the illusion of understanding by
others, the presence of potential listeners, and vocalization. In all cases the procedure was to create a relatively
high incidence of egocentric speech in children and then to
change the context such that one of the requirements for
such speech was no longer present.

In the first study Vygotsky placed children whose coefficient of egocentric speech had already been established
in a situation where the illusion of being understood by
others was no longer tenable. He did this by putting individual subjects in a group of deaf mute children or in a
group of children who spoke a foreign language (one not
known by the subject). In all other respects the child collective and the activity remained the same as it had been in
the setting in which the baseline of egocentric speech had
been established. Vygotsky reported that when the illusion
of understanding by others was removed the coefficient of
egocentric speech fell drastically.    In the majority of
cases egocentric speech disappeared altogether, and in the
remaining cases its mean coefficient was only one-eighth
what it had been under normal conditions.

In his second critical experiment Vygotsky examined the
effect of removing potential listeners on the incidence of
egocentric speech. In this case, after having established a
baseline coefficient of egocentric speech for each subject
in a collective monologue, he removed the subject from the
setting that included the potential listeners. He did this
either by placing the subject in a group of children with
whom the subject was not acquainted and with whom he did not
converse before, during, or after the study, or he excluded
the possibility of collective monologue altogether by placing the child alone at a table separated from others or in a
separate room. He reported that as in the first study the
coefficient of egocentric speech fell in the experimental
condition. The decrease was not as drastic as in the first
study, but it still was striking, falling to one-sixth the

level it had been in the control condition.   In   the  third
study  Vygotsky  varied conditions such that the requirement
of vocalization in egocentric speech could not be met. After
establishing a baseline level of egocentric  speech   in   his
subjects  Vygotsky  placed them in a setting in which vocalization was difficult or impossible.  For example, he  placed
them  in  a  room  in which the noise level was so high that
subjects could not hear their own speech or  the   speech   of
others around the table at which they were sitting.  (In one
case this involved an orchestra playing behind a wall of the
laboratory  in which the study was being conducted.)  In another case all the children in the collective were forbidden
to speak loudly and had to converse in whispers.  As in  the
first  two  studies,  Vygotsky found that the coefficient of
egocentric speech in the   experimental   condition   was   much
lower (5.4 times lower) than in the control condition.   Thus
the  difference  was somewhat less (but still striking) than
in either of the first two studies.

     Vygotsky summarized the results of these three  experi-
ments as follows:
     In  all three studies we were pursuing the same goal.  We
     focused on the three phenomena that  appear  with  almost
     all  egocentric  speech  by  the   child:   the illusion of
     understanding, collective  monologue,  and  vocalization.
     All  three  pheomena  are common to egocentric and social
     speech. We experimentally compared situations  in  which
     these  phenomena  were  present  with situations in which
     they were absent and found that the  exclusion  of  these
     features...inevitably results in the dying out of egocen-
     tric speech. On the basis of these findings we can legit-
     imately  conclude  that  although  the child's egocentric
     speech is already becoming distinguished in function  and
     structure  it  is  not definitively separated from social
     speech, in whose depths it is all  the  while  developing
     and maturing.  (1956, pp. 353-354)

     Thus we see the kind of evidence that Vygotsky provided
in  support of his claim that egocentric speech grows out of
social speech.  The other aspect of his analysis of the origin and fate of egocentric speech concerns the way in  which
this  speech  form diverges more and more from social speech
as it develops toward inner speech.  Vygotsky did  not  disagree with Piaget that the quantity of egocentric speech de-
creases  with  age.   However, he argued that the quality of
egocentric speech changes in a way that supports his  inter-

pretation and not Piaget's. Specifically, he argued that if the development of egocentric speech represents the progressive differentiation of speech for oneself from speech for others, it should become less intelligible to others with age. Such a prediction contradicts the prediction that would be generated by Piaget's interpretation of egocentric speech. On the basis of Piaget's contention that egocentric speech is a manifestation of egocentric thinking, one would expect its intelligibility to increase with age. This follows from the fact that progressive socialization should result in the child's being more likely to take others' perspectives into account, therefore producing egocentric speech (and speech in general) that is more intelligible to others. Vygotsky found that his analysis of egocentric speech supported his prediction and not Piaget's.

> One of the most important and decisive factual results from our research is that we established that the structural characteristics of egocentric speech which distinguish it from social speech and make it incomprehensible to others do not decrease, but increase with age. These characteristics are at a minimum at the age of three and are at a maximum at the age of seven. Thus they do not die away but evolve, their development is inversely related to the coefficient of egocentric speech. (1956, p. 345)

The research of Kohlberg et al. (1968) also supports Vygotsky's claim on this point. These researchers reported that with an increase in age there is an increase in the proportion of egocentric speech that is unintelligible. Specifically, they found that the proportion of all egocentric speech that fell in the category of "Inaudible Muttering" ("Statements uttered in such a low voice that they are undecipherable to an auditor close by") rose from about .24 to about .50 between the ages of five and nine. This increase in the proportion of unintelligible egocentric speech was offset by a decrease in the proportion of egocentric speech in categories such as "Describing Own Activity," categories made up of speech that could be understood by others.

Motivated by his deep interest in semiotic issues, Vygotsky carried his analysis of speech forms a step beyond arguing that the function, origin, and fate of egocentric speech indicate that it culminates in inner speech. In addition, he argued that the study of the specific properties of social and egocentric speech can provide insight into such properties in inner speech.

In this connection it should be noted that Vygotsky's concern for creating an objective, scientific psychology led him to reject the use of introspective reports and other subjective techniques for analyzing inner speech. Instead, he insisted that the objective properties and developmental tendencies of egocentric speech must serve as our main source of information. For him:

> ...the study of egocentric speech and of its dynamic tendencies toward the emergence of certain structural and functional properties and the weakening of others is the key to the investigation of the psychological nature of inner speech. (1956, p. 354)

This statement was qualified by the assumption that the study of egocentric speech cannot provide direct insight into the mature form of inner speech. According to Vygotsky it can only reveal the direction that development is taking. He assumed that the continued development of speech after it has gone underground results in a form of inner speech that bears little or no resemblance to external speech. Hence his statement that:

> ...even if we could record inner speech on a phonograph it would be condensed, fragmentary, disconnected, unrecognizable, and incomprehensible in comparison to external speech. (1956, pp. 355-356)

In reviewing the conclusions that Vygotsky reached about the characteristics of inner speech it should be pointed out that he provided virtually no concrete examples of how the characteristics he mentioned are manifested in egocentric speech. Instead, he described the characteristics of inner speech by drawing from theoretical literature in semiotics and poetics and by using illustrations from everyday and artistic texts. Therefore, in reviewing his account we do not have access to the specific egocentric speech data on which he based his claims.

With this limitation in mind, let us turn to Vygotsky's comments about the characteristics of inner speech. He divided these characteristics into two broad categories: "syntactic" and "semantic". He stated that the "first and most important" (1956, p. 355) property of inner speech is its unique abbreviated syntax. According to him we can gain insight into the "fragmentary, abbreviated nature of inner speech as compared to external speech" (1956, p. 355) by examining egocentric speech. Vygotsky's analysis of egocentric speech led him to conclude that this abbreviation takes

a specific form, a form that he described in terms of "predicativity" (<u>predikativnost'</u>).

As it develops, egocentric speech does not manifest a
simple tendency toward abbreviation and the omission of
words; it does not manifest a simple transition toward a
telegraphic style. Rather, it shows a quite unique tendency toward abbreviating phrases and sentences by preserving the predicate and associated parts of the sentence at the expense of deleting the subject and other
words associated with it. (1956, p. 356)

As we have pointed out elsewhere (Wertsch 1979b) Vygotsky's notion of subject and predicate here is actually concerned with a functional level of analysis of the utterance
rather than the formal analysis of sentence types or "system
sentences" (cf. Lyons, 1977). Although these levels of analysis were not clearly elaborated at the time Vygotsky was
writing, he apparently understood this distinction as is
evidenced by some of his comments about the difference between the "phasic" and "semantic" aspects of speech. In this
connection he wrote of the difference between the "grammatical" subject and predicate on the one hand and the "psychological" subject on the other. According to him the psychological subject (<u>psikhologicheskoe podlezhashchee</u>) is "what
is being talked about in a given phrase" or what is "in the
consciousness of the listener first" (1956, p. 333), whereas
the psychological predicate (<u>psikhologichskoe skazuemoe</u>) is
"what is new, what is said about the subject" (1956, p.
333). Units at this level of analysis have recently been
studied under rubrics such as "given and new information"
(Chafe 1974, 1976; Clark & Clark 1977; Clark & Haviland
1977), "given and new information" (Halliday 1967), and
"theme" and "rheme" (Firbas 1966). The fact that Vygotsky
saw the independence of the functional level of analysis
concerned with utterances and the formal level of analysis
concerned with system sentences is reflected in his statement that "any member of the sentence can become the psychological predicate, in which case it carries the logical emphasis..." (1956, p. 333).

Vygotsky developed his account of the predicative syntax of inner speech by tracing it back to its precursors in
egocentric and social speech. Borrowing heavily from Yakubinskii (1923), he noted that in social speech, pure predicativity occurs primarily in two types of situations. The
first is the situation in which the topic of conversation is

equally well known to all interlocutors.  In this connection
he provided the following example:

> Suppose  that several people are waiting for the "B" tram
> at a tramway stop in order to go in a certain  direction.
> Upon  seeing  the tram approaching, none of these ppeople
> would ever say in expanded form, "The 'B' tram for  which
> we are waiting to go to a certain point is coming".  Rath-
> er,  the expression would be abbreviated to the predicate
> alone: "It's coming," or "The 'B'".  (1956, p. 356)

The second type of situation in which  pure  predicativity
occurs  in  external social  speech is in response to a question.  In this connection Vygotsky pointed out that:

> In response to the question of whether you want a cup  of
> tea,  no  one answers with the expanded phrase: "No, I do
> not want a cup of tea".  The answer will be purely predi-
> cative: "No".  It will consist solely of  the  predicate.
> (1956, pp. 356-357)

In  developing his account of the factors that contribute to
predicativity Vygotsky argued that the end points of a  continuum  that  extends  from minimal to maximal predicativity
are represented by written language  and  inner  speech  respectively.  The nature of the dialogue and the level of the
shared or common "apperceptual mass" a la Yakubinskii guided
Vygotsky's argument throughout this discussion.

Thus,  according  to Vygotsky the first general characteristic of inner speech that makes possible  its  abbreviation  is  the  (functional) syntactic fact of predicativity.
The second general characteristic is  "semantic"  in  nature
according  to Vygotsky.  Under the semantic characterization
of inner speech he identified three interrelated properties:
the predominance of "sense"  over  "meaning,"  the  tendency
toward  "agglutination,"  and  the  "infusion of sense into a
word".  The key to understanding all of these properties  is
Vygotsky's distinction between meaning (znachenie) and sense
(smysl).  Vygotsky outlined this distinction as follows:

> The  sense of a word, as Paulhan has demonstrated, is the
> aggregate of all the psychological facts emerging in  our
> consciousness because of this word.  Therefore, the sense
> of a word always turns out to be a dynamic, flowing, com-
> plex  formation  which  has several zones of differential
> stability.  Meaning is only one of the zones of the sense
> that a word acquires in the context of speaking.  Further-
> more, it is the most stable, unified, and  precise  zone.
> As we know, a word changes its sense in various contexts.
> Conversely,  its  meaning is that fixed, unchanging point

which remains stable during all the changes of sense in various contexts. This change in a word's sense is a basic fact to be accounted for in the semantic analysis of speech. The real meaning of a word is not constant. In one operation a word emerges in one meaning and in another it takes on another meaning. This dynamism of meaning leads us to Paulhan's problem, the problem of the relationship of meaning and sense. The word considered in isolation and in the lexicon has only one meaning. But this meaning is nothing more than a potential that is realized in living speech. In living speech this meaning is only a stone in the edifice of sense. (1956, pp. 369-370)

For Vygotsky, the issue of sense is not a matter of lexical ambiguity. Such ambiguity is entirely a matter of different meanings and is commonly studied in linguistics. In contrast, the notion of sense is concerned with many aspects of signification that are not included in what is normally considered linguistic analysis.

If Vygotsky's notion of sense is not analyzable within the boundaries of linguistics, then one might be tempted to assume that it properly belongs in psychology. After all, he defined the sense of a word as "the aggregate of all the psychological facts emerging in our consciousness," and he went on to speak of "intellectual and affective sense" (1956, p. 370). However, it is just as mistaken to try to locate the notion of sense solely within psychology as it is to consider it to be a purely linguistic issue. This is because Vygotsky's analysis of it was based on the same line of reasoning that he began to develop in The Psychology of Art (1973). In this early work Vygotsky criticized Potebnya and others for their failure to ground aesthetic analyses in concrete objective texts while also criticizing the Formalists for focusing solely on text form and failing to appreciate the psychological significance of all aspects of the word. His proposed solution to these shortcomings was to examine how the psychological aspects of aesthetic reaction are related to the objective semiotic properties of artistic texts. For this reason his account of the aesthetic reaction and his account of sense do not fall neatly within the boundaries of either linguistics or psychology as these disciplines are normally defined.

In truth, however, the claim that sense is a notion be-

longing neither in psychology nor in linguistics is a by-
product of the starting point in Vygotsky's analysis rather
than the starting point itself. This starting point was
Vygotsky's concern with "living language". His use of this
term reflects the fact that his analysis focused on how
psychological processes are affected by signs that occur in
real time and space. While it is obviously important in his
account to recognize that actual utterances instantiate particular linguistic types, we see here that the instantiation
itself was also important for Vygotsky. The fact that his
analysis was concerned with the properties of the sign that
derive from its unique spatio-temporal properties, we see
that this concern with living speech also translates into a
concern with the inherent indexicality of signs.

For our present purpose of reviewing Vygotsky's ideas
it is important to note that at the time that he made his
distinction between sense and meaning, this contrast between
focusing on living speech and focusing on linguistic units
isolated from their context was also being discussed by other Soviet semioticians, especially Bakhtin (1929; Medvedev/
Bakhtin, 1928). For example, in his analysis of Dostoevsky's writing Bakhtin spoke of the difference between linguistics and "metalinguistics". According to him, in linguistics language "quite legitimately and necessarily, is
detached from certain aspects of the concrete life of the
word" (1929/1973, p. 150), whereas metalinguistics involves
"the study of those aspects of the life of the word which--
quite legitimately--fall outside the bounds of linguistics,
and which have not as yet taken their places within specific, individual disciplines" (1929/1973, p. 150). Bakhtin's
statement that metalinguistics "cannot, of course, ignore
linguistics, and must utilize its results" (1929/1973, p.
150) is consistent with Vygotsky's comments about the relationship between a word's sense and meaning. Both authors
argue that we must go beyond linguistics as it is traditionally constituted and deal with many of the critical issues
in the analysis of living speech. Vygotsky did this by
examining the aesthetic reaction produced by a word as it
appears in a particular artistic text, but he did not produce any systematic account of the semiotic mechanisms that
made this reaction possible. Rather, he worked at the level
of anecdotal illustration. Bakhtin on the other hand, extended this type of analysis in an important way by developing a systematic account of one of these semiotic mechanisms, the dialogical interrelations among parts of a text.

Given these general points about Vygotsky's notion of sense, let us turn to what he called the "semantic" properties of inner speech. The "first and most fundamental" of these properties is the "predominance of a word's sense over its meaning". Thus Vygotsky saw the context specific, indexical aspects of signification as predominating over the cross-contextual, stable aspects; the signification of the word is more a function of the stream of consciousness and the context in which it appears than it is a function of a stable, cross-contextual meaning system. This contrasts with the state of affairs in most interpsychological communication since for Vygotsky social interaction is necessarily linked with stable generalized meanings. Of course this is not to say that the emergence of inner speech destroys the capacity for using conceptual thinking as epitomized in scientific concepts. Vygotsky's point was simply that different aspects of signification predominate in different types of semiotic functioning. This is the point of his comment that:

> In spoken language as a rule we go from the most stable and permanent element of sense, from its most constant zone, i.e., the meaning of the word, to its more fluctuating zones, to its sense in general. In contrast in inner speech this predominance of sense over meaning that we observe in spoken language in various cases as a more or less weakly expressed tendency approaches its mathematical limit and occurs in an absolute form. Here the prevalence of sense over meaning, the phrase over the word, and the entire context over the phrase is not the exception but the general rule. (1956, p. 371)

Vygotsky (1956) and his followers (e.g., Luria 1975/1976, 1982; Akhutina 1975, 1978) have elaborated on this account of the relationship between sense and meaning in their analysis of speech production. A major point in these analyses has been that inner speech plays a role in speech production, but that it occupies a relatively early position in the series of steps that results in external speech. The fact that several steps are required to convert inner speech to external speech is a reflection of these investigators' understanding of the relative predominance of sense and meaning in the two speech forms.

Vygotsky's claim that the predominance of sense over meaning is the most important and fundamental semantic property of inner speech is reflected in the fact that he viewed the two remaining properties as following from it. Both of

these properties involve the ways in which words combine and
blend with one another. The first of these two remaining
semantic properties of inner speech is "agglutination".
Vygotsky's comments about this semantic property of inner
speech reveal that for him it was no more than an analogy
with the notion of agglutination as it is used in studies of
language typology. Thus he cited Buhler's observations about
Amerindian languages such as Delaware that have a tendency
toward agglutination. In this connection he noted that two
features of agglutination are of particular interest:

> ...first, in entering into the composition of a complex
> word, separate words often undergo phonic abbreviation
> such that only part of them becomes part of the complex
> word, second, the resulting complex word that expresses
> an extremely complex concept emerges as a structurally
> and functionally unified word, not as a combination of
> independent words. (1956, p. 372)

After noting the features of agglutinative languages that
were of most interest to him Vygotsky went on to argue that
analogous features characterize inner speech and that they
begin to make their appearance in egocentric speech.

> We have observed something analogous in the child's ego-
> centric speech. To the degree that this speech form
> approximates inner speech, agglutination as a means for
> forming unified complex words for the expression of com-
> plex concepts emerges more and more frequently and more
> and more distinctly. In his egocentric speech utterances
> the child displays a tendency toward an asyntactic com-
> bination of words that parallels the decrease in the co-
> efficient of egocentric speech. (1956, p. 372)

The third semantic property of inner speech is what
Vygotsky termed the "infusion of sense". He argued that as
a result of its being more dynamic than word meaning, word
sense follows different laws of combination. The basic difference between the two, a difference that reflects the relative predominance of either decontextual and contextual aspects of signification, is that a word's sense is influenced
and changed as a function of its entering into combinations
with other words, whereas a word's meaning is not. Vygotsky
again turned to artistic texts to illustrate and develop
this point.

> With regard to external speech we observe analogous phe-
> nomena particularly often in artistic speech. By passing
> through an artistic work, a word absorbs all the diver-
> sity contained in its sense units and becomes as it were

equivalent to the entire work. This is especially clear
in the case of titles of artistic works. In artistic
literature the title stands in a different relation to
the work than, for example, in painting or music. To a
much greater degree it expresses and crowns the entire
sense content of the work than, say, the title of a picture. Words such as Don Quixote and Hamlet, Eugene Onegin
and Anna Karenina reflect this law of the infusion of
sense in the purest form. The sense content of the entire work is really contained in one word. The title of
Gogol's poem <u>Dead Souls</u> is a particularly clear example
of the law of the infusion of sense. The initial meaning
of this title signifies dead serfs who have not yet been
taken off the census lists and therefore can be bought
and sold as living peasants can. They are dead but still
counted as live peasants. This expression is used in this
sense throughout the poem that is built on the buying up
of dead souls. But in passing like a red seam throughout
the fabric of this poem these two words take on a completely new, immeasurably richer sense like sponges absorb moisture, they absorb the deep sense message of the
various parts and images of the poem and they turn out to
be fully saturated by the sense only toward the very end
of the poem. At this point these words signify something
quite unlike their initial meaning....We observe something analogous...in inner speech. In inner speech the
word as it were absorbs the sense of preceding and subsequent words, thereby extending almost without limit the
boundaries of its meaning. (1956, pp. 372-373)

In summary, we see that Vygotsky's account of social,
egocentric, and inner speech includes several implicit and
explicit claims. First, egocentric and inner speech function
to control and regulate human activity. Second, a genetic
analysis of semiotic regulation must begin with social
speech. It cannot begin with intrapsychological forms of
verbal regulation, i.e., egocentric and inner speech. Third,
the earliest (interpsychological) forms of verbal regulation
involve the transformation of existing natural actions rather than the creation of something solely from the social
line of development. Fourth, intrapsychological forms of
verbal regulation will reflect the structural and functional
properties (e.g., dialogicality) of their interpsychological
precursor. Fifth, <u>contra</u> Piaget, egocentric speech does not
simply reflect egocentric thinking; rather, it plays an
important role in the planning and regulation of action.

Sixth, <u>contra</u> Piaget, the origins of egocentric speech are to be found in social speech (through the differentiation of speech functions), and the fate of egocentric speech is inner speech. Seventh, it is possible to identify specific structural and functional characteristics of inner speech through the study of egocentric speech. These include the functional syntactic characteristic of predicativity and semantic characteristics deriving from the relatively greater importance of sense as compared to meaning in inner speech.

## CONCLUSION

Vygotsky's work on inner speech is a "microcosm" of his theory and method. A new form of consciousness is created through the child's differentiation of the functional structure of a social institution, language, and the child's psychological processes take on some of the properties that structure social institutions in general. The uniqueness of language is that it systematically links generality, communication, and multifunctionality, and thus becomes the perfect "tool" for establishing new functional connections between thought, affect, and motivation. When the child differentiates the representational function, he can then use this function to represent these other functions. In so doing, he transforms these functions by "infusing" them with a new level of generality and also creates new interfunctional connections through their mutual regimentation by the representational function.

The development of inner speech is also an example of Vygotsky's application of Marx's dialectic to psychology. At first, the child is not aware of the representational function of speech, although his parents and other adults interpret him as referring and predicating. His system of communicative tools contains a potential level of consciousness which he will become aware of only through his interactions with others. He uses signs which have the interpreted <u>effect</u> of reference and predication; however, when the child differentiates this function from other speech functions and uses it to represent or mediate them, the representational function becomes a cause for his actions. This dialectical reversal of cause and effect in the development of inner speech is a key genetic law in his psychology. Vygotsky views this process of differentiation through action and the child's becoming aware of that differentiation as a confirmation of Marx's dictum that individual consciousness is produced through the subject's becoming aware of the structures underlying social relations. The same process that guides the development of inner speech, when applied to other functions that language can serve, forms the basis for the growth of all the higher mental functions and the development of the uniquely human forms of rule governed and plan guided human action.

## REFERENCES

Akhutina, T.V. 1975. The Neurolinguistic Analysis of Dynamic Aphasia. Moscow: Izdatel'stov Moskovskogo Universiteta. (in Russian)

Akhutina, T.V. 1978. "The Role of Inner Speech in the Construction of an Utterance". Soviet Psychology, Spring, XVI(3), 3-30.

Avineri, S. 1968. The Social and Political Thought of Karl Marx. Cambridge: Cambridge University Press.

Bakhtin, M.M. 1973. Problems of Dostoevsky's Poetics. Ann Arbor: Ardis Press. (Originally published in Russian in 1929).

Bibler, V.S. 1975. Thinking as Creation: An Introduction to the Logic of Mental Dialogue. Moscow: Political Literature Press. (in Russian)

Chafe, W.C. 1974. "Language and Consciousness". Language, L, p. 111-13.

Chafe, W.C. 1976. "Givenness, Contrastiveness, Definiteness, Subjects, Topics, and Point of View". In C.N.L.I. ed., Subject and Topic. New York: Academic Press.

Clark, H.H. & Clark, E.V. 1977. Psychology and Language: an Introduction to Psycholinguistics. New York: Harcourt, Brace and Jovanovich, Inc.

Clark, H.H. & S.E. Haviland. 1977. "Comprehension and the Given-new Contract". In Discourse Production and Comprehension (ed. by R.O. Freedle). Norwood, New Jersey: Ablex Publishing.

Firbas, C.J. 1966. "On Defining the Theme in Functional Sentence Analysis". Travaux Linguistiques de Prague, I, pp. 267-80.

Halliday, M.A.K. 1967. "Notes on Transitivity and Theme in English, II". Journal of Linguistics, III, pp. 199-244.

Ivanov, V.V. 1977. "The Role of Semiotics in the the Cybernetic Study of Man and Collective. In Soviet Semiotics: An Anthology, (ed. by D.P. Lucid). Baltimore: John Hopkins University Press.

Kohlberg, L., J. Yaeger, & E. Hjerthalm. 1968. "Private Speech: Four Studies and a Review of Theories. In Child Development, XXXIX, pp. 691-736.

Luria, A.R. 1976. Cognitive Development: its Cultural and Social Foundations. Cambridge, Mass.: Harvard University Press.

Medvedev, P.N. & M.M. Bakhtin. 1978. The Formal Method in Literary Scholarship: A Critical Introduction to Sociological Poetics. Baltimore: the John Hopkins University Press.

Smirnov, A.A. 1975. The Development and Present Status of Psychology in the USSR. Moscow: Pedagogika. (in Russian)

Vygotsky, L.S. 1956. Collected Psychological Research. Moscow: Idatel'sto Akademii Pedagogicheskikh Nauk. (in Russian)

Vygotsky, L.S. 1962. Thought and Language. Cambridge, Mass.: MIT Press.

Vygotsky, L.S. 1973. Psychology of Art. Cambridge, Mass.: MIT Press.

Vygotsky, L.S. 1977. "From the Notebooks of L.S. Vygotsky." Moscow University Record. Psychology Series. XV, April-June, pp. 89-95. (in Russian)

Vygotsky, L.S. 1978. Mind in Society: the Development of Higher Psychological Processes. (ed. by M. Cole, V. John-Steiner, S. Scribner, and E. Souberman). Cambridge, Mass.: Harvard University Press.

Vygotsky, L.S. 1979. "Consciousness as a Problem in the Psychology of Behavior". Soviet Psychology, XVII, 4:3-35.

Vygotsky, L.S. 1981. "The Genesis of Higher Mental Functions". In The Concept of Activity in Soviet Psychology, (ed. by J.V. Wertsch). Armonk, N.Y.: M.E. Sharpe.

Yakubinskii, L.P. 1923. On Dialogic Speech. Petrograd: Works of the Phonetics Institute of the Practical Study of Language. (in Russian)

# LANGUAGE, THOUGHT, AND SELF IN VYGOTSKY'S DEVELOPMENTAL THEORY

Benjamin Lee and Maya Hickmann

Center for Psychosocial Studies and
Max-Planck Institut für Psycholinguistik

## INTRODUCTION

Since Vygotsky never wrote about the self or its development, his contributions to such issues lies in the way he formulates the relationships between consciousness, language, cognition, and emotions. We will show how his semiotic and functionalist psychology is the starting point for a unified and dialectical theory of subjectivity. This theory is about how language creates new functional connections between psychological processes and thereby changes consciousness. Therefore one cannot understand any particular aspect of development, be it play, motivation, or egocentric speech and inner speech, without seeing its place in his overall developmental theory.

This paper is divided into two parts. The first section introduces Vygotsky's theory of consciousness and his psychological functionalism, and then discusses the role language plays in the development of the higher mental functions. Vygotsky sees the differentiation of the representational function of language from its communicative and social functions during the development of egocentric speech as founding two lines of development—scientific concepts and inner speech. Each line has its own organizing principles. Scientific concepts are structured along principles of generality and abstraction, while inner speech works along principles of deletion, presupposition, and condensation. This latter line of development is also the foundation for Vygotsky's theory of the mediation of affect and motivation. The motivational consequences of this mediation are clearest in his account of play and its relation to inner speech; through play the child acquires a new system of socially constructed generalized desires.

343

In the second section, we show how Vygotsky's theory can be developed into a more general theory of how language mediates cognitive and affective development. We interpret the development of egocentric speech as the child's differentiation of a "metapragmatic" level of sign use whereby he can use linguistic signs to refer to and represent different aspects of language use. This self-reflexive property is unique to language among all semiotic systems; only language seems to be a system of signs organized along principles of generality and self-reflexiveness. The differentiation of a metapragmatic level makes it possible for the child to represent the referential aspects of linguistic use; this development also makes it possible for him to think in propositions which is the foundation for "scientific" or formal thought. It also allows the child to represent the interpersonal and pragmatic aspects of language use which contain the whole range of rhetorical devices such as condensation, ellipsis, and metaphor, and thus founds a linguistically mediated level of unconscious "inner speech" processes.

## VYGOTSKY'S THEORY OF CONSCIOUSNESS

Vygotsky began his analysis of consciousness with a critique of the counterparts to materialism and idealism in the psychological theories of his day. He rejected the radical empiricist reductionism espoused by Bekhterev, who insisted that all behavior could be analyzed as combinations of reflexes and that consciousness was therefore an unnecessary concept. On the other hand, Vygotsky believed that the subjective-idealist approach was not able to support a science of psychology because of its use of introspection as its main source of data: "vulgar behaviorism" reduces consciousness to an attribute of the physical aspects of behavior, while subjective-idealism reifies it into some form of mental substance. Both positions, in some sense, "objectify" consciousness by not seeing it as a relation between subject and object.

Vygotsky's solution at the psychological level was like Marx's at the societal level: consciousness is neither reducible to behavior, nor separate from it, but is instead an attribute of the organization of practical activity. Vygotsky also insisted that the proper psychology for the study of consciousness would have to be a functionalist one in which the definition of all psychological states and processes presuppose one another.

A functional analysis of a given item consists in showing what role or effect that item has in some system of which it is a part. Although many types of functional systems exist, the most interesting ones for Vygotsky are those complex systems which are made up of interfunctional connections among their various subsystems. Each such subsystem depends upon the effect or relation it has with some other subsystem, and it is precisely because of these interconnections that each system exists. The overall system must be analyzed in terms of how its various subsystems are functionally interconnected. Vygotsky maintained that all psychological states are functionally related by consciousness.

> Memory necessarily presupposes the activity of attention, perception, and comprehension. Perception necessarily includes the function of attention, recognition or memory, and understanding. However, in previous, as well as contempory psychology, this obviously correct idea of the functional unity of consciousness and the indissoluble connection of the various forms of its activity has remained on the periphery. (1956, p. 243)

Vygotsky recognized the implications of a functional psychology. He criticized much of the psychological research of his time for its "atomistic" and "piecemeal" analysis of psychological processes. Such work did not investigate the functional connections between psychological processes and states, and thus completely bypassed the study of consciousness.

> The atomistic and functional analysis that has dominated scientific psychology in recent decades has resulted in the examination of individual psychological functions in isolation. The problem of the connections among these functions, the problem of their organization in the overall structure of consciousness has remained outside the field of the investigators' attention. (ibid., p. 43)

By "functional analysis", Vygotsky meant Thorndike's faculty psychology which he was contrasting with his own version. Consciousness is not an attribute of any particular state or process, such as attention or memory, but rather an attribute of the way in which such states are organized and functionally related both to behavior and to each other.

The atomistic study of psychological functions not only does not investigate consciousness, but it also fails to see that the very processes it studies depend upon the integrative characteristics of consciousness. Consciousness establishes the connections between the various processes, thereby giving them a certain unity and continuity--in organizing such processes, consciousness transforms them.

The unity of consciousness and the interrelation of all psychological functions, were, it is true, accepted by all. The single functions were assumed to operate inseparably in an uninterrupted connection with one another, but in the old psychology, the unchallengable premise of unity was combined with a set of tacit assumptions that nullified it for all practical purposes. It was taken for granted that the relation between two given functions never varied, that perception, for example, was always connected in an identical way with attention, memory with perception, thought with memory. As constants, these relationships could be, and were, factored out and ignored in the study of separate functions. Because the relationships remained, in fact, inconsequential, the development of consciousness was seen as determined by the autonomous development of single functions. All that is known about psychic development indicates that its very essence lies in the change of the interfunctional structure of consciousness. (Vygotsky 1962, pp. 1-2)

In applying Marxist principles to psychological development, Vygotsky had to face the problem of how higher mental functions involving abstract thought develop. Among his contemporaries much was being made of the primitive tool using capabilities of chimpanzees and apes which indicated that tool use, the psychological counterpart to labor at the social level, was not a distinctly human characteristic. Something else had to occur to lift the essentially noninstitutional, ahistorical and asocial nature of animal tool use into its human form. Vygotsky's great contribution to the formulation of a Marxist psychology is his proposal that it is the semiotic mediation of tool use that creates the truly human forms of labor activity.

In Vygotsky's later work, portions of which appeared as Thought and Language and Mind in Society, these themes are developed much more deeply and supported with empirical research. Although he does discuss isolated semiotic devices such as knots used as memory aids, Vygotsky focuses upon language as a mediating device because it is a <u>system</u> of <u>reversible</u> signs organized in terms of principles of multi-functionality and self-reflexivity. According to Vygotsky, consciousness, is "the experiencing of experiences" (1979, p. 19), and to be conscious of one's own experiences "means nothing less than to possess them in object form (stimulus) for other experiences" (idem.). Speech is reversible in that it can be both a stimulus and a response--"A heard word is the stimulus, and a word pronounced is a reflex producing the same stimulus" (1979, p. 78-79). This property of reversibility allows signs to feed back or control their users. It also implies that consciousness is basically social in origin since it uses reversible signs which simultaneously allow the child to take his own experience as an object for other experiences and at the same time form a system, one of whose structuring principles is communication.

Two other properties of language, its multifunctionality and its self-reflexivity, play a vital role in determining how the linguistic mediation of consciousness produces both more abstract cognitive structures and more complicated forms of motivation. The property of multifunctionality lies in language's potential for use in many types of goal-directed activity. According to Vygotsky, any action, whether it is carried out internally or externally, must be analyzed in terms of its goal(s) and the means whereby this goal is achieved. Since in any goal directed system, the means has the function of contributing to the goal of the action, and is selected at least in part because of this effect, the multifunctionality of language derives from its use as a system of means in a variety of different interactive settings, such as getting someone to do something, providing information, promising, etc. In particular, speech forms are structured by two major functions, communication ("social contact") and representation the former function ties speech to its social roots and contextual determination, while the representational function links language to its unique property of self-reflexiveness.

In addition to its multifunctional nature, human language is also unique among sign systems in that it can represent not only other actions, but also itself. This self-reflexive property allows language to have systematic levels of generality linked together by principles of hierarchical organization which all depend upon language's ability to take itself as its own object and context (or, in more contemporary terms, natural language contains its own metalanguage).  The systematic linkage of different levels of generality in language allows it to be the perfect mediator for the creation of more abstract forms of thought.  Word meanings have several levels of generality, from early complexes to scientific concepts.  Abstract, propositional or formal thought depends upon differentiating the different levels of generality already embodied in words.

The development of higher mental functions, both cognitive and motivational, is made possible by the multifunctional and self-reflexive nature of language.  The systematic multifunctionality of language, embodied in the fact that a given linguistic form can serve several functions and a given function be effected by different forms, allows it to be the semiotic mediator of consciousness par excellence. Because it is already multifunctional, language is the perfect tool for linking functions.  Speech is a system "of reflexes of consciousness, a system for reflecting other systems" (idem). At the same time, since language is also self-reflexive and contains systematically linked levels of generality, the linguistic mediation of psychological processes also entails that all processes, including non-representational ones, are "regimented" through its representational function. By being regimented in this way, the non-representational functions are transformed and acquire a new structure of generality. As will become clearer below, the application of the self-reflexive property of language to these non-representational functions becomes the foundation for Vygotsky's theory of "inner speech" and the development of "second order" motivations, while the application of self-reflexiveness to the representational function of language leads to the development of scientific concepts and formal thought.

Vygotsky's work on the development of the higher mental functions is a constant interweaving of the Marxist theme that the means mediating practical activity shape consciousness, and his focus on the unique properties of language as a mediator. Once the lines of practical activity and speech cross and begin to interact, speech becomes a way of integrating and creating interfunctional connections between such basic psychological processes as attention, perception, memory, and emotions, creating the distinctly human forms of motivation and thought.

When speech and thought begin to interweave in action, a dialectic is set up in which these three properties of linguistic signs allow the child to "bootstrap" himself "up" through the various levels of abstraction present in language. At first, the child is guided by these semiotic signs without being aware of their effects. They are silent means in his goal directed activity. Yet these very activities, because they involve the use of means structured along principles of reversibility and generalization, create effects which later will become causes of the child's behavior. For example, for Vygotsky, the young child's thinking is in many respects determined by memory. Definitions given by young children are based on their memories of the impressions objects and events make upon them. When asked to define some concept, their definitions are determined not so much by the logical structure of the concept (as it would be in an analytic definition) as by their concrete recollections. In adolescence, a change occurs--to recall means to think. Thinking, at first a product of memory, now becomes its guide: the act of thought controls recall (meta-memory). Memory becomes logicalized and is concerned with establishing and finding logical connections. In the experiments reported in Mind and Society the major shift occurs when the child consciously uses mnemonic devices to help himself solve problems. What earlier had simply been an effect of these devices, in that they helped the child solve the problem without his being aware of their efficacy, now becomes a conscious choice of behavior.

This reversal of cause and effect is one of Vygotsky's interpretations of Marx's dialectic and is a key genetic law in his psychology. Vygotsky views the process of differentiation through action and the child's becoming aware of the differentiation as a confirmation of Marx's thesis that individual consciousness is produced through the subject's becoming aware of the structures underlying social relations. His particular addition is to show how the duplex structure of language as both a multifunctional tool and a self-reflexive device allows it to mediate and represent action. This mediation and representation of practical activity forms the basis for the uniquely human forms of rule governed and plan-guided action. The child's conscious appropriation of a culturally and socially established means of practical activity radically restructures his consciousness. Nowhere is this dialectic more graphically described than in Vygotsky's accounts of inner speech and play.

SEMIOTIC MEDIATION AND THE DEVELOPMENT
OF INNER SPEECH AND PLAY

Vygotsky's overall approach to a theory of motivation
was influenced by his reading of Marx. In Mind and Society
he self-admittedly sets out to write a "Kapital" for psychology. Vygotsky's starting point is the Marxist position
that human tool use differs from animal instrumental activity in that man plans his activities beforehand, creating in
his mind a plan or rule to guide himself. These rules, of
course, are produced in a given society according to processes of historical evolution governed by the laws of dia-
letical materialism. What Vygotsky adds to Marx is to show
how this planning ability develops through the semiotic
mediation of action.

The child plans how to solve the problem through speech
and then carries out the prepared solution through overt
activity. Direct manipulation is replaced by a complex
psychological process through which inner motivation and
intentions, postponed in time, stimulate their own development and realization. This new kind of psychological
structure is absent in apes, even in rudimentary form.
(1978, p. 28)

The semiotic mediation of activity allows the child to plan
his activities according to what will eventually be a socially created and transmitted set of rules.

According to Vygotsky, acting and speaking are, for the
very young child, undifferentiated parts of the same psychological function which is directed to fulfilling some ongoing, context-specific and goal-directed activity. Speech is
a mere component of the means to instrumental ends. Since
speech and activity are initially undifferentiated in the
immediate context of behavior and are thus part of the same
overall perceptual field, the gradual differentiation and
internalization of speech allows language to mediate between
direct perception and action. This restructuring of the perceptual field gives the child's operations a greater freedom
from the concrete, sensory aspects of the situation. Speech
mediates and supplants the immediacy of natural perception--
the child begins to perceive the world through his speech,
as well as through sensory perception. With the help of
words, the child masters his attention, creating new structural centers in the perceived situation. The change in the
relation between perception and attention, allows the child
to shift his attention away from the immediate situation and

makes possible the development of a new kind of motivation.
Instead of a preoccupation with the outcome of an interaction, the emotional thrust can now be shifted to the nature
of the solution and of the very problem at hand. The use of
an auxiliary sign-system such as language dissolves the fusion of the sensory and the motor system, making new kinds
of behavior possible. The semiotic mediation of activity
allows the planning function to develop by restructuring the
decision-making process on a totally new basis.

> [A] functional barrier is created between the initial and
> final moments of the choice process; the direct impulse
> is 'shunted' by preliminary circuits. This change is a
> mandatory condition for the development of all uniquely
> human, higher psychological functions. The child who
> formerly solved the problem impulsively, now solves it
> through an internally established connection between the
> stimulus and the corresponding auxiliary sign; the movement which previously had been the choice now serves only
> as a system which fulfills the prepared operation. The
> system of symbols restructures the whole psychological
> process enabling the child to master his environment.
> (ibid., p. 35)

The child uses artificially created external signs as the
immediate causes of his behavior. These signs have the property of reversibility: they can act upon the agent using
them in the same way as they act upon the environment or
others.

> In this new process the direct impulse to react is inhibited and an auxiliary stimulus which facilitates the completion of the operation by indirect means is incorporated. Careful studies demonstrate that this type of organization is basic to all higher psychological processes, although in much more sophisticated forms than that
> shown above. The intermediate link in this formula is not
> simply a method of improving and perfecting the previously existing operation. Because this auxiliary stimulus
> possesses the specific function of reverse action, it
> transfers the operation to higher and qualitatively new
> forms, permitting humans by the aid of outer stimuli to
> control their behavior from the outside. The use of signs
> leads humans to a completely new and specific structure
> of behavior, breaking away from the traditions of biological development and creating for the first time a new
> form of a culturally based psychological process. (ibid.,
> p. 40)

The key to Vygotsky's account of the development of distinctly human forms of motivation lies in the relationship between what, at first glance, seem to be two unrelated areas: the development of play and inner speech. We will show that the former line of development illuminates some of Vygotsky's cryptic remarks about the development of inner speech at the end of <u>Thought and Language</u>, while the latter line of development provides the linguistic foundation for the development of play into its more adult forms.


## Inner Speech

The role of inner speech in Vygotsky's account of the development of the higher mental functions can be understood only in light of his overall theory of how speech creates new interfunctional connections among psychological processes, thereby radically restructuring consciousness. Vygotsky's work on concept development shows his conviction that language provides a foundation for abstract thought through its systematic interweaving of different levels of generality and reversibility. His work on inner speech shows the influence of language in mediating affect and motivation.

Before the child uses speech, his development is characterized by what Vygotsky calls "practical intelligence," or the ability to use tools to mediate his actions in order to achieve goals. This ability is also present in some forms of animal behavior but a fundamental change occurs in human ontogenetic development as the child begins to use speech as a particular sign-system to mediate his actions. Uses of speech eventually take on an organizing function which "penetrates the process of tool use and produces fundamentally new forms of behaviors" (Vygotsky 1978, p.24), in fact characteristically human forms of behavior.

As pointed out earlier, Vygotsky believed that language eventually comes to serve two major well-differentiated functions for the adult. These may be called the social-communicative and the representational functions. For the adults, these two functions are constantly intertwined. Ontogenetically, the representational function grows out of the social-communicative function which is primary. The major body of Vygotsky's work on language development focuses on showing the long and gradual differentiation between these two functions and the internalization of speech. This development takes the child from the earliest phases, where speech, ongoing actions, and perceptions are completely fused and the different functions of speech completely undifferentiated, to the adult phase. At this point speech is distinguished as a particular kind of action and forms an "internalized" system which allows the adult to communicate with others in social interaction, as well as to regulate and reflect internally on any action.

The gradual changes in the form/function relationships of child speech can be briefly outlined as a three-phase development: 1) external speech, 2) egocentric speech, and 3) inner speech. During the first phase, speech tokens seem to accompany the child's ongoing activity: there is always a relationship of copresence between the speech produced by the child and the context of ongoing activity. The speech tokens then bear an existential (indexical) relationship to the ongoing activity. Furthermore, the context of ongoing activity is for the most part an interpersonal one: the child carries out activities mainly through his interaction with adults. "Social contact" is the essential characteristic of child speech during this phase, although in its earliest (unreflective) use by the child, it does not constitute a separate communicative function. The functions of speech during this phase are undifferentiated for the child: the social contact created between the child and the adult through speech in the context of ongoing activity allows the adult to interact with the child and to regulate, monitor and organize his actions. As far as the child is concerned, speech is a global, functionally undifferentiated mass which simply creates interpersonal contact.

The first evidence for a differentiation of the functions of speech appears with "egocentric speech" (at approximately 3 to 4 years). During this phase, the child begins to use external speech to mediate and organize his own activity. Unlike many interpretations of "egocentric" speech in other developmental theories, Vygotsky does not attribute the quality of the child's speech during this phase to his "non-communicative" or "unsocialized" nature. Rather, it is the product of his experimenting with a new function of speech, its representational function, using it to regulate (to plan, guide, organize, and monitor) his own activity. This new function, then, emerges from the global uses of language for social contact, although the child produces

speech externally, regardless of whether he is using it to
regulate his own activity or addressing it to others.

It is not until the third developmental phase (from approximately 8 years on in our culture) that uses of the representational function of speech become clearly differentiated from uses of "external" speech to communicate with others.  The gradual disappearance of egocentric speech, which
accompanies the full-fledged differentiation of speech
functions, coincides with the process whereby the child internalizes the interpersonal processes characteristic of external speech into an inner system guiding his own intraper-
sonal processes. This phase completes a gradual process of
differentiating and internalizing speech: speech now forms
an inner system of signs (verbal thought), which has two
well-differentiated functions: the internalized regulatory
function of speech is differentiated from the uses of external speech to communicate with others. The resulting multifunctional system essentially contains the basic division
present in the adult system between thought processes and
communicative processes, both of which are mediated by language.

Thought and Language records both the child's development of abstract scientific concepts and complex inter- and
intrapsychological patterns of motivation from their beginnings in early child language. First the child uses speech
to accomplish certain goals. Although his speech can be interpreted by adults as referring and predicating, speech and
action are still undifferentiated for the child. When this
function emerges, the child can use language not only to get
things done in the world, but also to represent them. As
representation is differentiated as a distinct function of
language, speech can be used to represent anything, including itself.  Vygotsky's insight was to see that the use of
language to represent language functions results in two dia-
lectically related but contrasting "vectors" of development.
The use of language to represent or refer to the referential
-and-predicational aspects of language eventually results in
the development of logic and abstract thought. The use of
language to represent the means-ends and interpersonal aspects of language leads to the development of egocentric and
inner speech, as well as semiotically mediated motivation.

In the former "vector" which leads to the development
of abstract concepts, once the referential function is dif-

ferentiated, the development of "thinking in complexes" is the result of the child's gradual formation of stable denotational class criteria for word meanings. For example, a given word stands for some set of objects because it possesses some perceptual property in common across referential uses of the word. At this early stage of cognitive development, these denotational equivalences are determined only by stable perceptual equivalences among the denotata themselves. Later on, the equivalences become less perceptual and more conceptual.

If we call the way a word denotes its objects the "mode of presentation", then Vygotsky's theory of concept development shows how the child's grasp of the mode of presentation of words changes from perceptual criteria to purely symbolic or conceptual features. Vygotsky describes the levels of generality and their relation to reference in terms of a geographical metaphor.

If we imagine the totality of concepts as distributed over the surface of a globe, the location of every concept may be defined by means of a system of coordinates, corresponding to longitude and latitude in geography. One of these coordinates will indicate the location of a concept between the extremes of maximally generalized abstract conceptualization and the immediate sensory grasp of an object—i.e., its degree of concreteness and abstraction. The second coordinate will represent the objective reference of the concept, the locus within reality to which it applies. Two concepts applying to different areas of reality but comparable in degree of abstractness—e.g., plants and animals—could be conceived of as varying in latitude but having the same longitude. (1962, p. 112)

The first coordinate is the degree of generality possessed by a given "mode of presentation" of a word. The second coordinate is the objective referent, the object(s) to which the term applies. The mode of presentation can vary from concrete, perceptual properties to abstract relationships defined only by conceptual equivalences, not by any perceptual qualities of the denotata.

The higher levels in the development of word meanings are governed by the law of equivalence of concepts, according to which any concept can be formulated in terms of other concepts in a countless number of ways....The manifold mutual relations of concepts on which the law of equivalence is based are determined by their respective measures of generality. (ibid., p. 112)

An example of this dialectical interplay between mode of presentation and reference as mediated by grammar is Vygotsky's discussion of the relation between "spontaneous" and "scientific" concepts. With everyday, spontaneous concepts (which are really "pseudo-concepts" in Vygotsky's other terminology), the child "has" a concept only in the sense that he knows to which objects the concept refers, but he is not conscious of his act of thought. The mode of presentation of the concept is not the object of thought—what is referred to is language-external reality, not language internal concepts. Scientific concepts begin with verbal definitions and are used in nonspontaneous operations. They are definitional forms which talk about words as referring to concepts which are related to each other in systems of concepts. Each concept's relation to objects is mediated by some other concept. Spontaneous concepts are related to what they denote by qualities of the external world, and scientific concepts are related to language external reality through a language internal system of concepts.

> Though scientific and spontaneous concepts develop in reverse directions, the two processes are closely connected. The development of a spontaneous concept must have reached a certain level for the child to be able to absorb a related scientific concept. Working its slow way upward, an everyday concept clears a path for a scientific concept and its downward development. It creates a series of structures necessary for the evolution of a concept's more primitive, elementary aspects which give it body and vitality. Scientific concepts in turn supply structures for the upward development of the child's spontaneous concepts toward consciousness and deliberate use. Scientific concepts grow down through spontaneous concepts; spontaneous concepts grow upward through scientific concepts. (ibid., p. 109)

The child's acquisition of scientific concepts provides a system of generality which changes the psychological structure of everyday concepts.

In the second vector of development, the differentiation of the representational function in egocentric speech and its subsequent regimentation of the regulative and interpersonal aspects of language use begins the slow develop-

ment of inner speech. Language can be used to represent the
highly contextualized and interpersonal aspects of speech
events.   As spontaneous concepts move upward (become relatively more decontextualized), egocentric speech "goes underground", becoming "inner speech".  "Inner speech" obeys
principles that are the inverse of grammar, logic, and scientific discourse.  As Vygotsky repeatedly points out, psychological notions of topicality and  given-new  are  structured along pragmatic discourse principles which are different  from  those that structure traditional notions of grammar.  Whereas discourse using "scientific concepts" tends to
have maximal elaboration of surface distinctions and  explicit  presuppositional structure (precisely because such discourse tends  to  use  "decontextualized"  concepts),  inner
speech  obeys a principle of minimal surface elaboration and
maximal presupposition and condensation.  This   reaches  the
point  where all subjects are eliminated and thinking occurs
only in predicates since the thinker already knows and  thus
can presuppose what the topics of his thoughts are.  What is
maximally presupposed in the context can be deleted, and the
efficiency of  the system depends upon knowing what to omit
as well as what to say.

     The devices of inner speech parallel the repertoire  of
mechanisms Freud postulates for the unconscious.  This would
not  surprise Vygotsky, since inner speech is the asymptote
of the process whereby egocentric speech "goes underground",
and egocentric speech itself is the mediation of  motivation
by  speech because it involves the use of language to represent and guide action.  Inner speech becomes  the  means  by
which  motivation  and  non-verbal  thought can become regimented into external speech forms communicable  to  others.
Although  external  and  inner  speech serve different functions, the former for communication with others, the  latter
self-regulation and "communication for one's self", inner
speech is still speech ("thought connected with words").  In
external speech, thoughts are expressed in  words  so  that
they  can be communicated; in inner speech, words are merely
routes to their meanings, "words die  as  they  bring  forth
thought"  and inner speech becomes "to a large extent thinking in pure meanings".  This  peculiar  structure  of  inner
speech allows it to be a mediator between non-linguistic motivation, thought and external speech.  It is connected with
non-verbal  thought because all thought, according to Vygotsky, involves generalization; as <u>verbal</u> thought, it is  connected with external speech. Thus the psychological analysis

of another's external speech is complete only when one understands its motivation.

We come now to the last step in our analysis of verbal thought. Thought itself is engendered by motivation, i.e., by our desires and needs, our interests and emotions. Behind every thought there is an affective-volitional tendency, which holds the answer to the last "why" in the analysis of thinking. A true and full understanding of another's thought is possible only when we understand its affective-volitional basis. (ibid., p. 150)

In examining Vygotsky's account of inner speech, the question arises as to its relation to motivation. If inner speech mediates motivation and external speech, is it merely a transparent means of motivational expression, or is inner speech in some sense constitutive--does the development of inner speech change the nature of motivation? An examination of Vygotsky's theory would suggest the latter since egocentric speech is the precursor of inner speech and is the use of language to plan and guide action. By representing and guiding action, egocentric speech introduces a new level of generality into means-ends relationships, thereby creating a new motivational structure. Unfortunately, Vygotsky's discussion of motivation in Thought and Language is rather truncated; his most explicit discussion of motivation is in his work on play, which he sees as the foundation for the development of "second-order" motivations to which we will now turn.

## Play

The importance of play in Vygotsky's theory of development is that it allows a "zone of proximal development" for the child, through which a new kind of desire as well as a new kind of attitude towards reality are created. These developments are critical in preparing the child for his later position in society.

At school age play does not die away but permeates the attitude to reality. It has its own inner continuation in school instruction and work (compulsory activity based on rule). All examination of the essence of play have shown that in play a new relationship is created between the semantic and visible fields--that is, between situations in thought and real situations. (Vygotsky 1976, p. 554)

The critical point about play is that it completely reorganizes the child's motivational structure through the semiotic
creation of an imaginary situation. The process through
which this transformation is accomplished involves the
child's gradual differentiation of the various semiotic
functions of play, his gradual awareness of the effects of
these differentiations, and his subsequent ability to use
them to guide his own activity.

Although Vygotsky links play directly to motivation and
feelings, he does not equate them--for example, he does not
believe that all pleasure is play since there are many pleasurable activities such as eating which are not definable as
play. Any delay in fulfilling a desire leads to frustration
on the part of the young child. By the age of two or three,
the child begins to develop long-term desires, some of which
cannot be fulfilled at once. These unfulfilled desires, instead of being immediately abandoned, remain as generalized
unfulfilled wishes which thereby acquire a generalized motivational force. A tension is set up between these unsatisfied desires and the child's earlier tendency to immediately gratify his desires. In the case of the young child,
if a desire is not immediately gratified, then he may have a
temper tantrum unless, for example, the adult distracts him.
In the case of the older child, the desire for immediate
gratification exists in a different form: he develops long
term desires and needs which cannot be fulfilled at once;
yet they maintain a firm grip upon him and cannot be easily
put aside.

Play resolves this tension by embedding a generalized
rule for behavior in an imaginary situation. Play is not
random: the child plays at being a fireman, a father, or a
teacher; he is the father, the doll is the child, his sister
is the nurse, etc. Playful actions fit the rules of behavior for these models. What is unnoticed in real life,
because it is automatically responded to or enacted, becomes
a rule, a model for action. Within play there is an interesting dialectical progression. Early play consists of the
construction of an imaginary situation in which there are
guiding rules for behavior. In later play with games, the
rules set up the imaginary situation. There is a progression from immediate gratification and its eventual incorporation within an unfulfilled desire, to the setting up of a
rule of behavior as a means for an imaginary gratification;
in other words, there is a fusion of the general but unreal-

izable with the particular and realizable. Eventually, the
rule itself becomes a starting point for play in a game with
rules.    The  rule  now  sets up the imaginary situation, and
direct gratification becomes transformed into  a  gratification  over  the course of the game.  The child's ontogenetic
history mediates desire through the establishment  of  rules
to be followed.

The   uniqueness  of rules in the play situation is that
unlike rules for conduct, they are  self-created  and  self-
applied.    In a game, it is not simply a matter of one thing
being allowed and another not.  Rather, the child chooses to
play and to obey rules: his imagination creates these rules,
and he also chooses to act in accordance with them.

Play creates a leading edge in the child's  development
because  through it he begins to sever the direct connection
between a thing, a situation, and an action.  For the  young
child, a given situation dictates what he can do, given what
he  wants.  A doorbell is to be rung, a rattle to be shaken.
For the child, the meaning of an object is  the  interaction
between  the history of its effects upon his actions and his
desires in the immediate moment. In play, the child sees one
thing but acts differently in relation to what he sees. Both
the situation and the action are imaginary; the  child  constructs  a  meaning  to guide his action, and a symbolically
mediated desire governs reality.  A stick becomes a horse, a
doll becomes a mother.  Reality bends to the  will  of  conception and action is dictated not by the object, but by the
idea  represented by it.  However, a complete tie to reality
is never severed in play--the arbitrariness of true  symbolism  is  not present.  In play, it is not the case that anything can stand for anything--a horse is a stick because  it
can  be  ridden  on.  In play, meaning is separated from the
objects or sign vehicles that normally embody them,  but  in
real life, this relation is unchanged.

Once  meanings  and  ideas  begin  to determine action,
rather than vice versa, a critical shift has occurred in the
child's relation to reality.
But all the same  the  basic  structure determining  the
child's relationship  to reality is radically changed at
this crucial point, for his perceptual structure changes.
The special feature of human perception--which arises  at
a  very early ago--is so-called reality perception.  This
is something for which there is no analogy in animal per-

ception. Essentially it lies in the fact that I do not
see the world simply in colour and shape, but also as a
world with sense and meaning. I do not merely see something round and black with two hands; I see a clock and I
can distinguish one thing from another. (ibid., p. 546)

In play, a child treats an object as standing for something and as embodying a meaning. The meaning of the word
'horse' supplants the actual horse, the child foregrounds
the meaning and backgrounds the object, thus experimenting
with the link between object and meaning. In early play,
this experimental foregrounding of the meaning over the
object requires a "pivot" or mediational device. The stick
severs the meaning of the word 'horse' from a real horse.
The child transfers the meaning of the word from a real
horse to a stick and then by treating the stick as if it
were a horse he "acts out" a meaning. This realization in
action, in praxis, is a necessary precursor for the child's
realization of the meaning behind the object, an epistemological shift which is comparable to a shift from "knowing
how" to "knowing that". In play, the child unconsciously and
spontaneously separates meaning from object and meaning from
action without knowing that he is doing so. The effect of
his actions, however, form the basis for his later conscious
appropriation. This enactment and creation of divisions between reality and meaning forms the basis for all the higher
mental functions.

Play involves several paradoxes which are the source
for the dialectical movements which will form new motivational structures. The first is that the child uses an imaginary or "alienated" meaning in a real situation. The second
is that play is a combination of two opposed lines of resistance. The child plays because it gives him pleasure, yet
at the same time in order to have maximum pleasure in play
he must regulate his actions with rules. Play continuously
demands that the child control his immediate impulses,
thereby creating a second order desire, an affect which
overcomes another affect not merely by supplanting it, but
by incorporating it. In real life, obeying a rule and refraining from action often occur as renunciations, perhaps
stirred by fear of punishment. In play, the reverse is
true--voluntary subordination and restraint lead to pleasure. Play is "a rule which has become an affect" (ibid.,
p. 549), and is the foundation for self-restraint and
self-determination. This gives the child a new self which

is the locus of the desires of the "I" who is the subject-role of the game.


### Inner Speech, Play, and Motivation

In a recent article, Vandenberg (1981) draws out some interesting parallels between play and inner speech. The child's early speech is primarily social in function. With the development of egocentric speech, the child uses language to represent reality and guide his actions, although the forms that he uses are still primarily social forms. Gradually the child differentiates external speech for others and internal speech for himself which eventually becomes inner speech.

Symbolic play also seems to start with an undifferentiated social phase where the play items stand for social role categories in the real world (nurse, doctor, mother, father, etc.), but there is a lack of differentiation of the child's own actions from those of his fellow participants. In such social play, children often seem egocentric, each child playing in accordance with his own perspective upon the rules without any awareness of their social nature. However, the growth of this early "play egocentrism" should not be overinterpreted as a kind of egocentric individualism. As is the case with egocentric speech, the play categories are ultimately social in origin, but they are used by the child to guide his own activities. The external social form of play develops into play with rules, while the "egocentric" aspect "goes underground" to become an inner world of imagination, subjectivity, and fantasy. As Vandenberg points out, games with rules develop into their adult forms of "arbitrary but mutually agreed upon systems of interaction, and fantasy becomes more realistic and tied to adult concerns." (ibid., p. 362).

If Vandenberg's extrapolation of Vygotsky's ideas is correct, then the parallel between the development of play and inner speech is too striking to be overlooked. In each, we have a dialectical opposition between an external, social form, and an internal, individual form, with each developing out of an originally undifferentiated social matrix through the differentiation of the representational function of signs. In each case, the external, social form becomes more abstract and generalized while the inner forms move toward

condensation and personal symbolism. Vygotsky makes the connection between play and inner speech explicit.

> Play is converted to internal processes at school age,
> going over to internal speech, logical memory, and ab-
> stract thought. In play, a child operates with meanings
> severed from objects, but not in real action with real
> things. A child first acts with meanings as with objects
> and later realizes them consciously and begins to think
> just as a child, before he has acquired grammatical and
> written speech, knows how to do things but does not know
> that he knows, i.e., he does not realize or master them
> voluntarily. (ibid., p. 548)

How is play related to language? The key to Vygotsky's
analysis of symbolic play is that meaning and object
separate, but that meanings of things are given by words.
Play develops through the linguistic mediation of the
meaning-object relation. Words stand for their objects, and
when the child uses a stick to stand for a horse, he is
really replacing the word 'horse' with another sign--a semiotic substitution of one sign vehicle for another. The stick
does not stand for a real, existent horse, but rather for
the meaning attached to the word 'horse'.

> Separating words from things requires a pivot in the form
> of other things. But the moment the stick--i.e., the
> thing--becomes the pivot for severing the meaning of
> 'horse' from a real horse, the child makes one thing
> influence another in the semantic sphere....Transfer of
> meanings is facilitated by the fact that the child ac-
> cepts a word as the property of a thing; he does not see
> the word but the thing it designates. For a child the
> word 'horse' applied to the stick means, 'There is a
> horse'; i.e., mentally he sees the object standing behind
> the word.(ibid., p.  547-548)

Since play is also the basis for the child's development of
second-order desires, Vygotsky is thereby proposing a theory
of the linguistic creation of a new form of motivation.

Vygotsky's accounts of the development of inner speech
and play both involve linguistic mediation. In each case,
the representational function of language is differentiated
from other functions, and then is used by the child to guide
his actions. In both cases, the differentiation of the representational function of words sets up two lines of development, one towards external forms of generalization and abstraction, the other towards internal forms of "subjectiv-

ity" and motivation. The "second order" motivation that lies
"behind" inner speech is the product of the history of the
development of play, an effect of an earlier differentiation
of speech functions that, in the adult, becomes a motivating
cause for his actions.


## UPDATING VYGOTSKY

In order to place Vygotsky's work in a more comprehensive semiotic and linguistic framework, we will draw out
some parallels between speech and non-linguistic actions.
All theories of child development have to address this problem, whether implicitly or explicitly, and regardless of any
other difference among them. Thus, for example, both Piaget's and Vygotsky's developmental theories discuss extensively the child's early "sensori-motor" period or the
development of "practical intelligence" before the uses of
speech and then show the relation between speech and these
early cognitive processes.

In general, both nonlinguistic and linguistic actions
are goal directed, so that speech is an action which can
accomplish something in the same way as any non-linguistic
action can. Every action involves a means used to accomplish
some end or set of ends. The means are inherently context
sensitive, and any particular action will be the product of
adjusting the means used to the dual demands of goal and
context. Speech and nonlinguistic actions are simultaneously
distinct and continuous with one another: speech is a kind
of behavior, but it also has properties which make it a
priviledged kind of action, with great consequences for the
child's development. First, language is multifunctional,
i.e., the same sign can contribute to more than one function
and the same function can be effected by more than one
sign. Although various linguists and psychologists have dif-
fered in their conception of the term "function" (Silverstein 1978), there has been a gradually more consistent realization that referring and predicating is but one purpose
for which language can be used. Second, speech can be used
to refer to both linguistic and nonlinguistic actions and to
predicate something about them. In this way, language has
within it its own metasystem, a property which can take different forms and is not present in any other sign systems.
Finally, linguistic signs are structured not only according
to the general principles which all action systems must fol-

low, but also with respect to the parameters of language use itself, particularly the speech situation. For example, Jakobson (1971) proposes that there are at least six foci of the speech event itself which form the foundation for the structure of language--speaker, addressee, channel, code (grammar), message or signal form, and context (what the message is about).

The multifunctional nature of speech requires that it be analyzed in terms of the interplay between three "macro-functions"--the pragmatic, metapragmatic, and metasemantic levels of linguistic forms (Silverstein 1980). Pragmatics is the examination of the relationship between signs and their contexts of use. This aspect of speech is essential for any analysis of discourse relationships and much of the discussion below involves the indexical nature of signs, as defined within Peirce's framework (Peirce 1932), i.e., mini-mally a relationship of direct copresence between linguistic signs and some aspect of the context of utterance. This sign-context relationship can take many different forms and signs can have an indexical relationship with a number of nonlinguistic and/or linguistic aspects of the context of utterance (Jakobson 1971; Silverstein 1976). Linguistic indices, by signalling parameters of the ongoing speech event, link language to the "here and now" of social action. In this respect, they are closely tied to the use of lan-guage as a means in action, and are subject to the same con-straints as any means.

Linguistic indices vary in terms of two major dimen-sions. First, some are interpretable in relation to pre-ex-isting parameters of the speech event, i.e., parameters which exist independently of the use of the index, while some transform or bring into being some aspect of the speech situation through their very use. The former Silverstein (1976) has called "presupposing" indices, the latter "crea-tive". Since most indices have both properties, this dimen-sion is a continuum ranging from "relatively presupposing" to "relatively creative". To this continuum, Silverstein adds the distinction between referential and non-referential indices. For example, the first person pronoun is a rela-tively presupposing referential index because it refers to the person uttering it who exists independently of the use of the sign. It is also creative in that its use indexes the interpersonal roles of speaker and hearer, creating a minia-ture social system of discourse role relations by indicating who the personae of the speech event are. Examples of non-referential, relatively creative indices would be such status/solidarity markers as the 'tu-vous' distinction, which in holding referential content constant (i.e., singular addressee) index but do not refer to or talk about the closeness or familiarity between speaker and hearer. An example of a relatively presupposing, non-referential index would be a social sex marker where the choice of a given lexical item would indicate the sex of the speaker.

Whereas pragmatic uses of speech can be characterized in terms of the indexical relationships between signs and aspects of the context of utterance, metapragmatic uses of speech refer to or represent the indexical relationships of pragmatic uses. As such, metapragmatic forms are inherently part of the referential apparatus of language. They can refer to and describe any of the indexical foci of the speech event (for example, Jakobson's six foci mentioned earlier), and any kind of index (presupposing or creative, referential or nonreferential), but thereby "regiment" such indexical relationships into propositional form. For example, reported speech is a metapragmatic use of speech: speech is used to refer to speech, i.e., a speaker uses speech in one situation in order to report speech which was uttered in another situation. Thus, both of the following examples constitute reported speech. In the first one, the exclamation point graphically indicates an intonation contour.

1. John said "No! I don't want to do it!"
2. John angrily exclaimed that he didn't want to do it.

Although both examples can be used to report the same utterance, they differ in a number of ways. In the first case, all aspects of the original utterance are preserved, both what was said and how it was said. Secondly, a non-referential creative index in the original speech situation is put into referential and predicational form and represented through lexemes ("angrily", "exclaimed"); in addition, the personal pronoun "I" shifts into a third person form, and other markings shift in the verb, thereby indicating that the reported speech event occurred in another speech situation. In this case, then, the reporting speech has transformed some interpersonal aspects of the reported event into content. Metapragmatics regiment the pragmatics of language and our capacity to understand them by molding them into propositional form through reference-and-predication, and in

so doing,   transforming the pragmatic effect of the regimented forms (Silverstein 1978).

The   last   type  of  linguistic function is that involved
in making definitions or equivalence statements (e.g.,   the
word  "optometrist" means the same as the word "eyedoctor").
This "meta-semantic" function of language  is   the   starting
point  for  most modern treatments of grammar.  Grammar consists of the formal devices in language which  are  used  to
create propositions analyzable in terms of the modern predicate  calculus.   Frege's discovery of quantification theory
showed how  the  truth  value  of  sentences  regimented into
propositional  form  was  a function of systematic relationships of "multiple generality".  In order  to  make  clearer
some  of the epistemological implications of his work, Frege
distinguishes between the "sense" and "reference"  of  words
and sentences. This distinction hinges upon the relationship
between  equivalence  and generality on the one hand and the
self-reflexive use of language to talk about its own  referential  capacity  on  the other.  Frege's classic example is
that of the two names "the evening star"  and  "the  morning
star"  which both refer to the planet Venus.  An equivalence
statement such as "the evening star  is  the  morning  star"
states  that  these  two names  have the same referent, but
different modes of specification or "senses" are  associated
with  each  name.   The sense of "the evening star" is something like "the brightest starlike object visible after  the
sun  sets"  and that of "the morning star" is "the brightest
star visible before the sun rises".  Frege insists that  the
sense of a sentence is determined by the senses of the words
which make it up, and that the sense of a word is its general capacity to determine the senses of sentences of which it
is  a  part.   Sense thus determines reference, but not vice
versa.  The sense of an expression is what we know  when  we
know  the meaning of an expression, or at least that part of
the meaning of an expression that determines the truth value
of the sentence.  Notice that Frege's distinction of  sense
and  reference  hinges upon the fact that two linguistic expressions are  equivalent  in  referential  value  ("morning
star"  and  "evening star") in all truth functionally evalu-
able contexts.  Such uses of language are  metapragmatic  in
that  uses of language are being talked about and defined as
equivalent, but they form an unusual subclass in  that  the
context  of  usage  itself  is  irrelevant--signs  in a true
identity relationship are intersubstitutable  in  all  truth
functional  contexts, salvae veritate as Leibniz put it.  If

something holds in all contexts, then context is irrelevant. The use of language to talk about the denotational value of signs independent of context of use is thus a particular type of metapragmatic activity which we shall call "meta-semantics". If grammar is the linguistic means of encoding propositions, grammar becomes systematic metasemantics.

These distinctions are not mutually exclusive, and a given linguistic form can involve all three functions. For example, the use of a token of the pronoun "I" can be characterized in three ways: 1) it signals copresence with the person who utters it and is thus pragmatic, 2) it represents that person as "speaker" and is thus metapragmatic, and 3) it has certain grammatical, hence metasemantic, properties such as usually being the head of a noun phrase in subject position with nominative case marking. It seems that all lexical items, including proper names, involve, to various degrees, all three functions (Silverstein 1978). Some of these items, however, indicate one of these functions more clearly than others. For example, performative verbs of speaking display their metapragmatic aspects "transparent-ly"—their very denotational meaning is to classify events as speech events. Indeed, these constructions are unique in combining all three functions in certain instances of use. Because of certain grammatical (hence, metasemantic) proper-ties, these verbs, when put in the present nonprogressive tense and aspect, and with a first person agentive construc-tion, (i.e., "I promise to do x".), act as metapragmatic indexicals. They not only refer to and describe the ongoing speech event as a token of the type named by the predicate (and are therefore presupposing referential indices), but they also bring about the very speech event they name (and are thus creative). To say "I promise to go" is not only to describe oneself as promising, but to make a promise.

In light of these distinctions we may now return to Vygotsky's theory of development. Since all words contain pragmatic, metapragmatic, and metasemantic aspects, Vygotsky views language as influencing psychological development by providing systematic grooves which thought can differenti-ate. The pragmatic, metapragmatic, and metasemantic func-tions structure the surface forms of language, providing systematic form-function distinctions of which children will eventually become aware. If language is structured in terms of these three functions, and if the higher mental functions develop through the child's awareness of the systematic structure of the means used in social activity, then the use of language and the subsequent differentiation of these functions molds thought according to the principles that govern the structuring of language.

### The Implications of Vygotsky's Theory for the Development of the Self

The development of scientific concepts from its roots in egocentric speech is the child's gradual differentiation of the referential function of speech and its self-reflexive application to the representational function of language itself. In the earliest uses of language, speech is an accompaniment to practical activity. Although the indexical qualities of speech may focus the child's attention on different aspects of the immediate context, for the child at this point, speech functions are not separate from other means-end behaviors, despite adult overinterpretations. When the child can use language to both represent action and guide it, he begins to differentiate among two major dimensions of speech—language as presupposing and referential, and language as creative. The differentiation of the referential function in egocentric speech is the development of the child's awareness of both the presupposing indexical qualities of words—they name things which are independent of speech usage—and the creative aspects—speech not only reflects reality, but can be used to transform it. When a child names a painting and then draws the object named, he is treating the name as both a referential and a creative index, and is displaying a metapragmatic awareness of the indexical qualities of speech—"This word used here and now stands for something which I will now bring into being", a kind of microperformative.

Once the referential function is differentiated, the child can use words to classify objects. In what Vygotsky called "complexes", words are treated as presupposing referential indices which have denotational consistency because of perceptual similarities among the objects named. Such classifications are inherently metapragmatic. They state a word is being used to refer to an object that is similar to the object referred to by the same word in a previous utterance or instance of classification. At first, these equivalences across acts of referring are determined by perceptual

features of the objects themselves. At this stage of development, each token usage of a word is a presupposing referential index. The child names an entity (in Vygotsky's examples, colored blocks) which exists independent of the use of
the sign, but exists in a relationship of copresence with
the sign, making it an index. A given word type has a general meaning which is a classification of these token usages
as equivalent--they all refer to objects which share some
perceptual similarity and it is this perceptual similarity
which grounds the mode of presentation of the word.

Words, however, contain not only pragmatic and metapragmatic aspects that determine the mode of presentation of
the objects they stand for, but also metasemantic properties
defined in terms of their position in a system of sense relations. As pointed out earlier, the clearest example of a
system of such language internal relationships would be the
concepts defined by a grammar. Concepts expressed by grammatical relations are not determined by perceptual qualities
of the denotata of such concepts, but rather by their position in language internal "sense" systems defined by propor-
tional differences and equivalences among signifiers and
signifieds (in the Saussurean sense). Grammar is the way in
which words are arranged to form sentences, with each mor-
pheme or word containing a conceptual value determined by
its ability to form sentences which express propositions. As
the child differentiates the denotational level of word
meaning, he simultaneously discovers the language internal
equivalences in which words can participate and he gradually
begins to take certain concepts created by language internal
grammatical relations as "objects" of cognition.

Vygotsky saw grammar as the mediating device between
the upward growth of spontaneous concepts and the downward
growth of scientific concepts because the child's everyday
uses of language follow certain grammatical regularities
(they express them and are guided by them) without talking
about (or taking as an object of thought) the concepts described by those regularities. On the other hand, scientific
concepts are introduced by definitional equivalences established through the regimentation of sentences into full
fledged propositional or logical form. The mode of presentation of a scientific concept is its place in a system of
such concepts. Such definitions require that sentences be
treated as totally decontextualized. Since grammar is a system of language internal, decontextualized equivalences, all

scientific definitions presuppose a grammatical regimentation of language. The regimentation of speech by grammar allows grammar to be the link between indexical reference and logic.

Piaget demonstrated that the child uses subordinate clauses with "because", "although", etc., long before he grasps the structures of meaning corresponding to these syntactic forms. Grammar precedes logic. Here, too, as in our previous example, the discrepancy does not exclude union, but is, in fact, necessary for union. (1962, p. 126-127)

If the metapragmatic regimentation of the presupposing and referential aspects of signs and their use founds the development of classification, then it is not surprising to discover that the metapragmatic regimentation of the relatively non-referential and/or creative aspects of speech is the development of egocentric and inner speech. Scientific concepts push towards a maximal differentiation of form and function--each word should name one and only one object or class. This demand occurs because scientific concepts involve the regimentation of presupposing referential indices by language internal conceptual equivalences, a process which cannot be fulfilled unless there is a preexisting object for each index. The regimentation of the creative and/or non-referential aspects of speech moves along opposite principles, precisely because there is no presupposed "object" which exists independent of the use of the index. Instead, such indices signal the copresence of some speech parameter which is brought into being by the use of the index. These would all be included in what Halliday calls the "interpersonal function" of language and which is concerned with:

the social expressive and conative functions of language, with expressing the speaker's angle: his attitudes and judgments, his encoding of the role relationships in the situation, and his motives in saying anything at all. (Halliday & Hasan 1976, pp. 26-27)

This interpersonal function of language would also include uses of language which others have called style or rhetoric. These particular devices are relatively creative and non-referential--although they do use referential speech, they function to produce certain effects in the audience which are not described by the referential content of the utterances, but play off of it. For example, meta-

phors often look like definitional equivalences--"Girls are
faded lilies"--but are really metapragmatic uses of metasemantic forms for non-definitional purposes. The clearest
example in Vygotsky is Stanislavsky's instructions to his
actors as reported in <u>Thought and Language</u> (pp. 150-151)
where the underlying motives are given for the dialogues of
each of their parts. As Benveniste has pointed out, this
dimension of speech is the linguistic expression and constitution of unconscious motivation.

We thus come back to "discourse." By following this comparison, one would be put on the way to productive comparisons between the symbolism of the unconscious and
certain typical procedures of the subjectivity manifested
in discourse. On the level of speech, one can be precise:
these are the <u>stylistic</u> devices of discourse. For it is
style rather than language that we would take as term of
comparison with the properties that Freud has disclosed
as indicative of oneiric "language." One is struck by the
analogies which suggest themselves here. The unconscious
uses a veritable "rhetoric" which, like style, has its
"figures," and the old catalogue of tropes would supply
an inventory appropriate to the two types of expression.
One finds in both all the devices of substitution engendered by taboo: euphemism, allusion, antiphrasis, preter-
ition, litotes. The nature of the content makes all the
varieties of metaphor appear, for symbols of the unconscious take both their meaning and their difficulty from
metaphoric conversion. They also employ what traditional
rhetoric calls metonymy (the container for the contents)
and synechdoche (the part for the whole), and if the
"syntax" of the symbolic sequences calls forth one device
of style more than any other, it is ellipsis. In short,
to the extent that symbolic images in myths and dreams,
etc., will be listed, one will probably see more clearly
into the dynamic structures of style and their affective
components. What is intentional in motivation obscurely
controls the manner in which the inventor of a style
fashions common material and, in his own way, releases
himself therein. For what is called unconscious is responsible for the way in which the individual constructs
his persona, and for what he accepts and what he rejects
or fails to recognize, the former being motivated by the
latter. (1971, p. 74-75)

It is not surprising, therefore, that in Vygotsky's
concluding section of <u>Thought and Language</u> he compares the

actual dialogues in a play with their underlying motivations as mediated by "inner speech". Inner speech becomes the linguistic encoding and transformation of motivation.

> The development of verbal thought proceeds...from the motive which engenders a thought to the shaping of the thought, first in inner speech, then in meanings of words, and finally in words. (ibid., p. 152)

Metapragmatics is also linked to motivation through symbolic play. The symbolic substitute (a stick for a horse) is a substitute for the lexical item which refers to the object. Children view words as directly naming their objects, as part of them. They are treated as pure presupposing referential indices. In symbolic play, the child uses something to stand for the name, severing the direct bond between name and object. This separation is effected by symbolic vehicle which acts as a "pivot", allowing the child to work in the realm of meanings.

> He cannot sever meaning from an object, or a word from an object except by finding a pivot in something else, i.e., by the power of one object to steal another's name.(1976, p. 547)

Instead of having a name that stands indexically for its object, in symbolic play he uses the play vehicle to stand for the meaning of the object, its "mode of presentation", and treats this mode as if it were the object normally designated by the replaced word. Thus symbolic play is metapragmatic--the play vehicle replaces the name-object nexus, and causes the child to focus upon the meaning that the word embodies.

The metapragmatic differentiation of the representational function of words in the ongoing play situation allows the child to bend his will in accord with the world of representation and meaning, to choose to use a meaning to guide his action, just as with egocentric speech, where the child's naming something guides his action. Since we have seen that symbolic play involves a linguistic dimension for Vygotsky, it should not be surprising to see how the regimentation of the presupposing and creative aspects of the signs used in play lead to the development of adult games with rules on the one hand, and internal processes of fantasy and imagination on the other. The external forms of play, as reported in Piaget's work on children's games, develop toward an increasingly explicit awareness of the social nature of rules and their creation or maintenance by

mutual agreement. This can be traced back to the referential
function of words in designating the social role categories
that guide play and the child's growing awareness of their
social constitution. The development of the external aspects
of play into games with rules is the metapragmatic regimentation of the presupposing and referential aspects of play,
i.e., when a stick stands for what the word 'horse' stands
for. The development of a social epistemology parallels the
development of external speech towards its asymptote, the
development of scientific concepts which are the epitome of
propositional thought and discourse. The creative indexical
aspect of play is the child's use of a symbolic vehicle to
stand for the meaning of the word, thereby bringing about a
separation of word from object, and at the same time using
the "freed" meaning to guide his actions. The choice by the
child to control his desires with respect to a higher rule
creates a second order, linguistically mediated and distinctly human form of motivation, leading to the formation
of the will, and the higher forms of imagination, fantasy,
and internal subjectivity. Inner speech joins with symbolic
play in becoming the creative mediator of a new level of
motivation, as well as its expression in ongoing social
interaction.


## CONCLUSION

Although Vygotsky never developed a theory of the self,
his approach provides the foundation for such a theory
through its linking of a social institution, language, with
the development of the higher mental functions. Speech, as
a multi-functional and self-reflexive means in social interaction, shapes consciousness by creating new functional connections among psychological processes. Since language is
both a social institution and a communication system, its
development socializes thought at two levels--the child
joins a socio-historical system and a world of others to
whom he speaks. This point is critical for any truly social
theory of the self. Writers such as George Herbert Mead,
James Mark Baldwin, and present day cognitive-developmental
psychologists have tended to focus on dyadic interactions as
the foundation for the social origins of the self. Vygotsky
goes one step further by insisting that the dyad is merely
part of a larger system. This is particularly clear with the
personal pronouns which Mead, for example, uses as a critical part of his analysis of the self. Mead treats these pro-

nouns as providing the foundations for a theory of the social origins of the self, since 'I' and 'you' are reversible
and presuppose each other. In an early paper on consciousness, Vygotsky makes the same point:

> We are aware of ourselves in that we are aware of others;
> and in an analogous manner, we are aware of others be-
> cause in our relationship to ourselves we are the same as
> others in their relationship to us. (1979, p. 29)

There is a critical difference, however. Mead never
points out that part of the meaning of 'I' and 'you' is
their unique position in the system of language, as indexical anchors of the speech event. If the personal pronouns
are the origins of the self, they owe their uniqueness to
their determination by the system of language and discourse.
Benveniste makes this point clearly:

> Consciousness of self is only possible if it is experi-
> enced by contrast. I use 'I' only when I am speaking to
> someone who will be a you in my address. It is this con-
> dition of dialogue that is constitutive of person, for it
> implies reciprocally 'I' becomes 'you' in the address of
> the one who in his turn designates himself as 'I'. Here
> we see a principle whose consequences are to spread in
> all directions. Language is possible only because each
> speaker sets himself up as a subject by referring to him-
> self as 'I' in his discourse. Because of this, 'I' posits
> another person, the one who, being, as he is, completely
> exterior to "me", becomes my echo to whom I say 'you' and
> who says 'you' to me. This polarity of persons is the
> fundamental condition in language, of which the process
> of communication, in which we share, is only a mere prag-
> matic consequence. It is a polarity, moreover, very pecu-
> liar in itself, as it offers a type of ' opposition whose
> equivalent is encountered nowhere else outside of lan-
> guage. (Benveniste, 1971, p. 225)

If we take Benveniste's point seriously, Vygotsky's position
about the social origins of psychological processes take on
a renewed force. Language sets up the possibility of a dialectic between the individuality of every act of discourse
and the linguistic system, and this dialectic is constitutive of the development of subjectivity. This implies that
our consciousness of self and other emerges as the product
of the use of language in concrete situations of interpersonal communication. The self is social in two senses.
First, it is one pole of a dyadic relation constituted repeatedly in the ongoing praxis of discourse. Second, the

dyad itself is structured by its position within a social
system, language, which has a structure that systematically
links social indexicality and grammatical structure in an
ongoing socio-historical dialectic. Both abstract thought
and motivation are inextricably altered when speech begins
to mediate practical activity, and this alteration occurs
not only at the level of the content of the categories, but
also in terms of the structure of the processes. All the
higher mental functions take on a basically social, communicative, and dialectical cast. The application of the representational function to indexical aspects of speech gives
them a new level of generality. Second order motivations
are generalized desires whose particular "rhetorical" structure is due to both the representation and internalization
of pragmatic stylistic devices which are inherently social
and communicative in origin. For Vygotsky, the higher forms
of motivation are social in terms of their structure because
they originate from the internalization of these interpersonal components of discourse, even if they are no longer
social in function. Inner speech, for example, is no longer
social in function, but its basic properties of condensation
and ellipsis derive from the generalization and internalization of discourse pragmatic devices such as anaphora and
ellipsis. More importantly, the structure of speech itself
is the product of the demands of representation, context
specificity, and social history. As the child acquires language, so too does he become a member of society.

Vygotsky saw that any linguistic sign was part of a
larger socially created system of grammatical and discourse
relationships. In our extension of Vygotsky, we briefly indicated how the pragmatic, metapragmatic, and metasemantic
functions determine the meaning of words, and how Vygotsky's
work could be viewed as showing that development involves
the differentiation of a metapragmatic level and its subsequent regimentation of other functions of speech and action. The development of the higher mental functions, be
they abstract thought or second order motivations, all depend upon this dialectical interplay of speech and action.
The principles that structure speech are not merely biological, psychological, or even social psychological. Rather,
they form a socio-historical system created by the use of
language in a socially constituted world and thereby creating language-internal and language specific principles of
structure and change. Vygotsky, of course, saw these principles as basically historical and dialectical, and acting

only at the level of social institutions. To the extent that consciousness is dialectical, its development results from the socializing effect of language on the child's thought. The unity of self and consciousness is the unity of the social world refracted by language.

We cannot close our survey without mentioning the perspectives that our investigation opens up. We studied the inward aspects of speech, which were as unknown to science as the other side of the moon. We showed that a generalized reflection of reality is the basic characteristic of words. This aspect of the word brings us to the threshold of a wider and deeper subject--the general problem of consciousness. Thought and language, which reflect reality in a way different from that of perception, are the key to the nature of human consciousness. Words play a central part not only in the development of thought but in the historical growth of consciousness as a whole. A word is a microcosm of human consciousness. (Vygotsky 1962, p. 153)

## ACKNOWLEDGMENTS

This paper is the result of joint work while the second author was a research fellow at the Center for Psychosocial Studies in Chicago.

## REFERENCES

Benveniste, E. 1971. Problems in General Linguistics. Trans. by M.E. Meek. Coral Gables, Florida: University of Miami Press.

Halliday, M.A.K. & R. Hasan. 1976. Cohesion in English. London: Longman.

Jakobson, R. 1971. "Concluding Statement: Linguistics and Poetics". In Style in Language (ed. by T.A. Sebeok, Cambridge, Mass.: M.I.T. Press.

Silverstein, M. 1976. "Shifters, Linguistic Categories, Cultural Description". In Meaning in Anthropology (ed. by K.H. Basso & M.A. Selby), Albuquerque, New Mexico: University of New Mexico Press.

Silverstein, M. 1978. "The Three Faces of 'Function': Preliminaries to a Psychology of Language". In Proceedings of a Working Conference on the Social Foundations of Language and Thought (ed. by M. Hickmann), Chicago: Center for Psychosocial Studies.

Vandenberg, B. 1981. "Play: Dormant Issues and New Perspectives". Human Development, XXIV, 6:357-65.

Vygotsky, L.S. 1956. Selected Psychological Research. (In Russian). Moscow: Academy of Pedagogical Sciences Press.

Vygotsky, L.S. 1962. Thought and Language. Cambridge, Mass.: M.I.T. Press.

Vygotsky, L.S. 1976. "Play and Its Role in the Mental Development of the Child". In Play (ed. by J.S. Bruner, A. Jolly, K. Sylva), New York: Penguin Books.

Vygotsky, L.S. 1978. Mind and Society. (ed. by M. Cole, V. John-Steiner, S. Scribner, E. Souberman), Cambridge, Mass.: Harvard University Press.

Vygotsky, L.S. 1979. "Consciousness as a Problem in the Psychology of Behavior". Soviet Psychology, XVII, 4:3-35.

# CONTRIBUTORS

Michael Basch
> Training and Supervising Analyst, Chicago Institute for
> Psychoanalysis, Chicago, Illinois
> Attending Psychiatrist, Michael Reese Hospital, Chicago,
> Illinois
> Clinical Professor of Psychiatry, The Pritzker School of
> Medicine, The University of Chicago, Chicago, Illinois

Gil G. Noam
> Lecturer, Harvard Medical School
> Assistant Psychologist, McClean Hospital
> Co-director, Clinical Developmental Institute, Belmont,
> Massachusetts

Lawrence Kohlberg
> Professor of Education and Social Psychology, Harvard
> University

John Snarey
> Post Doctoral Research Fellow, Harvard Medical School,
> Massachusetts Mental Health Center
> Co-director, Clinical Developmental Institute, Belmont
> Massachusetts

Werner van der Voort
> Max-Planck Institute for Social Sciences, Starnberg, West
> Germany

Augusto Blasi
> Professor of Psychology, University of Massachusetts/Boston
> Boston, Massachusetts

John Broughton
> Associate Professor of Psychology, Columbia Teacher's
> College, Columbia University

Robert Kegan
> Lecturer in Human Development, Counseling, and Consulting
> Psychology, Harvard University
> Co-director, Clinical Developmental Institute, Belmont,
> Massachusetts

Benjamin Lee
     Research Fellow and Project Coordinator, Center for
          Psychosocial Studies, Chicago, Illinois

James V. Wertsch
     Chairperson, Department of Linguistics, Northwestern
          University, Evanston, Illinois
     Associate Professor, joint appointments in Department
          of Linguistics, Department of Psychology, and
          Program on Learning Disabilities, Northwestern
          University, Evanston, Illinois
     Research Fellow, Center for Psychosocial Studies,
          Chicago, Illinois

C. Addison Stone
     Assistant Professor, Program on Learning Disabilities,
          Department of Communicative Disorders, Northwestern
          University, Evanston, Illinois

Maya Hickmann
     Max-Planck Institute for Psycholinguistics, Nijmengen,
          Holland